LEADERSHIP IN DISASTER

LEADERSHIP IN DISASTER

Learning for a Future with Global Climate Change

Raymond Murphy

McGill-Queen's University Press
MONTREAL & KINGSTON • LONDON • ITHACA

ISBN 978-0-7735-3524-4

Legal deposit second quarter 2009
Bibliothèque nationale du Québec

Printed in Canada on acid-free paper that is 100% ancient forest free (100% post-consumer recycled), processed chlorine free

McGill-Queen's University Press acknowledges the support of the Canada Council for the Arts for our publishing program. We also acknowledge the financial support of the Government of Canada through the Book Publishing Industry Development Program (BPIDP) for our publishing activities.

Library and Archives Canada Cataloguing in Publication

Murphy, Raymond, 1943–
Leadership in disaster : learning for a future with global climate change / Raymond Murphy.

Includes bibliographical references and index.
ISBN 978-0-7735-3524-4

1. Ice storms – Canada, Eastern. 2. Ice storms – New England. 3. Emergency management – Planning. 4. Climatic changes – Canada. 5. Climatic changes – United States. I. Title.

QC926.45.C2M87 2009 363.34'92609713 C2008-907795-4

This book was designed and typeset by Pamela Woodland in Sabon 10.5/13

To Ruth,
Patricia, Lorrie,
Kiernan, and Maya

CONTENTS

ACKNOWLEDGMENTS

This research would not have been possible without a grant from the Social Sciences and Humanities Research Council of Canada, for which I am particularly grateful. I would like to thank *Canadian Geographic* for permission to reprint its map of the extreme weather event of 1998 that is the focus of this study, which shows its scope and consequences. My appreciation goes also to Elsevier JAI Press for allowing me to reprint on pages 27 to 33 a slightly revised version of what I had written about disasters in Raymond Murphy, "The Challenge of Disaster Reduction," *Community and Ecology*, edited by A.M. McCright and T.N. Clark (Oxford: Elsevier JAI press, 2006), 93–103.

I would like to thank the leaders interviewed in the research behind this book for their insider information and insightful analyses of both this disastrous weather event and the broader environmental problem of global climate change. Their names are given in appendix 1. I also want to acknowledge the invaluable help of Dr Karen Johnson-Weiner, an anthropologist at the State University of New York in Potsdam, for introducing me to Amish families and making me more sensitive to their values. Mr David Luthy, an Amish archivist in Aylmer, Ontario, deserves much praise not only for his help in finding materials but also for his warm hospitality. The book has benefited greatly from the valuable comments of two anonymous readers chosen by McGill-Queen's University Press, and it is a pity that the rules do not permit their names to be disclosed. Most of all, I want to thank my wife, Ruth Marfurt, for her editorial advice, astute comments, and love in this journey of writing a book.

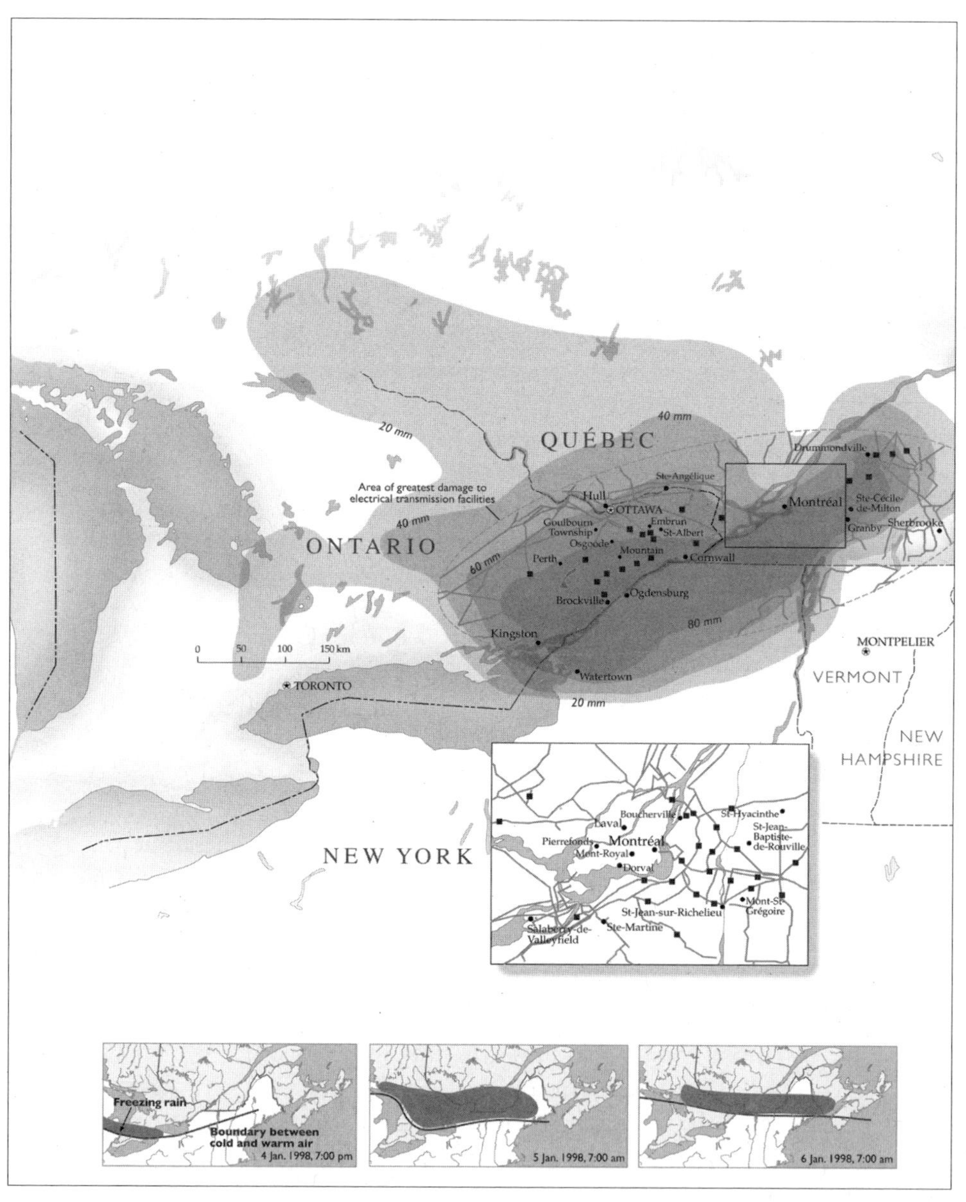

Map of the area affected by the ice storm of 1998. Canadian Geographic.

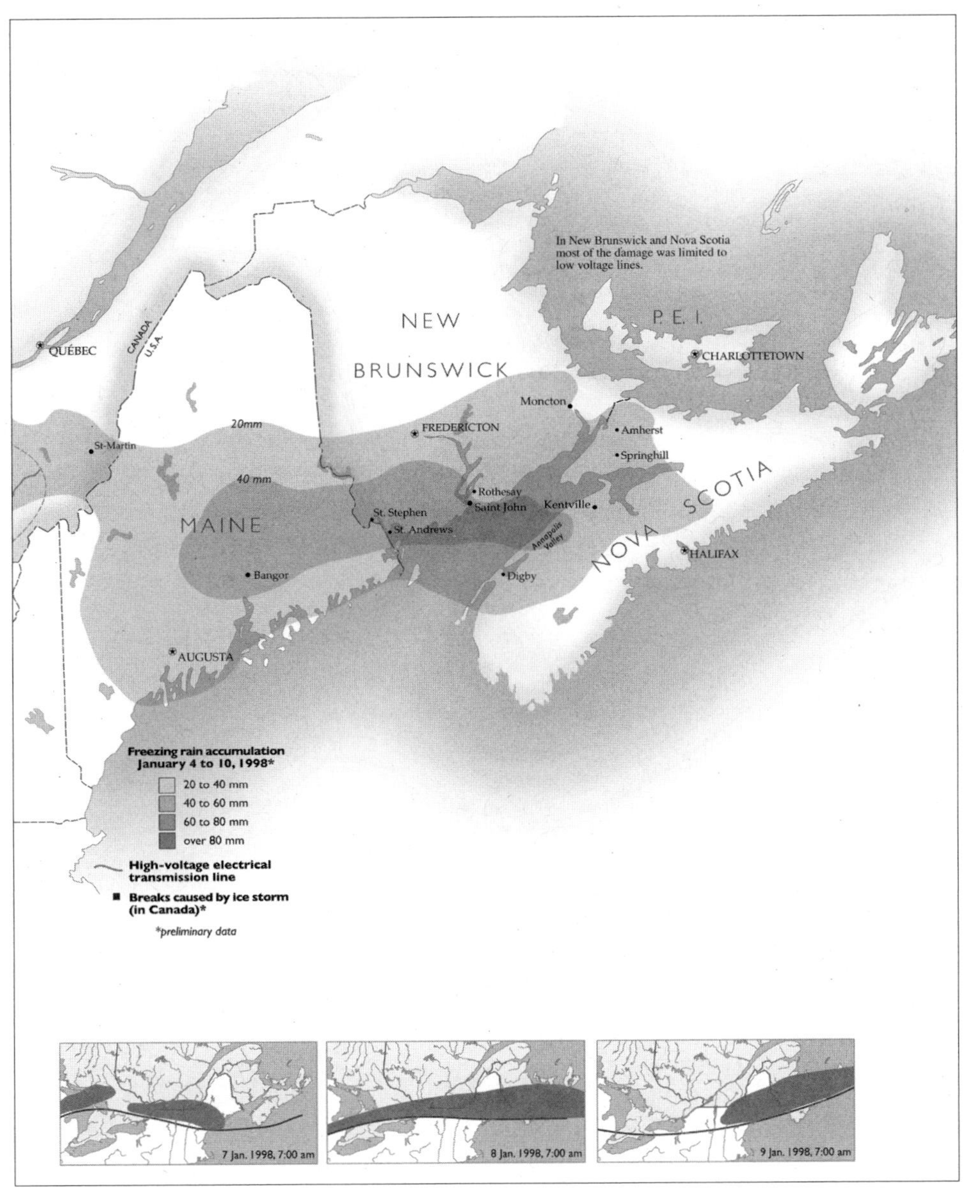

In New Brunswick and Nova Scotia most of the damage was limited to low voltage lines.
NEW BRUNSWICK
P. E. I.
CHARLOTTETOWN
QUÉBEC
CANADA
U.S.A.
Moncton
FREDERICTON
Amherst
Springhill
St-Martin
20mm
40 mm
Rothesay
Saint John
Kentville
St. Stephen
St. Andrews
MAINE
NOVA SCOTIA
Annapolis Valley
HALIFAX
Bangor
Digby
AUGUSTA
Freezing rain accumulation
January 4 to 10, 1998*
20 to 40 mm
40 to 60 mm
60 to 80 mm
over 80 mm
High-voltage electrical transmission line
Breaks caused by ice storm (in Canada)*
*preliminary data
7 Jan. 1998, 7:00 am
8 Jan. 1998, 7:00 am
9 Jan. 1998, 7:00 am

LEADERSHIP IN DISASTER

INTRODUCTION

Disastrous ice storms are rare, especially when caused by warming. Disasters are, however, not so rare, and ice storms are part of that category. Both natural and technological disasters, even terrorist-induced ones, involve the interaction of nature's hazards with vulnerabilities that have been constructed inadvertently or recklessly. We usually think of disasters as sudden events, but they can also be slow-onset, such as droughts. There is scientific consensus, albeit not unanimity, that global warming provoked by human activities is likely to produce a variety of disastrous results: more intense extreme weather, hundred-year storms becoming fifty-year storms, flooding from the rise of ocean levels, extensive drought, wildfires, abnormal insect infestations, and so on. The perils are too numerous to list. Changes are predicted to be slow at first until positive feedback loops and tipping points are encountered; then changes will become rapid, irreversible, and perhaps disastrous. Mitigation is easiest when carried out early, whereas if it is delayed, it will require greater sacrifice because the problems will be cumulative. And global warming is but one effect of human activities on the environment in which we live. The chronic and the acute, the slow-onset and the sudden, environmental problems and disasters, nature's hazards and socio-technological vulnerability are now coupled together in what Kai Erikson calls "a new species of trouble."[1]

We need not worry about nature per se. It is robust and will persist no matter what humans do. The issue is the kind of environment that will result from the interaction of human activities with nature's dynamics. Will global environmental change induced by human activities produce a material context for humans and other species that is harmful? For such deep and broad issues that involve the future, science provides valuable indications but not certainty. Yet leaders and the population must make sense of these issues and decide what to do. In these contested matters, even ignoring the risk and pursuing business as usual constitutes a choice. It is commonly assumed that the success of modern society in manipulating nature's forces makes it more robust and resilient. In some ways that is true, but we need also to explore the ways in which those recombinations of nature's processes to satisfy human desires render society more vulnerable.

This study deals with the consequences of an extreme weather event in an enormous region and the response of the modern societies of Canada and the United States to it. The event consisted of wave after wave of intense freezing rain that fell on a large part of northeastern North America and lasted for five days in early 1998. It was caused by warm, moist air, resulting from the El Niño phenomenon in the Pacific Ocean, that was carried by wind currents across to the other side of the continent.

The first part of the book examines general issues concerning the modernization of the risk of environmental calamities and disasters. It begins with a critical assessment and synthesis of the literature drawn from general social theory, environmental research, and disaster research. These theories and studies offer insight from different and in some cases opposed perspectives concerning the creation of risk, its mitigation, and preparation for danger. This analysis is followed by my own contribution to the elaboration of concepts and a framework to help bridge the culture/nature divide, which has left much of the social sciences restricted solely to the culture side. Thus Part One presents analyses of risk and vulnerability that are significant for understanding the concrete interaction between constructions of nature and social constructions of humans during a disaster and for understanding socio-political questions involved in global warming. It provides a framework for the empirical investigation that follows.

I propose the metaphor of a dance to help us to comprehend the interaction between movements of the material world of nature and the social practices of humans. In the dance examined here, nature's constructions of weather prompted actions, expectations, and even emergent beliefs on the part of humans. Part Two provides a comprehensive description of that dance between nature's movements and the associated movements of humans – a dance that was partly choreographed by humans in anticipation of hazards and partly improvised. The dance was at times adroitly performed but at other times awkwardly executed. The description seeks to give the feel of what it is like to live through a disaster of massive scope, intensity, and duration as the freezing rain crept up on leaders and the population unexpectedly, leaving them, as it dragged on, with the question, "will it never end?" – so hard to imagine years after it ended. Our understanding of social practices can be deepened by closely examining how human groups responded to the extreme movements of their dance partner, nature. Understanding practices in their biophysical context is particularly significant at a time when human actions are provoking the emergence of movements of nature that create the challenge of a response by humans.

Underlying the description of the disaster in Part Two is a theoretical argument. That documentation calls into question mastery-of-nature and end-of-nature hypotheses, showing, on the contrary, that technological development internalizes new dynamics of autonomous nature in modern societies. It demonstrates that the advance of science, the market, and rational organization has led to dependence on a centralized electrical grid, which enabled freezing rain to have disastrous consequences for society. Vulnerability to nature's forces was constructed, and a natural disaster was manufactured by humans because of that dependence. Extreme forces of nature that result in disasters make visible social-natural relations which are often ignored in normal dynamics of nature, much as we ignore breathing until we no longer do it well.

All citizens are equal in a democracy, but some are more equal than others. Modern societies have a hierarchical structure in which the choices, both good and bad, of a leader are more consequential than decisions by an ordinary citizen. Part Three investigates the overall performance of key leaders in this disaster, as well as specific aspects of leadership – transparency transformed into withholding information when the crisis became grave, conflict between leaders, and other behaviour. That section is based on interviews with major leaders in the management of the disaster, and they do not fail to provide insider knowledge as the lead dancers in society's response to the movements of extreme weather and as actors in performance with other important actors. It examines the ways these leaders made sense of the crisis and arrived at decisions and actions. Jean-Bernard Guindon, director of the Civil Security Centre (Centre de sécurité civile) for the Montreal Urban Community, concluded that this disaster was "one of the most extraordinary life experiences that I have ever lived. It taught me a great deal about everything: about people, about life, about organizations." Learning from the strengths and weaknesses of leadership can contribute to preparing better for difficult-to-foresee hazards of the future and reducing vulnerability.

Learning is especially important if global warming turns out as predicted by consensus science to be a huge environmental problem affecting all others, with extreme weather disasters and slow-onset calamities the outcome. Leaders must decide whether to continue business as usual, to prepare for disasters (adaptation), or to prevent disasters from occurring by changing practices that exacerbate global warming (mitigation). Part Four examines how these leaders who experienced an extreme weather disaster made sense of the contested issue of global warming. Though not caused by global warming, the weather event examined here could well be a harbinger of life under it. The two countries studied, Canada and the United States, are the leading greenhouse-gas emitters internationally. Do these leaders have insights that can help in mitigating global climate change, adapting to it, or at least in understanding why mitigation and/or adaptation is not occurring? The answer is in the affirmative. In the interviews they increase our understanding of not only the management of risks but also the politics, economics, ethics, and cultural habitus involved. They reveal how modern societies unleash risks and then adapt to and/or mitigate them. Thus Part Four shifts from the level of managing disaster to that of policy to prevent future disasters. Chapter 13 investigates communities that were struck by intense, persistent freezing rain without suffering a disaster. Our comprehension of modern societies in which the same construction of nature proved disastrous is deepened by this unusual comparison. It provides an opportunity for modern societies to learn how to avoid disaster and environmental calamities.

In light of the empirical investigation in the book, the final chapter returns to an analysis of general issues concerning risk and survival in the new frontier of global environmental change resulting from human activities – that is, the emerging new dance between humans and nature. To give a preview of the conclusions, this study calls into question two misleading oversimplifications common in the cultures of modern societies: an idealized representation of nature and wishful thinking concerning the outcome of technological development.

Climatologists contend that ice storms are likely to occur from central Missouri across central Illinois to central Indiana and northern Ohio, extending into upper New York State and southern Canada. This prediction has led the North American insurance industry, through the Institute for Catastrophic Loss Reduction in Canada and the Institute for Business and Home Safety in the United States, to conclude as follows: "Experts concur that freezing losses similar to those which devastated Montreal in 1998 could impact Toronto, Boston, New York City, Buffalo, Detroit, Cleveland, Chicago, Minneapolis and/or St. Paul."[2] This ice storm just missed Boston. Ice storms are but one of many hazards of nature, some of which humans are unleashing.

PART ONE

Social Action in Its Biophysical Context

CHAPTER 1

The Modernization of Risk

The interaction between human activities and nature's dynamics has become one of the key social issues of our time. Humans are consuming resources, polluting, and recombining nature's processes and materials at a frenetic pace. The activities of a six-billion population, which is still rapidly growing, are unleashing new forces of nature such as climate change which are so massive that some geologists argue our epoch should be called the "anthropocene" period to distinguish it from the "halocene" era prior to the industrial revolution. Humans and their machines have become the principal forces modifying the planet. What this activity will produce is uncertain, involves novel risks for humans, and is a principal subject of public debate.

Environmental issues are problematic because they bring the risk of disasters, whether sudden or slow-onset. Even natural disasters, which, as we shall see in this study, are not as natural as they seem, incite a socially constructed response in both citizens and leaders. The effects on a society of a calamity involving nature's constructions reveal the strength or weakness of preparations for danger, as well as improvised, emergent social constructions. The central element in the theoretical framework used here to elucidate both environmental problems and disasters is based on Ulrich Beck's theory of reflexive modernization and the "risk society" because it remains the most general and penetrating conceptualization of the potential material threats unleashed by modern society and the possible mitigation of those dangers. Beck's theory has an affinity with Max Weber's concept of the paradoxical magnification of irrationality under rationalized modernity.[1] Other important theories are relevant as well:

the theory of the social reconstruction of nature by market forces and modern technology to achieve cornucopia; ecological modernization theory; the theory of the treadmill of production; and discourse analysis. These theories and perspectives in many ways contradict one another, but they each provide plausible accounts from different angles that are important for situating in a broader context the detailed empirical investigation that will be presented in subsequent sections. They give, on an abstract level, the basis for arguments that could have been used by the leaders who were interviewed. Disaster research will be integrated with these social theories, which are central to environmental research in the social sciences.

WEBER'S CONCEPTION OF RATIONALIZED MODERNITY

Weber, in his well-known theory, contended that modern capitalism and formal rationalization were launched by the inner logic of Protestant religious asceticism: "since asceticism undertook to remodel the world and to work out its ideals in the world, material goods have gained an increasing and finally an inexorable power over the lives of men as at no previous period in history."[2] This is a truly paradoxical result with enormous consequences in terms of consumption of resources, production of waste and pollution, and other environmental problems. Economic rationality and consumerism have become a force with its own inner logic: "victorious capitalism, since it rests on mechanical foundations, needs its support [of religious asceticism] no longer."[3] Weber argued that rational structures contained an inherent dynamic: they provided technically better solutions to problems and induced a motivation for the further development of those structures.[4] Specialized occupations are required by the logic of machine production, whether we like it or not. "The Puritan wanted to work in a calling; we are forced to do so."[5] Weber argued that rationalization – in terms of searching for the best means to accomplished goals through science and technology, the market, formal organization, and the legal system – was the key distinguishing feature of the modern world. It enabled an extraordinary economic development, in particular of consumption. But he also pointed out the unintended harmful consequences of this development. The intensification of formal rationality brought with it a magnification of substantive irrationalities judged according to particular values. A century after Weber wrote, irrationalities that provoke environmental calamities and new vulnerabilities to disasters have confirmed his analysis.[6]

Weber characterized the intellectualized culture in the modern, rationalized world as the following belief: "there are no mysterious incalculable forces that come into play, but rather that one can, in principle, master all things by calculation ... Technical means and calculations perform the service. This above all is what intellectualization means."[7] He contended that the modern economic "order is now bound to the technical and economic conditions of machine production which to-day determine the lives of all the individuals who are born into this mechanism, not only those directly concerned with economic acquisition, with irresistible force. Perhaps it will so determine them until the last ton of fossilized coal is burnt. In Baxter's view the care for external goods should only lie on the shoulders of the 'saint like a light cloak, which can be thrown aside at any moment.' But fate decreed that the cloak should become an iron cage."[8] Here Weber comes close to a theory of a treadmill of production and consumption. But he did not draw the conclusion that destiny is predetermined by these technical and economic conditions. For him, the future is open-ended and to be constructed. In his critique, Weber gave a glimpse of how the door of the iron cage could be opened. "No one knows who will live in this cage in the future, or whether at the end of this tremendous development entirely new prophets will arise, or there will be a great rebirth of old ideas and ideals, or, if neither, mechanized petrification, embellished with a sort of convulsive self-importance."[9] New ideals and the rebirth of old ones can help mitigate this mechanized petrification, unlock the gate of the iron cage, and prevent the depletion of resources.

Weber ended his theory of the importance of Protestant asceticism in the rise of modern capitalism with the following words: it is "not my aim to substitute for a one-sided materialistic an equally one-sided spiritualistic causal interpretation of culture and of history."[10] He had a nuanced, complex view of the interaction between material interests and ideas: "not ideas, but material and ideal interests, directly govern men's conduct. Yet very frequently the 'world images' that have been created by 'ideas' have, like switchmen, determined the tracks along which action has been pushed by the dynamic of interest."[11] Human agents constructing the switchmen of world images determine whether rationalization occurs and, if so, whether it will interact with the dynamic forces of nature in a perilous or harmonious way. Economic rationalization was not the only social dynamic that emerged from Protestant asceticism. There was an intervening switchman of world images that resulted in a bifurcation of the tracks leading from Protestant asceticism. Much can be learned about modern, rationalized society, its interaction with nature's constructions, and the new risks this interaction engenders by comparing it with another social construction that emerged from Protestant asceticism, namely, Amish communities. These two results of Protestant asceticism pushed the dynamic of interest along very different tracks with dissimilar environmental and disaster outcomes. That comparison will be made in a later chapter.

Weber also argued that "the various value spheres of the world stand in irreconcilable conflict with each other ... here too, different gods struggle with one another, now and for all times to come."[12] By "value sphere" he is referring to a domain of activities that has its own requirements: the economic sphere, the political sphere, the religious sphere, the ecological sphere, and so on. The switchmen of dominant images determine which value sphere will have precedence in this conflict. Technological rationalization varies according to the value sphere that steers it. Developing and implementing costly disaster-mitigation practices and rationalization in an ecological sense requires a world image that gives high priority to these spheres in their conflict with the economic sphere and hence makes more secure and environmentally benign the tracks along which interest is pursued. W.J. Mommsen argued that Weber's dire predictions should be seen, not as an irresistible deterministic force leaving humans no choice for their fate, but rather as a "self-denying prophecy" to foresee and avoid the harmful consequences of proceeding along the path of economic rationalization.[13] In other words, they can be interpreted as a precursor to Beck's analysis of the transformation of primary modernity.

REFLECTIVE MODERNIZATION AND THE RISK SOCIETY

Ulrich Beck argues that the structuring of the future is occurring surreptitiously and indirectly in laboratories and boardrooms, rather than in parliaments and in political parties.[14] He contends that there has been a profound change in the past half-century from primary modernization to reflexive modernization. In the early stage of modernization, societies perceived only progress in their development of science, technology, rational organizations, and the market. Pollution occurred, but this was a local problem near factories and could be avoided by fleeing to the many pristine areas that remained. Moreover, many of the adverse side effects of the technological manipulation of nature's dynamics involve delayed-action consequences and were not visible, so people were unaware that they were occurring in this primary stage. There was a time lag before the reactor at Three Mile Island produced a near disaster and the reactor at Chernobyl generated a complete one. Those experiences educated technical experts and the population concerning the risks of nuclear energy and modernization in general. It was only after antibiotic-resistant bacteria emerged because of the overuse of antibiotics that awareness developed of the threatening side effects of that innovation. It took time for fossil fuels to have an impact on something as vast as the world's climate, for the ozone layer to become depleted as a result of the use of CFCs, for DDT to produce cancer, and for all of these consequences to become visible through science or/and experience. When cumulative consequences involve something as massive as the oceans, the atmosphere, and all the land of the planet, the delayed-action harm can take decades or centuries. This is brief on a geological or evolutionary time scale, but it appears lengthy compared to a human lifetime. The long, intensifying effect unfortunately implies that it will also take a very long time for humans or nature to undo degradation and to restore the environment to its previous state.

According to Beck, the time lag is over. Now, in this new period of reflexive modernization, the harmful secondary effects of the successes of science and modernization are turning back against the very societies that have developed them, creating the "risk society."[15] Human societies have always faced risks from nature's dynamics, but the type of risk is different under reflexive modernization: "the risks emerging today are distinguished firstly from all the earlier ones by their *society-changing scope*, and secondly by their particular *scientific constitution*."[16] The present period of modernization has become reflexive in that the very successes of science, the market, and rationalized organization, not their failures, are turning back to threaten society. Hence the distribution of "bads" becomes as important for society as the distribution of goods. The exceptional human capacity to manipulate nature's processes, combined with our unexceptional incapacity to master nature, have unleashed threatening new dynamics of nature that have hitherto been held in check by nature itself. At first these new risks were legitimated as latent side effects, but they have come out of the closet and are subject to scientific investigation and public criticism. Beck contends that now the "*latency phase of risk threats is coming to an end.* The invisible hazards are becoming visible. Damage to and destruction of nature no longer occur outside our personal experience in the sphere of chemical, physical or biological chains of effects; instead they strike more and more clearly our eyes, ears and noses."[17] Risk threats that previously operated in a latent phase are becoming manifest to our senses.

Beck proposes a solution to these self-inflicted problems, namely, modernization that anticipates environmental problems and possible disasters and intentionally takes measures to avoid them in advance. He refers to this approach as reflective modernization and ecological enlightenment. The important issue is whether risks are scientifically displayed and objectively interpreted or are instead concealed and dismissed. Science is the creator of new, emergent problems in human interaction with nature's constructions, but it provides, as well, the capacity to be the potential solution. It makes available the cognitive resources to diagnose problems as well as the means for overcoming the dangers. Whether those capacities will be used to deal with the threats is, however, not automatic and becomes a source of conflict and a socio-political issue: as the destructions of nature "are universalized by industry, they become social, political, economic and cultural contradictions inherent in the system."[18] Beck himself admits that there is what I would call a "treadmill of risk production" that makes it difficult to foresee risks: "under the surface of risk calculation new kinds of *industrialized, decision-produced incalculabilities and threats* are spreading within the globalization of high-risk industries ... *Along with the growing capacity of technical options grows the incalculability of their consequences.*"[19] This is an extension of Weber's thesis that the intensification of rationality (in particular, through calculations) leads paradoxically to a magnification of irrationalities and incalculabilities. Reflective modernization requires risk calculability, but reflexive modernization spreads new incalculabilities. Therein lies a major contradiction of modernization.

The distinctive feature of industrialization and modernization is that "risks depend on decisions ... society today is *confronted by itself* through its dealings with risks. Risks are the reflection of human actions and omissions."[20] People, businesses, state agencies, and politicians are responsible for risks. Hence human agents can be blamed for environmental problems and disasters. Accepting risk is not compelled by technology and is instead the result of "the system of organized non-liability"[21] that turns the institutions which have the mandate to control the production of hazards into accomplices. Beck gives the legal system as an example of organized irresponsibility. The legal requirement of proof beyond a reasonable doubt and the built-in bias that the accused is innocent until proven guilty place the burden of proof on potential victims. In practice, they translate "difficulty to prove danger" into assumptions of "safety." These requirements and biases of the legal system, understandable though they may be in their own sphere, are very different from the scientific requirement of eliminating bias and of conclusions based on the weight of the evidence.

Beck concludes that societies need to "install *brakes and a steering wheel* into the 'non-steering' of the racing techno-scientific development that is setting explosive powers free."[22] He argues that safety will be enhanced only if consequences are debated before major decisions are taken, if the injured do not have an impossible burden of proof, and if perpetrators are required to prove that their commodities and production are not hazardous. The debate will have to involve "the inclusion of experts and counter-experts, finely balancing a variety of disciplines, so that their systematic errors throw one another into relief."[23] Although Beck's solution of debate between experts and counter-experts has a pleasant ring to it, such debate can in practice throw the population and decision-makers into a state of confusion and indecision. Despite all the risks and obstacles under reflective modernization, he draws an optimistic conclusion: "the enlightenment is beginning anew ... from the industrial stone age of the past to an enlightened, future industrialism of actions where the basic questions of 'progress' are extricated from the anonymity of organized non-responsibility, and new institutions of attribution, responsibility and participation are created."[24]

Socially constructed discourse claiming safety must not be confused with its material referent. If discourse concerning safety is erroneous, "there remains only the social construction of non-toxicity. It does not, admittedly, inhibit the effect, but only its designation ... That might be a momentary consolation, but it is no help against poisoning."[25] Beck adds that hazard forces people to rediscover humans as beings embedded in nature's dynamics, implying that in periods without visible hazards, humans tend to assume they are free-floating socio-cultural entities. A disaster or even a threat can teach humans, through the pressure of necessity, about the contradictions of managing hazards in risk society. He argues that risk awareness is time and again shored up by "the objective counter-force of hazards: it is constant, enduring, not tied to the interpretations that deny it, present even where the demonstrations have long since weakened."[26] Denouncing talk about hazards as alarmist or hysteria does not make the hazards go away. Biophysical events undermine assumptions of safety and mastery of nature. Public relations campaigns and skilled rhetorical claims of safety run up against unintended, unwanted biophysical consequences. These strategies can suppress "the *perception* of risks, but only the perception, not their reality or their effects; risks denied grow especially quickly and well."[27] Material consequences send shock waves through institutions as well: "Disasters, near-disasters and suspected disasters expose to public view, and thus render fragile, the technological backwardness of policy and law."[28] Disaster and the risk of disaster inspire the mass media with the possibility of high circulation levels. Thus they create the basis for reflection and the potential to change. Anthony Giddens also argues that "bads generated by industrialism provide an impetus to change in and of themselves," giving as an example traffic congestion that leads city authorities to create traffic-free city centres.[29]

Usually, consciousness precedes action: we think, then we act. But this order is reversed under modernization's manipulation of nature's processes: action occurs, danger changes the world, and then people become conscious of it; this awareness provokes further social change. Similarly, "manufacture precedes research ... Test-tube babies must be produced, genetically manipulated beings engineered, reactors built, before and in order that their characteristics and safety may be studied."[30] Repercussions are examined in a field setting. The planet has become a real-life testing grounds where few, if any, experimental controls are possible. Beck argues that political action gains influence through the detection and perception of risks. Hence social structural change involves "the creation of awareness of the autonomous revolution of hazard that industrialism has turned into in its phase of technological self-creation."[31]

There are, nevertheless, major problems in perceiving risks because everyday life is culturally blinded: culture promotes the belief in normalcy even where threats lurk. Much of the construction of modernity is faith-based, but not on religion. Faith in progress leads to "the continuous changing of society into the unknown, without a program or a vote. We assume that things will go well, that in the end everything we have brought down upon ourselves can be turned back into progressiveness ... Consent without knowledge of wherefore is the prerequisite ... The productive forces, along with those who develop and administer them, science and business, have taken the place of God and the Church."[32] More than ethics is required to steer the development of applied science. Beck argues that an ethical renewal of the sciences would be no more effective than a bicycle brake on a jetliner because of the power of technological development and its connections with economic interests. He suggests a hypothesis: "as the hazards increase in extent, and the situation is subjectively perceived as hopeless, there is a growing tendency not merely to accept the hazard, but to deny it by every means at one's disposal."[33] The state may go into a mode of defensive aggression in which talking about hazards is discouraged. The hypotheses may be correct in some cases, but in other cases disasters and threats impel insight and action, rather than denial and despair. As described above, Beck himself contends that the experience of biophysical consequences can jolt the socio-cultural into change. Thus his denial hypothesis is contingent and can be modified by becoming aware of the tendency to deny hazards that seriously threaten ways of life.

Many of the dangers created by applied science are knowable only through science. These hazards are "invisible and yet all too present – and they now call for experts as sources of answers to the questions they loudly raise."[34] Science is required to increase awareness of the problems it has itself created and to have the cognitive means to deal with them. "Perhaps there will be a variety of alternative forms of science, of which we have as yet no conception, in the future of scientific-technological civilization, but not an alternative to science."[35] Beck's critique of science leads, not to the rejection of science, but instead to its expansion. Paradoxically, faith in science is a necessary part of the critique of modernization. Thus it is not the proponents of a new Stone Age culture who are warning of perils but often people who are themselves scientists. An anti-science ideology leads to perverse consequences, such as the obfuscation of risks. The complexity of risks under reflexive modernization in the risk society, however, leads Beck to specify a social danger. "The democratic institutions sign their declaration of surrender, and in the splendour of their formal responsibility they delegate power over matters of safety to the technocratic 'alternative government' of corporately organized groups."[36] He has a nuanced, complex assessment of science in the modern world, whereby science is becoming more necessary but less sufficient.[37] Democratic institutions are needed to check and balance science. They have the difficult task of mitigating the risk of becoming subservient to scientific institutions that commit their own errors and have their own vested interests.

Peter Dickens has elaborated a theory about the interaction between modern communities and nature that adds further important elements.[38] The development of modern technology has transformed nature, and this effect turns back and modifies humans psychically: efficient technology has prompted large groups to presume their invulnerability and omnipotence, to become unconcerned with the future and obsessed with the present. Industrial development and massive consumption have led to an egocentric, self-absorbed form of individualism that creates the illusion people are independent by obscuring their deep dependence on other people and on nature's dynamics. They are alienated from material reality. Industrial capitalism has created a personality type of passive selves engaged in pursuing commodities, which has led to serious environmental problems, including global climate change, and is not conducive to sustainable communities. Dickens argues that "it will take the breaking in of reality, in the form of, for example, a substantial transformation of weather systems, for the culture of narcissism to be transformed."[39] Thus he ominously implies that it will take a disaster to break the culture of consumerism based on a presumption of invulnerability.

THE SOCIAL RECONSTRUCTION OF NATURE UNDER MODERNITY

A very different theory has been and still is the basis of most practices in modern society. The theory of what I would call the social reconstruction of nature holds that there is nothing to worry about concerning the depletion of resources, pollution, and other environmental problems: future discoveries and inventions will solve them all. Why? Because human reason is the ultimate resource.[40] Scarcity has been made obsolete: when one resource is depleted, reason can in a timely fashion transform something else into a substitute, so that there will be no problem of depletion of resources. Unlimited economic growth is possible if the infinite substitutability of resources is assumed. Similarly, waste sinks will be invented as needed. Thus the economist Julian Simon argues that "within a century or two, all nations and most of humanity will be at or above today's Western living standards."[41] Rationality leads to cornucopia if one assumes that nature can be mastered, moulded, and recombined at will to satisfy the increasing desires for consumption by a still-growing human population. Extrapolations from the recent past and short-term time series data are given as proof that present trends are robust and will continue into the future. Economics, which previously was called the "dismal science" because it dampened aspirations by calculating the costs of projects, has in environmental matters become the euphoric science, predicting wealth for everyone forever.

This theory has been applied to disaster readiness as well: the wealthier the country, the greater capacity it has to be robust and resilient when confronted by nature's forces. Wealthy nations have the resources to protect their citizens, so they need not be concerned about environmental problems such as sea-level rise from global climate change. Bjorn Lomborg argues that "it seems likely that rich countries (as almost all countries will be by the end of this century) will protect their citizens at such a low price that virtually no one will be exposed to annual sea flooding."[42] He claims that the continued growth of economic wealth has occurred almost automatically because of our fundamental organization in a market economy, and that environmental problems will be solved more by the World Trade Organization than by the International Panel on Climate Change. Lomborg contends that the precautionary principle must be strictly circumscribed to allow the market to create wealth.[43] Progress consists of humans as anti-nature beings freeing themselves from the constraints of nature and of animality and ascending to the sphere of culture and reason.[44] The only danger seen in this theory is a scarcity of rationality if anti-market, anti-technology ideologies gain ascendancy.

Some versions of this theory frame the ethical dimension in terms of dealing with one problem (AIDS) rather than another (global warming), but not both. Other versions dismiss the ethical problem of high greenhouse-gas emitters harming low emitters. Both argue against policies that would limit the growth of consumption by high emitters, and instead place all their hope in the unregulated market producing technological innovations that will make clean energy cheap.

The above theory has been a continuing refrain in wealthy societies, with slight variations on the same theme being advanced by different authors, typically economists, business futurists, and philosophers.[45] Although the broader population does not articulate the theory in an abstract fashion, it too often expresses the same views. For example, the most popular talk-show host in the United States preaches it regularly.[46] Many, and in some societies most, politicians are predisposed to accept this cornucopian theory rather than the opposite message that present consumption in wealthy societies is unsustainable and needs to be reined in. Corporations, states, and the population act as if this theory were the basis for their social practices, even when their discourse appears concerned about environmental problems. Assumptions of nature's malleability and its mastery by science, the market, and rational organization have been and remain the central organizing creed of modernity and postmodernity. The premise is that nature's dynamics have been tamed and harnessed by reason to safely fulfill human desires. The belief that technological inventions will appear in a timely fashion to solve problems threatens, however, to make the population and leaders complacent about such issues and to discount risk. This is why disasters are so shocking: not only is there physical destruction but also, on the cultural level, the hubris of reason mastering nature is challenged.

THE MODERN TREADMILL OF PRODUCTION

The previous theory focuses on the way the modern market and applied science reconstruct nature in new technologies and in the broader environment, and it predicts that those institutions will generate an ever-increasing quantity of wealth. Other theorists, however, view this development as a source of danger. Allan Schnaiberg, Alan Gould, and Adam Weinberg use the metaphor of a "treadmill of production,"[47] which draws attention to the need of the market to produce a profit by continuously creating consumer demand for new commodities: for example. through advertising to enhance lifestyles. It "refers to a type of political economy that comprises a set of practices, assumptions, and structures which are geared toward economic growth, technological innovation, and diffusion and, therefore, continued ecological destruction."[48] They contend that a treadmill of production not only of commodities but also of pollution and emissions is an inherent feature of the modern market which is transforming human interaction with the environment from surplus to eventual scarcity. This treadmill is a complex mechanism that propels further expansion even when confronted by environmental problems such as pollution and resource depletion. Resource shortages are temporarily solved, not by conservation and reducing consumption, but instead by extracting resources from new geographical areas and substituting new materials. This apparent solution only displaces problems to new sites, depletes additional sets of resources, and shifts risk to a broader level: the global scale. Disaster mitigation and preparedness are also limited by requirements of profitability.

When environmental problems threaten, the state does what it can as regulator so as to avoid having to diminish economic growth.[49] There is a dialectical tension between the treadmill of production and demands for environmental protection, a contradiction that governments deal with through environmental managerialism.[50] Gould and colleagues contend that "the support of private capital, labor, and the state for economic growth implied conscious or unconscious support for ecological disruption and environmental degradation. This was true even when those social actors claimed an awareness of and concern for such negative impacts."[51] The conclusion of the treadmill perspective is that environmental problems cannot be solved or modern risks diminished by new technologies because increases in efficiency per commodity produced are more than offset by growth in the production of more commodities. "The driving force is the private sector's economic search for profitability, which has been expanded to a global quest for markets."[52] They refer to this process as an increased tilt or acceleration of the political-economic treadmill. Only by a sustained political mobilization against the treadmill of growth and hence a major structural change of the market, they argue, will environmental problems be solved.

ECOLOGICAL MODERNIZATION

Another theoretical perspective pitches itself between the previous two theoretical poles. The problems generated by technological innovation and the market are not automatically solved by increasing the wealth of nations. Nor is the treadmill of risk production a necessary feature of market forces, solved only by the elimination of the market. The modern world can, according to this theory, deal with environmental problems such as global climate change if and only if ecological goals are explicitly and intentionally integrated with economic goals in the market and in technological development so as to attain sustainable development.[53] Ecological modernization theory is related to the sustainable development theory of the Brundtland Commission.[54] Early versions of ecological modernization emphasized technological innovations, more efficient production, and the market. Later versions drew attention to the crucial role of the state in setting ecological rules (eco-taxes, recycling programs, etc.) for the market contest, to the involvement of environmental movements and NGOs in state decision-making, and to lifestyle changes. Thus Arthur Mol argues that "neo-liberals who would have us believe that we can leave the environment to the economic institutions and actors are wrong."[55] Ecological modernization theorists contend that environmental problems are too urgent to await the revolutionary overthrow of the market, hence constructing ecological reform is "perhaps the only alternative to counteract the devastating environmental consequences of today's economic globalization."[56] The market that has produced resource depletion and environmental degradation is nevertheless one of the most dynamic institutions of the modern world, and it will have to be redirected to solve those problems.

The practice of ecological modernization began in Germany, the Netherlands, and the United Kingdom around 1980, and so did its theorization. Mol argues that by the mid-1980s environmental reform had started to detach economic growth from intensifying ecological disruption, in some cases enabling economic growth despite an absolute decline in resource consumption and pollution. These developments also occurred in the Scandinavian countries, Japan, and the United States. It should be noted that disaster mitigation and prevention and the construction of robustness and resilience are logically part of ecological modernization, even if they are not always mentioned. The fact that disaster fatalities have been dramatically reduced in modern societies supports the disaster modernization component of the ecological modernization hypothesis.

DISCOURSE

Discourse and rhetoric are crucially important because they result in particular practices that are either benign or harmful in human interaction with biophysical dynamics. In particular, what the population and leaders define as safe or as risky determines the actions that will be taken. Discourse is key in shaping practices, even when winning rhetoric leads to disastrous consequences.

Discourse analysis focuses on "claims-making" by complaining groups. The key question is "how are claims presented so as to persuade their audiences?"[57] For example, how are claims of risk of disaster or environmental degradation assembled, presented, and contested? Where does the claim come from, who manages it, what resources do they have, and what interests do they represent? Storylines create meaning and mobilize action. Thus discourse analysis claims to deal not only with how power is expressed in language but also with "the actual creation of structures and fields of action by means of story-lines, positioning, and the selective employment of comprehensive discursive systems."[58] Consent of the population is internalized by framing the debate in a particular way and suppressing opposing framings, which both use and construct power.[59]

This type of analysis constitutes a reminder that science, too, is a socially constructed discourse, one that is not always convincing to non-scientists. New scientific findings can contradict previous ones; hence science can be an "unreliable friend"[60] for the environmentally dogmatic and the risk-adverse. Scientific claims of risk often do not result in remedial action because they are phrased in jargon and hedged in research caution. This practice leads them to be deconstructed in the public sphere and in a court of law, where definitive answers are demanded before polluters are found guilty and made to stop.[61] Outside the scientific sphere, findings have to be packaged and negotiated in a different way in a context of interest groups and lifestyle habitus. The scientific discovery of the depletion of the ozone layer by CFCs had to be sold to the public and decision-makers so that an international agreement could be reached to control them. That Montreal Protocol was an illustration of the successful selling of a scientific assessment of risk that resulted in an international agreement to mitigate risk. However, the Kyoto Protocol, constructed on the model of the Montreal Protocol, has not enjoyed similar success. Science is just one of many competing voices in the public sphere. Discourse analysts investigate rhetorical strategies that determine success or failure; for example, they argue that the risks resulting from ozone-layer depletion documented by scientists did not attract the attention of the media, the public, and decision-makers until it was depicted as a "hole."

One of the most quoted definitions of discourse views it as "a specific ensemble of ideas, concepts and categorizations that is produced, reproduced and transformed in a particular set of practices and through which meaning is given to physical and social realities."[62] What is noteworthy about this definition is that it defines discourse in terms of "ideas, concepts and categorizations." "Practices" constitute the context within which discourse is produced. Hence it is analytically distinct from practices such that the interaction of the two can be studied. A jumbo concept that conflates discourse and practices is thereby avoided. This leaves as an empirical question whether environmental discourse leads to environmental practices or to "greenwashing" that camouflages environmentally harmful practices, and whether rhetoric about safety is used to legitimate risky practices. There is, nevertheless, a weakness in the definition. Although the social context of practices is included, the material context (nature's dynamics) is not. Yet we know that ideas about risk and nature vary enormously between normal and extreme weather, during earth stability in contrast to earthquakes, and so on. The definition is incomplete and should be extended to situate discourse in its biophysical as well as sociocultural context by adding "in a particular set of biophysical dynamics" to the definition. Discourse analysis typically limits itself to the sociocultural, but there are exceptions, such as the conclusion that some claims-making processes are event driven by disasters[63] and by the state of the environment itself.[64]

Discourse analysis and social constructionism have documented how nature's dynamics are mediated by culture. However, they have been criticized for abstracting those dynamics out of the analysis, rather than studying how they are mediated. They have also been criticized for deconstructing the claims of environmentalists and of those who signal risk more than of those who claim safety and promote business as usual. They have tended to focus their research on "complaining groups" rather than contented groups who claim that the status quo is just fine, and the latter groups often have the most power and impact on the social construction of practices that interact with nature's constructions. Discourse analysts have been accused of lapsing into a relativism that is blind to whether environmental problems and risks exist, which they call agnosticism. They have bracketed out of their analysis the important issue of whether there is danger, and they examine only how discourse about it has been constructed. This emphasis focuses the analysis on how winning rhetoric is constructed but neglects the material consequences of whether prudent or reckless rhetoric wins. It also assumes that the presence or absence of danger has no influence on discourse. It is true that discourse analysts are not experts concerning the presence of biophysical risks, but they could take the best available knowledge from other sciences out of brackets and incorporate it into their analysis.

NECESSARY OR SUFFICIENT?

Do reflective modernization theory, ecological modernization theory, and sustainable development theory have referents in human interaction with nature's dynamics, or are they just talk? I want to argue here that there are different versions of those theories that all too often tend to be conflated. A distinction has to be made between what I would call the necessary and the sufficient versions. The necessary version contends that ecological goals have to be added to modernization to make it sustainable.

Modernization in ecological terms is a requirement because without it the planet will be stripped of its resources and degraded by pollution. Similarly, Beck's risk-society argument contends that reflection which anticipates environmental problems and disasters is needed in order to avoid or mitigate them. This first version is very convincing. It is necessary to make modernization ecological and to mitigate risk reflectively by anticipating risks in order to render development sustainable and avoid disaster.

The question then becomes, will this be done? The sufficient version implies that such reflective, ecological modernization and risk reduction will occur because they are needed to avoid environmental calamities and disasters. New technological inventions and superior forms of organization, markets, and institutions will appear in a timely manner. They constitute punctual magic bullets. Although risk-society theorists, ecological modernization theorists, and sustainable development theorists might not like the company, the sufficient version of their theories has logical similarities with theories of the social reconstruction of nature, which hypothesize that wealth suffices to solve environmental problems and mitigate disasters. The latter do admittedly go further by giving priority to economic growth, even at the expense of the environment, which can be restored later after wealth provides more finances. Even though ecological modernization theorists, risk-society theorists, and sustainable development theorists would not agree with that conclusion, it is an empirical question whether the sufficient version of their theories is being used to legitimate economic growth at the expense of a degraded environment.

These sufficient theories are based on nothing more than a primitive functionalist assumption: functional needs result in institutions and practices by which the needs are met. All these sufficient theories are implausible because needs are not always satisfied. Specific actions were needed to avoid a nuclear reactor explosion at Chernobyl, to protect New Orleans from hurricanes, and so on, but the actions were not taken and the needs were not satisfied. So the nuclear reactor blew up, New Orleans was flooded, and other disasters occurred. The depletion of fisheries in many areas of the world was not taken seriously in time, and the fisheries collapsed. Complete societies have collapsed when they chose activities that were not appropriate for the ecosystems upon which they were constructed.[65] Practices do not always change when confronted by indications of risk. The treadmill-of-production theory implies that ecological modernization will fail because its good intentions and environmental discourse run up against market forces developing new technologies that will prove dangerous, as has occurred in the past.

Economists who take environmental problems seriously[66] argue that problems such as global warming occur because harm caused by the production and use of commodities such as pollution and greenhouse-gas emissions are treated as externalities not included in their price. Producers escape responsibility for commodities once they sell them. The market contest is structured in a way that renders the production of pollution profitable and polluting commodities cheap. Consumers have no financial incentive to purchase environmentally benign products, and producers have an incentive to keep polluting and emitting because doing so is easier and more profitable than having to develop new, environmentally friendly technology and products. These economists refer to this perverse incentive structure as "market failure."[67] It consists of a failure to design commodities and technologies with the goal of avoiding environmental problems and disasters. These analyses go beyond simplistic notions that the market always succeeds or that it always fails. Instead, they focus on specific incentive structures in the market that lead it to either succeed or fail in terms of specific goals, such as mitigating environmental problems like global warming. I would argue that market failure constitutes, more broadly, a failure of modernization – namely, a failure to rationalize ecologically and to reflectively mitigate risk.

Furthermore, whether environmental problems and risks are even acknowledged is problematic, not automatic. The risks posed by fossil fuels, CFCs, and acid rain were at first denied by the producing industries and their allies. Danger was only acknowledged after much scientific documentation and political struggle. Hence it is an open question whether reflective anticipation of calamities will occur and social practices change to avoid them; that is, whether mitigation of problems or even adaptation will take place. Reflective, ecological modernization may be needed, but needs are not always met. Modernization in an ecological sense and reflection that produces safe and sustainable practices are a goal, not a given. Reflective, ecological modernization and the treadmill of production should be seen as two alternatives that can be constructed, with tendencies toward one or the other being a variable to explain. At times Beck admits as much. Risks force a transformation of society "either through the normality conspiracy of the side-effects, no longer unseen, of human genetic, nuclear, chemical and ecological hazards; or through an active policy of manufacturing attribution and liability in the system of organized non-liability."[68] In other words, society can change in one of two ways. It can deny the risks, practise normal business as usual, and suffer the consequences. Or it can acknowledge the danger and construct the new social practices necessary to mitigate the risks. Modern society's manipulation of nature's dynamics changes society. The question is, will it be with society's consent or without it through the unintended, unwanted consequences of that manipulation?

Not all environmental reforms are merely window dressing or public relations, but it is nevertheless imprudent to conclude that "ecological interests and criteria are slowly but steadily catching up with economic criteria" and that "we are moving beyond the era of a global treadmill of production that only further degrades the environment."[69] Such declarations are hasty because Mol himself admits that, globally, the overall trend is toward the opposite of ecological rationalization and dematerialization. Counterbalancing pessimism is important, but so is avoiding naive optimism. Whether ecological modernization will be sufficient to deal with environmental deterioration caused by modernization is best left an open question. "Serious global environmental governance implies that the central rules and institutions that govern human activities worldwide are redesigned from an ecological rational point of view."[70] If this is what ecological modernization means, then the road ahead will be long indeed before it is achieved. Ecological rationalization and disaster avoidance are not predestined just because they are necessary. Social constructions may or may not be made when they are needed in their interaction with nature's constructions. The construction could be risk acknowledgment or risk denial, prudence or recklessness.

Theories of stages, such as reflective modernization developing automatically out of primary modernization and ecological modernization evolving straightforwardly from economic modernization, with risks foreseen and action taken to avoid them, are characterized by an inherent naïveté. Such naïveté must be tempered by the findings of market failure and by the theory of the treadmill of production. Reflective, enlightened mitigation of risk and ecological modernization are variables rather than predetermined higher stages of development. Ecological rationalization could occur. Polluters could be made to pay for the consequences of their pollution. They could be given the total responsibility for a commodity (as has been begun in Europe with Extended Producer Responsibility regulations). A new design paradigm could replace the linear model (from resources extracted to waste dumped) with a cyclical model whereby commodities after use would become a treasured supply of resources.[71] To achieve this objective, ecological goals would have to be designed into commodities, products, and institutions, and market incentives would need to be restructured to promote environmentally friendly technologies and products. These changes would constitute a true ecological rationalization in practice rather than only ecological modernization in discourse. They would improve the chances of rendering development sustainable by resulting in an absolute decrease in pollution, in greenhouse-gas emissions, and in the depletion of resources. But all this does not necessarily happen. The problem is not that reflective, ecological modernization has been tried and found wanting; rather, the dilemma is that most societies have refused to try it because of lifestyle habitus and consumption aspirations. The significant question is, will risk be perceived and acknowledged,[72] expectations about nature's dynamics improved, foresight developed, and mistakes diminished? We turn now to the study of cases where that did not happen.

DISASTER

Charles Perrow has shown that the incongruence between socially constructed expectations of safety and nature's autonomous constructions has led to high-technology accidents: "we acted in terms of our own designs of a world that we expected to exist – but the world was different."[73] Diane Vaughan described her study of the Challenger Space Shuttle disaster as "the sociology of mistake" in arguably the world's foremost institution of scientific knowledge, NASA.[74] Disaster research has found that there has often been a "failure of foresight" during "the incubation of disasters" which has led to "man-made disasters."[75] Thus researchers argue that disasters occur when there is a divergence between socially constructed expectations about nature's energy and nature's movements resulting from that energy.

Conceptualizations of Disaster

It is important to note that "disaster" and "hazard" are defined by humans in terms of severe adverse consequences on human communities. For example, the extinction of the smallpox virus is not referred to as a disaster in human discourse. An earthquake, volcano, or extreme weather event that threatens land or ocean so remote from humans or so pristine that it has no effect on humans is not viewed as a hazard.[76] There are many perturbations of nature, as ecologists use the expression,[77] but only the fraction of them that threaten human communities constitute hazards.

Severe adverse consequences for human communities can occur in different forms. Some researchers focus on fatalities.[78] If disasters are indicated by huge numbers of fatalities (many tens of thousands), then disasters only occurred in the past or now in developing countries.[79] About 1,300 Americans died as a result of Hurricane Katrina, there were 2,941 fatalities from the terrorist attacks on the World Trade Center, and about 15,000 died as a result of the 2003 heat wave in France. Horrible though these statistics are, they constitute a different order of magnitude from the 220,000 killed by the 2004 tsunami in the Indian Ocean, the 87,000 who died in the 2005 Pakistani earthquake, the 242,000 fatalities in the 1976 earthquake in China, the 131,000 who died in the 1991 cyclone in Bangledesh, and the 300,000 who died there in the 1970 cyclone.[80] In Japan the number of fatalities from natural disasters has significantly decreased since modernization began. Modern technology and organization have dramatically reduced the death toll, making modern communities largely disaster-free according to the indicator of many tens of thousands of fatalities. That pattern may not remain true, however, if global environmental change occurs or if an epidemic escapes control. Modern mitigation of fatalities is impressive but tenuous.

If disasters are defined by social disruption and destruction of property, however, modern communities are severely afflicted by disasters – indeed, by ever-more-costly and highly disruptive ones. Hurricane Andrew caused enormous disruption and damage in Miami and bankrupted several insurance companies. Hurricane Katrina triggered even more serious consequences. The escalating costs of disasters are causing serious concern among global reinsurance companies, governments, and the United Nations. Modern technological and organizational methods used to avoid the acute problem of fatalities have also created chronic, painfully expensive problems of disaster preparedness and the monitoring of nature's dynamics, yet the costs of disasters are increasing.[81]

The most basic distinction concerning disasters has to do with their source, such as differentiating between natural disasters, technological disasters, and those having terrorism at their origin, all three affecting the local sustainability of communities. This distinction has explanatory usefulness: some researchers argue that natural disasters promote community solidarity, whereas technological disasters and terrorist attacks arouse condemnation and conflict.[82] However, these tendencies should not be exaggerated, and the opposite propensities occur as well. Natural disasters frequently result in governments being blamed for an inadequate response or preparedness. The consequences of Hurricane Katrina on New Orleans reminded everyone that the impact of a natural disaster depends not only on the force of a hurricane but also on the effectiveness of preparation and response. Furthermore, Barry Turner and Nick Pidgeon have shown that what is commonly referred to as a technological disaster often consists of nature's dynamics, which had been assumed to be technologically harnessed, slipping their leash.[83] Terrorists, too, use nature's dynamics embedded in technology (e.g., jet-fuel-filled airplanes and gravitational forces on skyscrapers) to cause a disaster. Thus there is some explanatory usefulness in the distinctions between natural, technological, and terrorist disasters, but it must not obscure the fact that the interaction of social constructions with nature's constructions is involved in all three and needs to be unpacked case by case.

A division[84] has arisen between "normal accident" theory[85] and high-reliability theory.[86] The former argues that interactive complexity and a tightly coupled system using dangerous materials of nature or in an environment where nature's dynamics are perilous result in the magnification of trivial operator errors into system disasters. Human error is normal, hence so too are disastrous accidents in these systems. High-reliability theorists contend, on the contrary, that interactively complex and tightly coupled systems operating with hazardous materials or in dangerous natural locations have few accidents and are highly reliable if they satisfy certain criteria. The difference between normal-accident theory and high-reliability theory should not, however, be overstated since much of normal-accident theory can be interpreted, not as claiming that high technology accidents are inevitable, but rather as a self-denying prophecy to promote greater reliability by making organizations and technologies less tightly coupled and less complex and by using less dangerous materials from nature.

A further distinction is between sudden disasters and slow-onset ones, such as drought.[87] However, the "new species of trouble" produced by intensified human activities blurs "the line we have been in the habit of drawing between the acute and the chronic."[88] Slow-onset processes of nature inadvertently unleashed by human activities, such as global climate change, have been predicted to bring more frequent extreme weather events, thereby fostering sudden disasters.[89] Evidence is now coming in to support this prediction.[90]

Another important difference is between post-disaster research and pre-disaster vulnerability investigations. This distinction must not, however, be overstated since post-disaster research involves learning lessons from a disaster in order to reduce vulnerability and improve preparations against the threat of future disasters. Post-disaster investigations[91] of the incubation of "man-made disasters" have yielded impressive analyses of their socio-cultural and organizational origins in mistaken expectations of control over nature's energy, failures of foresight, structural secrecy that inhibits communication about risk within organizations, and other factors.

Hazards research has examined disasters of specific origins (hurricanes, tornadoes, earthquakes, volcanoes, wildfires, nuclear reactors, toxic waste spills, etc.) and studied whether communities are robust and resilient to each.[92] This work is useful for technical and organizational defences that are hazard-specific. But the hazards approach has been criticized for being too focused on technology. Moreover, safety devices against one type of calamity can be vulnerable to other types.[93] Resilience and robustness are best examined, not in terms of a single hazard, but rather in a field of different hazards. Disaster preparedness is somewhat generalizable to a set of hazards: the same emergency measures organization can protect against multiple hazards. "By attending to vulnerability, the effects of all potential hazards can be accommodated to some degree."[94] Thus a broader biophysical vulnerability analysis of, for example, disaster-prone places has been developed.

It is important, nevertheless, to avoid over-naturalizing disasters: that is, seeing only nature's disturbance and failing to perceive socially constructed differences in the vulnerability of various social classes in a community. There has been a shift away from the study of technical defences toward the investigation of everyday vulnerabilities.[95] Joblessness, landlessness, homelessness, food insecurity, social disintegration, marginalization, and morbidity are ongoing sources of vulnerability to whatever force comes along in both developing countries and sectors of developed ones.[96] Eric Klinenberg's social autopsy of the 1995 heat wave in Chicago concluded that "extreme exogenous forces such as the heat will prove deadly again so long as extreme forms of vulnerability, isolation, and deprivation remain typical features of the urban environment."[97] Mike Davis went further by inferring that the state produced vulnerability to extreme weather events in modern Los Angeles[98] and in nineteenth-century Brazil, China, and India.[99] Key questions include asking which groups are most affected and which have most difficulty recovering from disaster.

The distinction between voluntary and involuntary risk is central: it is very different for a group or leaders to choose to take some risks themselves than for them to make decisions that place other people in jeopardy without their consent or leave them without defences and the means to cope.[100] Restoring the post-disaster situation to normality is not a vital goal if normality involves poverty, misery, and vulnerability to disaster. This observation points to the need to investigate the vulnerabilities of everyday life.[101] James Lewis contends that it is the policies and activities which occur without natural hazards in mind that prove to have the most impact on them.[102] K. Hewitt argues that "remote agencies, that only fire up in emergencies, are not to be relied upon."[103]

That the poor are typically most affected by disasters should not mislead us into assuming that wealthier groups always fare better. R. Dynes, E.L. Quarantelli, and D. Wenger found that middle socio-economic levels were more affected by the 1985 Mexico City earthquake than were poorer groups,[104] and J. Lewis documented that fatalities in the Antigua earthquake of 1843 were greater among groups that could afford masonry than among the poor who had to build their homes from timber and shingles.[105] Wildfires in Los Angeles and the French Riviera destroyed high-status homes of wealthy inhabitants who could afford to live in leafy communities. Quarantelli even speculates that "perhaps such future occasions might even impact more on the better off in social systems, contrary to what most research presently indicates."[106] Although that outcome is unlikely, the privileged cannot take their present comforts for granted if human activities unleash new dynamics of nature that have an adverse impact on human communities.

An exclusively social focus has arisen that rejects concern with a destructive agent (a disturbance of nature or technological danger) as cause. This approach has led to a formulation whereby the physical origins of disaster are intentionally not studied: "there is no reference to disaster agents which implies that all disasters are socially caused."[107] Quarantelli contends that the "more we get away from our hang-up of including an agent or the physical environment as part of our conceptual view and focus on the social behavior involved, the better off we will be."[108] In the interpretative current within this approach, the sociologist's task is "to observe how 'ordinary' language itself fashions and shapes human experiences of life's events and circumstances."[109] This results in a paradigm which argues that disaster is the result of the upsetting of the system of meaning.[110] The old Thomas premise – "If men define a situation as a crisis, it will be a crisis in its consequences"[111] – is used in this paradigm to dismiss the importance of the reality of the danger.

However, humans actively construct in a context not only of language but also of nature's autonomous dynamics. These dynamics are what A. Oliver-Smith refers to as exosemiotic, prediscursive processes.[112] Most dynamics of nature do not depend on human perceptions or interpretations yet affect both. Social-vulnerability analysis becomes weak if it neglects biophysical hazards analysis because then it lacks "an explanation of how one gets from very *widespread conditions* such as 'poverty' to very *particular vulnerabilities* that link the political economy to the actual hazards that people face."[113] There are many poor people in the United States, but it was in New Orleans that they died or suffered a stressful evacuation because of its geographical vulnerability to hurricanes. It is the interaction between the social and the biophysical that determines disastrousness, not one or the other taken separately. Vulnerability is a vacuous concept if it is not paired with an answer to the question "vulnerable to what?" "By ignoring hazard potential or by the assumption that development of any kind will make disasters go away, disasters could actually be made to increase."[114]

Neglecting non-social forces of nature and their effects on communities and downplaying the difficulty of solving material problems result in what W. Dombrowsky refers to as "a misleading sociologism. It is not only human interaction itself, or interaction with material culture and its autodynamics, that may generate failures, but also the interaction with nature and its own autodynamic and self-organizing processes."[115] For humans embedded in a material world of nature's own dynamics, the character of nature's precipitant has major consequences for society. Social life is premised on socially constructed expectations about nature's dynamics, whose independent actions at times contradict those expectations, with disastrous consequences, because essential material infrastructures depend on the premises being met. Naturalizing risks and disasters yields a weak analysis of them, but so does the opposite fallacy of sociologizing them: "disasters are totalizing events ... [that] bring about the conjunction of linkages in causal chains of such features as natural forces or agents, the intensification of production, population increase, environmental degradation, diminished adaptability and all their sociocultural constructions."[116] Disasters and environmental calamities are not just nature's constructions, nor are they just social constructions; rather, they consist of an extreme interaction of the two.

When the vulnerability approach is integrated with the hazards approach, analysis focuses on the conditions and structures that make a society vulnerable to both socio-economically and environmentally generated hazards. Modernization is a fallible attempt at adaptation to the environment, the success of which varies among communities and over time. "The same patterns of adaptation, while reasonably effective for some or many in the short run, may equally sow the seeds of future vulnerability and disasters in the long run."[117] Thus modern "development is the prime medium of vulnerability and its reduction."[118]

Disasters and Environmental Problems

Environmental problems and disasters are related. Hurricane Katrina's devastation along the coast of the Gulf of Mexico was so great because wetlands, which play a crucial role in absorbing storm surges from hurricanes, and barrier islands, which protect the coast, had been degraded by decades of development. Fossil-fuel emissions threaten to exacerbate the greenhouse effect, thereby causing sea-level rise and more energy in ocean waters, which in turn have been predicted to lead to more frequent and intense extreme weather and hence disasters in vulnerable places. For their part, disasters cause environmental problems. Hurricane Katrina overwhelmed technological controls, thereby releasing a cesspool of sewage and toxic materials in New Orleans.

Investigations of natural disasters have concluded that the intensified activities of industrialization have exacerbated vulnerability and will increase the frequency and cost of disasters in the twenty-first century.[119] Choices of development inappropriate for nature's dynamics lead to "disasters by design,"[120] "repeat disasters,"[121] and "unnatural disasters."[122]

Those studies documented that disasters are fostered when the state's role has been restricted to post-disaster reimbursement rather than the regulation of development that lies in harm's way. Environmental problems are catalysts of disasters; hence it is necessary to link protection against natural hazards[123] and disaster reduction[124] to sustainable development. As one researcher put it, "sustainable development is about disaster reduction."[125] Conversely, disaster reduction is about sustainable development. Disasters have been referred to as "the monitor of development ... Whether these processes [of development] have been planned or whether they have been fortuitous, whether they have caused or exacerbated vulnerability, or whether they have reduced vulnerability, will be exposed in the manifestation of natural hazards."[126] Disasters have been called "unpaid bills" and an externalized "debt of development"[127] because costly preventative measures were not implemented.

CHAPTER 2

The Internalization of Autonomous Nature into Society

In his examination of late modernity, Britain's best-known social scientist, Anthony Giddens, contended that the invasion of the natural world by abstract systems brings nature to an end as a domain external to human knowledge and involvement.[1] Although the celebrated German social scientist Ulrich Beck was somewhat more nuanced in his study of the "risk society," he too claimed that the "process of interaction with nature has consumed it, abolished it ... it no longer exists."[2] The French philosopher Luc Ferry argued that progress consists of humans as anti-nature beings, freeing themselves from the constraints of nature.[3] The historian of science Donald Worster has documented that "for generations technological development had progressed on the premise of transforming, even replacing, the natural world."[4] Some works in the social sciences theorize the "social construction of nature,"[5] others claim that nature has been abolished,[6] and still others contend that it is best to bracket nature's dynamics out of any analysis in the social sciences.[7] Even authors critical of the effects of human activities have written books about the "death of nature"[8] and the "end of nature" in a "post-natural world."[9] Unfortunately, there is massive conceptual slippage and confusion in these works between discourse concerning nature and the material dynamics referred to by the word "nature"; between interpretations, which are flexible, and material consequences, which are not; and between waning pristine nature as all of it on our planet becomes affected by human activities and the biophysical dynamics of nature, most of which humans do not control and which continue to affect social practices and influence beliefs.

THE UNINTENDED INCORPORATION OF AUTONOMOUS NATURE INTO SOCIETY

Here I will make the opposite argument, namely, that as societies have expanded to eliminate pristine nature and have manipulated and recombined its dynamics to attain human goals, autonomous nature has been internalized into those societies. Hurricanes, volcanoes, earthquakes, and tsunamis, which used to affect pristine wilderness, now have disruptive consequences for human communities constructed where they occur. Icebergs became dangerous when shipping lanes and oil-drilling rigs were put in their path. Previously harmless solar storms now disrupt societies dependent on long electrical transmission lines and satellite communication systems. Nature unleashes its independent, emergent dynamics within societies instead of within virgin wilderness precisely because human activities have expanded and affect all the biosphere.

I have suggested the concept of *primal nature* to avoid reducing nature to only its pristine form and to denote autonomous biophysical dynamics that continue to affect human societies.[10] Extreme weather events, climate dynamics, earthquakes, bacteria, and viruses are but a few examples of the unwanted forces of primal nature affecting societies. Of course, there are also beneficial dynamics of primal nature that provide the necessary infrastructure for human life, such as photosynthesis. The focus of this book will, however, be on its threatening side. Primal nature has until recently been *naturogenic*: most dynamics of nature were not significantly affected by human activities. But now *anthropogenic primal nature* is emerging, namely, processes of nature modified or unleashed by human activities, either inadvertently or recklessly. Ocean level rise and extreme weather events associated with global warming as a result of greenhouse-gas emissions by humans are examples.

Nature's constructions, including those unleashed by human activities, result in either material risk or material safety for humans. These constructions of nature do not necessarily determine perceptions of either risk or safety, which are social constructions that are developed from socio-cultural processes in a material context. *Perceived risk* or *perceived safety* can accurately capture material risk or safety, or it can be at odds with them. Hence there are different categories of conceptions of risk or safety that need to be distinguished, and these have divergent relations with the material world. There is a time dimension whereby perceptions of risk or safety consist of a projection of what will occur in the future. Risk involving nature's dynamics can be estimated before confronting those forces, but it is only known definitively after the fact. Subsequent events inform us whether there was material risk or safety at the earlier period. *Accurately perceived risk* results when there is material risk and it is perceived. The perception of danger from Hurricane Rita in 2005 and subsequent evacuation of the Houston area is an illustration.

If, on the other hand, there is material risk but socio-cultural processes or ignorance of physical dynamics lead to a perception of safety, then there is either *unacknowledged risk* or *unperceived risk*. The danger from Hurricane Katrina in 2005 and from the 2004 tsunami in the Indian Ocean are examples respectively. If the situation is one of material safety and that is what is perceived, then the outcome can be referred to as *accurately perceived safety*. This is the situation that usually prevails. If, instead, there is material safety but socially constructed dynamics lead to an erroneous perception of risk, then that situation could be called a *social scare* or *moral panic*. The prediction of a great earthquake off the coast of Peru in 1981 falls into this category.[11] A strong argument could be made that the millennium bug was also an unfounded social scare. These are ideal types, and in a concrete situation there is much shading from one to another. The categories are valuable in that they draw attention to the fact that material risk and perceptions of risk or safety are irreducible one to the other and that it is important to investigate the variable relations between constructions of nature and socially constructed perceptions.

Unperceived risk and unacknowledged risk are particularly dangerous because no action is taken to mitigate or prepare for the danger.[12] The dynamics of nature eventually produce occurrences or symptoms visible to sensory humans: the extreme weather event does or does not occur, the Challenger Space Shuttle is launched safely or explodes after being launched, tumours do or do not appear as a result of exposure to radiation, codfish remain abundant off the Grand Banks of Newfoundland or disappear, pregnant women who took thalidomide give birth to infants who are normal or visibly deformed, and so on. Visibility often occurs after a time lag or after a disaster, when even previously unperceived or unacknowledged risk becomes transformed into perceived risk. The material world is the testing ground where socially constructed expectations about nature's dynamics and about safety or risk prove to be appropriate or inappropriate.

Being risk-adverse is often portrayed in discourse as bad – unwilling to take the chance of changing and set in one's ways. However, whether being risk-adverse is being timid or prudent, and whether a risk-taker is wise or reckless, is determined by the particular set of hazards and vulnerability to them. American municipal, state, and federal authorities were not sufficiently risk-adverse when they rejected scientific advice on economic grounds and refused to make the dikes in New Orleans more robust. The consequences were devastating for the citizens of that city when Hurricane Katrina struck.

The assumption that greater knowledge of nature's dynamics results in increased control over nature is a dubious oversimplification. Some kinds of knowledge have not produced control over nature: for example, astronomy and meteorology. Even the premise that an increase in knowledge reduces ignorance may be unwarranted. It assumes that nature is a finite, closed system with no emergent properties. But nature is almost certainly an infinite, open system with emerging processes. Hence the counterintuitive idea that an increase in knowledge about nature does not necessarily reduce ignorance of nature must be envisaged. Knowledge of nature cannot be assumed to be mutually exclusive of ignorance of nature. Increases in scientific knowledge typically reveal new puzzles and show that the unknown is more vast than expected. Thus German social scientists argue that additional scientific knowledge produces additional "non-knowledge":[13] we now know that we do not know things about nature which we previously never imagined, as is the case with global climate change. Science consists of *partial knowledge*, which results not only in technological advance but also in unknown risks.

To remind us that technology is more than a human construction and that it consists of materials and dynamics of nature rearranged in ways not found in pristine nature in order to accomplish goals, I have referred to it as *recombinant nature*. This conception draws attention to the fact that the dynamics and materials of nature entrenched within technologies – whether these be hydroelectric dams, nuclear reactors, electrical transmission lines, antibiotics, or genetically modified organisms – retain their potential to burst through humanly constructed controls. Nature, though harnessed, sometimes slips its leash, with disastrous consequences.[14] Humans can stretch the elastic relationship with nature through the technological manipulation of its dynamics, but there is recoil whereby nature can strike back, and there are breaking points whereby nature can tip into a new steady state less advantageous to humans. The conception of technology as recombinant nature underscores the need for its prudent management and for wise leadership. The recombinant nature that will be examined in this study is the electrical grid, the naturogenic primal nature consists of extreme weather, and the anthropogenic primal nature involves global warming.

New forces of nature are being incorporated into societies as pristine nature is replaced by socially occupied primal nature and as recombinant nature is socially constructed. A particularly important example consists of global warming resulting from human activities, especially fossil-fuel emissions. By internalizing these dynamics of nature in an ever-expanding new frontier, modern societies have brought upon themselves new risks of recombinant nature, upon which they have become dependent, upsetting primal nature, upon which they depend as well. The assumption that humans have freed themselves from the constraints of nature is dubious, and acting as anti-nature beings only exacerbates the recoil effects and the biting back by nature.

NATURE SOCIETIZED AND SOCIETY NATURIZED

The above argument leads to the conclusion that, in the modern world of advanced technology, rationalized administration, and market competition, nature is being societized and society is being naturized. I am using these two admittedly awkward neologisms because I am reluctant to employ the expressions "nature socialized" and "society naturalized" since the terms "socialized" and "naturalized" have become burdened with so much ideological baggage. By *nature societized*, I am referring to the elimination of pristine nature on our planet as human activities affect all of the biosphere and to the manipulation and recombination of so many of nature's processes by science and technology. Even the biological evolution of non-human species is now based on selection by humans as well as on natural selection.

Whereas the hypothesis of nature societized is widely accepted in the social sciences, the idea of *society naturized* has not yet been recognized. By saying that society is being naturized, I am referring to the new, emergent dynamics of nature that have been internalized into society when communities and human constructions are built in locations exposed to perilous forces of nature, when technological recombinations of nature's hazardous processes render society dependent on their complete mastery for safety (as in nuclear reactors) or vulnerable to broader forces of nature (as when dependence on long electrical transmission lines enabled sun storms to black out electrical energy in Quebec), and when inadvertent or reckless human activities unleash new forces of nature into society such as global climate change. Society naturized is the consequence of "reflexive modernization"[15] in that the very successes of science, the market, and rationalized organization in manipulating nature's dynamics, not their failures, turn back to menace society. Far from modernity having ended nature, the anthropocene epoch of rationalized modernity is characterized by a uniquely intense interaction between human social constructions and nature's constructions. Reflection concerning which human constructions are appropriate for nature's constructions that modernity has internalized into society and which are not has become important for survival in this new modern frontier. Leadership is particularly crucial. Whether this need will be met is problematic.

TOWARD AN EPISTEMOLOGICAL EXPANSION OF THE SOCIAL SCIENCES

Many social scientists have argued that institutions in modern society, such as the market economy, medicine, and the educational system, have been based on a "great divide" between nature and society.[16] Social scientists cannot, however, presume that their own practices and institutions have somehow been left unaffected by this construction of modern society. The social and the natural sciences have themselves been based on this divide during the period of "primary modernization,"[17] with disciplines on each side of the divide bracketing the findings of the other out of their own studies. Social scientists who abstract biophysical dynamics out of their analyses claim they are not denying the existence of the natural world – which they could hardly do without undermining their credibility – but their publications even about risk and environmental issues are written as if the socio-cultural existed in a biophysical vacuum.[18] The great divide in modern culture between the biophysical and the socio-cultural has relegated the social sciences to taking into account only the socio-cultural as if humans were exempt from being affected by nature's dynamics.[19] This divide has now become particularly restrictive in a period when nature is being societized and society is being naturized.

Although bracketing out the dynamics of nature from social science analysis can be done in the abstract, nature's forces cannot concretely be suspended and should be taken into account in research even in the social sciences. Social science is weakened when it restricts itself to examining only discourse, such as claims of risk. It is strengthened when it studies discourse in its material context of nature's dynamics. Rhetorical struggles are important, but behind them lie the physical referents of words and material consequences such that the plausible discourse which wins the rhetorical contest can lead to the loss of the struggle with nature's forces. Neither social science nor modern society can dodge the difficult, chronic problem of the truth or falseness of risk claims, which will eventually be revealed in material consequences. Only a very incomplete analysis of someone crying wolf and of listeners heeding or ignoring the warning would fail to take into account the best available evidence on whether or not a wolf is present.

Social processes and nature's dynamics continually cross the great divide, and hence it is counterproductive for the social sciences to remain solely on one side. But how can they go beyond the restriction of studying only the socio-cultural? How can they analytically traverse the divide with the biophysical? They can be advantageously expanded by bringing material consequences to the centre of social science analysis, namely, by investigating how material consequences are socio-culturally *and* biophysically generated, how those consequences in turn affect cultural expectations and social practices, and how the failure of the prophecy of safety incites or fails to incite a change of practices. The analytical starting point of this expansion is a reconceptualization of what is meant by "action" in the social sciences so that it can capture the interaction between social constructions and nature's constructions.

ACTION REVISITED

Theory in the social sciences and, in particular, sociology tends to be heavily focused on authors, especially classical authors. There are two pillars in classical sociology that underpin the culture/nature divide and keep sociology solely on the cultural side. One is Émile Durkheim's famous counsel that social facts need to be explained by other social facts. That dictum was useful in sociology's early days to establish the new discipline and prevent it from being reduced to psychology, biology, or other branches of learning. A century later, however, when sociology is well-established and modern society is confronted with environmental problems and disasters related paradoxically to its technological, organizational, and demographic successes in its interaction with nature's dynamics, Durkheim's advice has become unduly restrictive. The challenge in the contemporary world is to expand sociology and the social sciences so that they can contribute to the study of that interaction.

The other pillar is Weber's conception of action.[20] "We shall speak of *Gemeinschaftshandeln* (social action) when human action is *meaningfully* related to the behavior of other persons ... Social action is not the only type that is pertinent for causal explanation. However, it is the primary object of interpretive sociology."[21] Note that here Weber accepts that types of action other than social are pertinent for causal explanation and that those types of action are secondary objects even of interpretive sociology. Nevertheless, according to this definition, "social action" and with it "interpretive sociology" are largely restricted to the study of meanings and expectations concerning the behaviour of other persons. An important "component of social action is its meaningful orientation to the *expectation* that others will act in a certain way, and to the presumable chances of success for one's own action resulting therefrom."[22] But even

Weber admits that there is more to the action of humans than that involving expectations about other persons: "Not every kind of action, even of overt action, is 'social' in the sense of the present discussion. Overt action is non-social if it is oriented solely to the behavior of inanimate objects."[23] The term "inanimate" is poorly chosen by Weber and could be misinterpreted as implying that objects are incapable of dynamics. It would be better to follow Bruno Latour and replace "inanimate objects" by "non-humans": action is non-social if it is oriented solely to the behaviour of non-humans. Non-social action is oriented to the expectation that non-humans will act in a certain way and to the *presumable* chances of success for one's own action resulting therefrom. I would argue that crucially important actions of humans are oriented to non-humans in the form of nature's dynamics. This is true for both the population and leaders. Human actions are predicated upon their expectations about how nature's dynamics will act. For example, practices concerning disaster preparation and mitigation and concerning climate change are based on expectations about how nature's dynamics will act and the risk that entails. Even omission of mitigation practices based on the subjective expectation of safety in this field of biophysical forces constitutes a type of human action: "We shall speak of 'action' insofar as the acting individual attaches a subjective meaning to his behavior – be it overt or covert, omission or acquiescence."[24] Hence humans act when oriented to nature's dynamics. This action is by definition (Weber's) not social, but that does not mean "not important." Humans attribute meaning to nature's movements and constructions and thereby orient their action. Of course, these are ideal types. Non-social action oriented to nature's dynamics is mixed with social action oriented to other persons during normal weather and benign constructions of nature, during preparations to mitigate disaster and avoid environmental problems, and during a disaster.

Weber's definition of "non-social action" in terms of orientation has, however, some absurd implications if that is the only type of non-social action taken into account by sociologists. The 2004 tsunami in the Indian Ocean – where the risk was unperceived, where no one was oriented to it as it developed, and hence it caught most humans by surprise – would not count as "action" even though it had devastating consequences for societies and their members. The tsunami of 2006, on the other hand, would count as "action" because humans were oriented to these forces of nature precisely because they had been struck two years earlier by a "non-action" catastrophic tsunami. This affirmation, which I heard advanced by a sociologist, makes no sense whatsoever. The absurdity is eliminated by realizing that Weber was referring only to action by purposeful human actors and agents, whether social or non-social. But that is only a subset of possible action. There is also non-social action by non-humans, which is highly consequential for human societies. This type of action is not oriented, not purposive, but it is still action, as the 2004 tsunami attests. Lack of purpose in nature's activity should not be equated with lack of action. Restricting the field of a definition does not restrict the field of action of nature's dynamics. Nature is active in a different sense, and it creates new, emergent phenomena. Non-social action of non-humans is not purposively oriented, but it does nevertheless constitute a type of action and movement that is particularly significant for humans and societies. These non-human actions and reactions have been theorized in scientific propositions, such as the laws of motion in physics. The human species itself has been constructed by the creative actions of nature's dynamics. Disaster research teaches us that this type of action by nature's actants is most dangerous precisely when it is not perceived or not acknowledged by human agents, that is, when humans are not oriented to it. Hence they remain unprepared for it, as in the case of the 2004 tsunami. I am following Latour in using the term "actants" to denote non-humans – namely, nature's dynamics – that engage in non-purposive actions and reserving the terms "actors" and "agents" for humans who take part in purposive action. The theoretical challenge now is to find a way to conceptualize the interaction of the *non-social* and *social actions* of humans with the *non-social actions* of nature.

THE DANCE OF HUMAN AGENTS WITH NATURE'S ACTANTS

Erving Goffman suggested a dramaturgical metaphor to help understand everyday interaction between humans.[25] Similarly, I propose the metaphor of a dance to comprehend the interaction between movements of the material world of nature and those of social practices while capturing the specificity of each, both affecting the other. Nature is our dance partner. Human agents dance with the moves of nature's actants to form hybrid constructions, with both agents and actants influencing the other and both having some autonomy.[26] The metaphor of a dance focuses on the joint movements of purposive human actions and non-oriented actions of nature. It transcends the limitations of a one-sided focus solely on nature's determinisms or only on human social constructions. Human agents and nature's actants are performers in movements influenced by the other's creative actions. The metaphor captures the autonomous movements of nature's dynamics without implying intentionality by the non-human partner, only movement. The autonomy is that of two dancers, entangled and affected by the other's movements yet independently making moves that may be either in harmony or out of step. Pushing and shoving by one partner or the other is not unknown. In some cases, nature's actants take the lead and human agents react and improvise after nature's moves in this dance. Disasters are examples of these cases. In other cases, humans take the lead by planning a choreography in anticipation of nature's moves and proacting. Disaster preparation and the Montreal Protocol to mitigate the depletion of the ozone layer are examples. Technological innovation is a stronger illustration of humans taking the lead by manipulating nature's actants, unless human agents mistakenly innovate maladroitly and are overwhelmed by nature's actants in a technological disaster.[27]

Sustainable social constructions in a world of nature's constructions require social practices in harmony with nature's dynamics. If human expectations about nature's dynamics are faulty, then human constructions can be tripped up by being out of step with those of nature, leading to disastrous results. These "failures of foresight"[28] and "mistakes"[29] in the incubation period of disaster can be documented retrospectively after a disaster or an environmental calamity. Such tragedies have typically resulted from erroneous expectations about nature's autonomous dynamics or their timing. In the dance of human agents with the moves of non-human actants of nature, the specific beliefs and actions of humans are social constructions shaped by cultural predispositions, economic interests, the power of decision-makers, and prior material experiences. They result in expectations about nature's dynamics that are particularly important in influencing the steps taken by humans. Assumptions about nature's actants are real in their consequences because they lead to social practices which, if correct and acted upon, result in safety but, if erroneous or if risk is unacknowledged, result in harm and, in the worst case, disaster. Whether moves by humans and nature are in harmony or out of step determines the material consequences. The dance can be adroitly or ineptly performed.

In this dance, there are different levels and units of human agents as dancers: individuals, leaders, communities, organizations, and states. Similarly, there are different levels and units of nature's actants as dancers with their own movements: individual organisms, species, weather systems, tectonic plates, the sun, and so on. The metaphor of dance underscores the significance of appropriate movements by the human dancers in order to achieve sustainable social constructions, as well as the adverse consequences of practices incompatible with nature's processes. It highlights the importance of accurate expectations by the human dancers and the adverse consequences of erroneous ones. It points to the significance of studying why learning to move in harmony with nature's dynamics does or does not occur. An analytic approach analyzing movements by human agents and nature's actants, like partners in a dance influenced by the other's creative movements, can bridge the nature/culture divide and transcend the limitations of a one-sided focus solely on nature's determinisms or on human social constructions that abstract the other out of the analysis. The social and non-social actions of humans must be understood in their social, cultural, *and* biophysical context, that is, as movements of human agents influenced by the movements of non-human actants in the dance they perform together.

Disasters are particularly propitious events for investigating the dance because extreme movements of nature that result in disasters make visible relations between social constructions and constructions of nature that are often ignored under normal dynamics of nature, much as we ignore breathing until we no longer do it well. It is precisely disaster research that can teach environmental research about errors of expectations concerning nature's dynamics, about the material consequences of such errors, and about social barriers to learning from the prompts of nature.

MATERIAL CONSEQUENCES: THE TESTING GROUND FOR EXPECTATIONS AND DISCOURSE

Socially constructed beliefs matter not only because they lead to social practices but also because those practices then interact with the dynamics of nature in ways that have material consequences. Social life is premised on expectations of safety concerning nature's dynamics. Such expectations are, however, trial balloons that can be punctured by nature's dynamics. When the terrorist-flown planes struck the World Trade Center, firefighters followed their prepared script and rushed up the 110 floors of the two towers to rescue people. They assumed the risk was minor and that they would succeed in bringing victims to safety. This action would have been appropriate had the fire had been less intense and not melted the skyscraper support beams. The degree of risk was unperceived; hence the false belief in acceptable risk led the firefighters to climb the stairs of the towers. An expectation of great danger would have been appropriate for the dynamics of jet fuel incinerating the skyscraper's supporting infrastructure and would have led to caution and to the safety of the firefighters. Instead, the expectation of acceptable risk led to their deaths. The belief that won the socio-cultural clash of contested ideas about safety (or risk) lost the biophysical contest. It was not just beliefs that determined safety or danger. Rather, the outcome was determined by the lack of correspondence of beliefs and practices with the dynamics of burning a planeload of jet fuel 90 floors up a skyscraper.

The fate of the 9/11 firefighters illustrates a general principle of wide applicability for numerous risks underlying disaster preparedness and environmental issues: namely, that appropriateness with biophysical processes of practices by the lay population, experts, and leaders is necessary for ensuring safety and avoiding danger. To many readers, this observation might seem obvious, but its obviousness is not a guarantee that prevents the principle from being breached. The interaction between nature's dynamics and expectations about them is the testing ground for the truth or error of assumptions about safety or risk, with the consequences of truth being enjoyed and the repercussions of error suffered. To the chagrin of relativists, there is a difference between truth and error which can be seen in the dissimilarity between the material consequences of true beliefs and false ones. If a discourse inappropriate for nature's constructions wins the rhetorical contest, then it can result in practices that produce a net loss for society on the material level, as attested in dramatic fashion by societies that have collapsed.[30]

HOW CAN NATURE INFLUENCE EXPECTATIONS AND BELIEFS?

Nature's dynamics have the capacity to construct and destroy, and hence are actants but not purposive agents. They have material consequences for humans, and these are prompts that can encourage the social construction of interpretations, conceptions, and practices by purposive, embodied human agents. *Prompt* is a particularly useful concept because it captures the influence of nature's constructions on human beliefs and practices without implying that nature determines them. Seeing the cliff's edge, humans can decide to jump into the abyss or to move carelessly; but typically, the sight of the precipice prompts them to take precautionary action. The fall of one person usually influences the next into more prudent measures. This is an ideal-type illustration. In complex situations, dangers are more or less evident, and there are incentives to business as usual from the economy and lifestyle habitus. Distinctions among different types of prompts can help us to understand the interaction between biophysical dynamics and beliefs about environmental problems, in particular, the risk of disaster and global warming. Beliefs are influenced by prompts from nature's actants in three fundamental forms.[31]

Scientific prompts result from ingenuous experiments and measurements by scientists that enable nature to object to what had been said about it, thereby leading to deeper scientific theories, new technologies, and recomposition of society at a different level.[32] These prompts consist of findings that arouse an awareness of nature's dynamics before they are otherwise visible: for example, forecasts of hurricanes and tsunamis. Imaginative scientific measurements made visible ozone-layer depletion and its cause, the use of CFCs, which gave rise to the social construction of the Montreal Protocol. That successful international agreement between nations to diminish ozone depletion was a remarkable achievement of scientific and political leadership to solve an environmental problem that was a side effect of the successful scientific development of CFCs. A potentially disastrous consequence of recombining nature's molecules was avoided by an innovative recombination of scientific and political forces.

There are two other types of prompts ensuing from nature's dynamics that influence everyone and can override scientific prompts. One involves *extreme prompts*, which occur through exceptionally severe disturbances of nature. These produce natural or technological disasters if societies are vulnerable, and their observation and/or experience can lead to a modification of assumptions and practices. It took the destruction of a space shuttle to change the assumptions of scientists and engineers at one of the world's leading technological institutions that launching was safe: "At NASA, the crisis that precipitated a transformation of worldview and resulted in a paradigm shift was not the teleconference, but the Challenger Disaster."[33] Extreme prompts of disaster are so visible they are a critical material contingency in the "closing of controversies, the non-negotiability of facts"[34] and in arousing new controversies. When nature's constructions, such as hurricanes (cyclones), floods, earthquakes, tsunamis, and volcanoes, strike human constructions such as cities, they result in experiences extraordinarily perceptible to the senses, and hence they constitute multi-dimensional challenges to leadership, social structure, and culture. They expose the vulnerabilities and dependencies of even modern societies. Assumptions of safety are falsified. Risk assessments and cost-benefit analyses are laid bare as fallible, at times ill-founded, projections. Biophysical experience of consequences and their visibility lead some interpretations to gain credibility and others to be seen as flawed. This is especially true of beliefs about safety and risk, even science-based ones. Hence robustness constructed into technological infrastructures is typically inspired by previously experienced calamities.

Cultures can be shaken to their roots by extreme prompts. The slow response of Pakistan to the devastation of a cyclone in 1970 fed the nationalism that led East Pakistan to secede and become Bangladesh a year later.[35] When an earthquake, a tsunami, and subsequent fires killed more than 50,000 people and destroyed the city of devoutly Catholic Lisbon, Portugal, on All Saints' Day in 1755, it prompted many to question how an all-powerful God could allow his closest followers to suffer so much. This experience accelerated the already developing secularization in Europe. The type of interpretation that is given to the prompt and gains prominence is, nevertheless, problematic. It is not rigidly determined by the disaster itself or by the pre-existing dominant culture. Nature's prompt constitutes a material eruption in society that can visibly give credence to new interpretations and novel cultural constructions through little more than lucky timing. On a Sunday in San Francisco in 1906 a small Christian movement was told by a flamboyant pastor there would be a sign that God was annoyed with the licentiousness of the city. An earthquake destroyed San Francisco the following Wednesday. Thousands showed up for his next service, and the Pentecostal movement in the United States was born.

The third type of prompt can be missed by social scientists because it is so ubiquitous for humans embedded in nature's processes, much as we fail to notice breathing. These *everyday prompts* consist of mundane observations and experiences of nature's routine dynamics by humans. Sensory human agents attribute characteristics to nature's actants, and the attribution tends to be influenced by the biophysical properties observed and experienced when these are easily visible. For example, humans tend to depict fish as having fins and swimming, rather than having wings and flying, even though humans could theoretically construct the representation that fish have wings and fly.[36]

In some cases the three types of prompts converge. The Netherlands has the highest disaster and environmental consciousness of any wealthy country because of its everyday experience of the potential for disaster, in that most of its citizens live below sea level, because of scientific evidence of risk, and because both rich and poor experienced disaster when a 1953 storm surge drowned 2,000 people.[37] This experience led Holland to invest $5 billion in a water security system, including a hydraulic seawall. There was a national consensus, rather than just a regional one, to make such an investment because most of the landmass of the Netherlands lies below sea level. The United States did not learn from the Dutch experience, in part because only a small proportion of Americans live below sea level. The director of science and technology at the Netherlands Delft Hydraulics concluded: "There is a difference in risk perceptions between our country and the United States. There is a continued threat from the sea in this country that makes people aware and willing to invest in flood defenses. That kind of awareness was obviously less apparent in New Orleans."[38] The relationship between nature and consciousness is not, nevertheless, deterministic. The Dutch could theoretically have dismissed these prompts by, for example, interpreting the 1953 storm surge as a thousand-year event requiring no action. But they did not; rather, they let nature's prompts inspire prudence and the construction of robust dikes. In Canada it took the great flood of the city of Winnipeg, on the Red River, in 1950, which put one-third of the city under water and forced the evacuation of 150,000 people, to undertake the construction of a floodway that cost $63 million and is estimated to have saved $8 billion in damage since its construction. A major flood in 1997, which came close to exceeding the capacity of the floodway, prompted a $665 million enhancement to be completed in 2010. New Orleans experienced near misses by hurricanes, including Ivan in 2004, yet an emergency response exercise in 2004 found that defences were not even equipped to protect against a category 3 hurricane, which they were supposedly designed for. This comparison has led one commentator to conclude: "You need a near-catastrophe to get political attention in Canada. In the United States, it looks like you need a total catastrophe to get that attention."[39]

In other cases the three types of prompts diverge. For example, the scientific warning by the International Panel on Climate Change about the dangers of global climate change resulting from human activities is at variance with the everyday prompt of well-being in wealthy societies. This divergence leads to contradictory discourse, a discrepancy between talk and action, and political conflict concerning environmental degradation if the Kyoto Protocol is not implemented versus the claim of economic degradation if it is. The experience of relative well-being over periods of time is typically extrapolated into expectations of safety by lay people and many leaders. Even learning from the material consequences of a disaster is problematic when normality returns, because such learning can imply economic cost, cultural change, and adjustments to lifestyle. So imagine how much more problematic risk acknowledgment is when there has been no disaster clearly attributable to climate change. Prompts can lead to change, but they can also be missed or dismissed. Some societies have ignored prompts indicating danger from nature's dynamics and suffered the consequences, even to the point of collapsing.[40] Hence, enlightened leadership faces a major challenge when there are scientific indications of danger but hitherto no disasters straightforwardly attributable to greenhouse-gas emissions, when there are no easy technical solutions, and when mitigation requires apparent sacrifice of consumption levels.

SENSE-MAKING AND DECISION-MAKING

In the social sciences literature a distinction has been made between decision-making, which refers to solving a well-defined problem, and sense-making, which involves the challenge of figuring out what an ill-defined problem is and defining it.[41] Making sense of the situation is particularly significant in unfamiliar, non-routine contexts where never-before-experienced conditions present themselves. Disasters and novel environmental problems impose challenges for sense-making in that they typically cannot be completely foreseen and planned for; hence they require improvisation and innovation rather than only standardized solutions.

Sense-making, even in hindsight, is heavily influenced by cultural predispositions. Values and beliefs constitute mental frames which, much like a lens, filter out some facts, accept others, and refract still others. These frames are used to make sense of the world, both social and physical. It is reassuring to have a predetermined mental frame to interpret new information, and it is demanding to restructure the frame itself. When new socio-cultural or biophysical contingencies are experienced, it is particularly challenging for leaders and the population to think "outside the box" because the box or frame into which they have been socialized facilitates their interpretations. Thinking outside the box is equivalent to improvising music without a score or finding one's way in a new land without a map. Mental frames can remain rigid, they can be gradually modified on the basis of socio-cultural and/or biophysical contingencies, or they can be suddenly changed when driven by extreme socio-cultural and/or biophysical prompts.

Mistaking non-routine biophysical conditions for routine ones and making sense of the situation in terms of habitual interpretations is typical of the incubation of disaster and environmental problems. It is, to continue the above analogy, like finding one's way in a new land using an erroneous old map. When a fire broke out in the cockpit of Swissair flight 111 shortly after takeoff from New York on 2 September 1998, the pilots followed their training and used a time-consuming checklist of troubleshooting procedures as the fire got worse. The plane crashed into the ocean, and 229 people died. If the pilots had made sense of the situation as one that required a rapid landing, they and their passengers might have survived. Coal miners in Sago, West Virginia, followed their training routines for an underground accident, built an alcove, and waited for help. They were then asphyxiated, even though they could have walked five hundred unobstructed metres to safety. The pilots and the miners were over-socialized into a response that was inappropriate for the specific dynamics of nature they faced, and thus they were rendered incapable of improvising when confronted with a new set of physical prompts.

These conclusions are admittedly based on hindsight. The victims are not being blamed for following their training. Rather, the lesson is to learn retrospectively from their tragedy of the need to improve training procedures by teaching people to expect the unexpected and hence become more sensitive to emerging prompts from nature and more flexible. In non-routine situations, going by the book or business as usual can lead to catastrophic error because the book has not foreseen emerging dynamics of nature. Different ways of making sense of a novel situation are not equal in their biophysical consequences: some definitions of the situation lead to disaster and environmental calamities, whereas others result in sustainable well-being. This is the case because some definitions are appropriate for their biophysical referents, whereas others are not.

Sense-making is itself influenced by biophysical prompts, but at times they are too late. When the first plane crashed into one of the World Trade Center towers in New York on 11 September 2001, experts and lay people alike made sense of this unusual situation by concluding that it was an accident. The crash of the second plane into the other tower a short time later prompted them to modify their sense-making into a conclusion of an act of terrorism. Similarly, when there was an incident on the London underground on 7 July 2005, experts and other passengers initially made sense of it in terms of one more fire in the London tube. But further explosions in the underground and on a bus led them to conclude that it made more sense to infer a terrorist attack.

Sense-making has to correctly interpret the prompts coming from nature's dynamics. In the case of disasters and non-routine environmental problems, a new choreography must be improvised rapidly to complement or even replace what was planned. As biophysical conditions change, leaders and/or the population have the challenge of sensing an escalating non-routine problem and therefore not clinging to the usual solutions. In the case of slow-onset disasters and irreversible environmental problems, an openness to being influenced by the emerging biophysical prompts and contingencies is necessary to modify business-as-usual predispositions in order to avoid mistakes, failures of foresight, and the incubation of disaster.[42]

Thus it is important to integrate into the analysis both the cultural lenses used to see the material world and the dynamic biophysical world of emerging properties that exists to see. Analyzing a situation in its context consists of both socio-cultural and biophysical components: meanings, matter, and especially the relations between them. Attempts to diminish environmental problems raise the difficult challenge shared with the mitigation of future disasters: namely, that of learning before suffering harm in a situation where there are routine prompts of well-being and where scientific prompts of danger are indicative but not certain. The social obstacles to learning in the cases of both disasters and environmental calamities are similar: ingrained cultural values, lifestyle habitus, vested economic interests, organizational turf wars, political partisanship, and so on.

THE DIFFERENCE BETWEEN THE PLAUSIBLE AND THE ACCURATE

The plausibility of ways of making sense of a situation is often assessed by extrapolating present experiences into the future. Actions taken from this way of appraising the available cues are sufficient if the situation is one of continuity, but they can be misleading if it is discontinuous.

The George W. Bush administration in the United States constructed assumptions about hurricane protection and an approach that made sense for that administration at a time when it was waging an expensive war in Iraq. The Federal Emergency Management Agency (FEMA) was subordinated to Homelands Security after 9/11, and a strong supporter of the president was named to direct the agency. The budget for reinforcing the levees in New Orleans was cut on the assumption that it was unlikely that a category 4 or 5 hurricane would strike the city. This decision, however, ran up against a category 4 hurricane that refuted those assumptions, rendered the approach inadequate, and, most importantly, devastated the vulnerable city and people of New Orleans. Sense-making leads to narratives that energize a particular type of action, but the important issue is whether that action results in benign or harmful consequences in its interaction with the dynamics of nature. The catastrophic consequences of the sense-making described above demonstrate that it was insufficient for making New Orleans robust and resilient in its biophysical context of hurricanes; such sense-making was plausible in its cultural, political, and socio-economic context, but it was not accurate in its depiction of the forces of nature.

Many stories are *plausible*, but only some are *accurate*. Whereas accurate narratives lead, plausible but inaccurate stories mislead. There is no escaping the importance of the accuracy or erroneousness of discourse that directs us into one set of practices rather than another. Meanings result in specific practices and activities that interact with nature's biophysical dynamics. The significant question is whether those meanings will lead to apt movements in this dance with nature's movements or to injurious movements.

Examples are numerous of the successful mitigation of environmental problems and disasters when leaders perceived risk accurately and showed foresight. Many industrial nations have faced up to the serious pollution of their lakes and rivers and have taken measures to deal with it. Deforestation in Brazil was on an ascendant curve in the early 1980s; then ecological warnings were taken seriously, and destruction of the country's tropical rain forest has decelerated. Buildings were reinforced in countries such as Japan, where the threat of earthquakes was acknowledged, and fatalities decreased. Nations recognized the threat of CFCs to the ozone layer and signed the Montreal Protocol, which reduced their production, thereby diminishing depletion. The Montreal Protocol remains a model of international cooperation in order to give nature the opportunity to heal itself by curtailing destructive human practices. Many risks have been mitigated precisely because prompts were acted upon and not dismissed, and warnings were heeded rather than denied. The risk was acknowledged as real, and this foresight led to action that was successful in diminishing the threat. Human activities underlying global climate change are admittedly much more complicated and expensive to change, but each of the above cases also began with apparent complexity and costliness, claims that social adjustments were impossible, little foresight, and a lack of leadership.

AN EXTREME WEATHER DISASTER IN MODERN SOCIETIES

The ideas and concepts presented in the previous chapter and in this one provide a general theoretical framework for the following empirical and detailed investigation of the experience of an extreme weather disaster in two modern societies, of its management by political and emergency leaders, and of their post-disaster insights into the broader environmental problem of global climate change. Does modernization produce new emergent risks? Are modern societies on a treadmill of risk production? This book will now examine what happens when modern societies make themselves dependent on the infrastructure of an electrical grid for essential needs such as heat, light, production, and communication, and when that infrastructure is confronted by an extreme disturbance of primal nature's broader forces. Is that infrastructure highly reliable, or is its collapse in that confrontation normal? Can these at-risk societies perceive and anticipate risk in a timely way that mitigates harm? Are modern societies robust or vulnerable when confronted by nature's hazards? Are they resilient once disaster occurs?

This book will investigate how modern societies cope when risk becomes harm, which could be called the actualization of risk. It will examine how the population and, in particular, leaders take decisions in the context of an extreme weather disaster and make sense of the situation. How do moderns respond to risk and its actualization, and how do leaders manage the actualization of risk in an extreme weather disaster and the risk of global warming? Is reflective modernization practised, or has it and ecological modernization been subordinated to economic modernization? This study will analyze what was learned from the disaster in terms of reflective modernization and how learning was accomplished. Do biophysical calamities play an important role in motivating learning by the population and by leaders? Do they provoke new assumptions, values, practices, and social structures, thereby stimulating anticipation, mitigation, and avoidance of calamities or at least preparation for them? In short, the remainder of the book will attempt to show how the great research divide between the socio-cultural and the biophysical can be bridged, and how their interaction can be described and analyzed from a broadened sociological perspective.

During and immediately following this extreme weather event, small variations in precipitation, temperature, wind, and other elements resulted in major changes in the biophysical context within which humans make their social constructions. Part Two will provide a detailed account not only of the interaction between human actors during a disaster but also and especially the interaction between multifarious constructions of nature and those of humans, both leaders and the population. It presents a "thick description"[43] of the dance between nature's movements and those of humans during and following a disaster, thereby elucidating the context, both biophysical and socio-cultural, within which sense-making, discourse, decision-making, and other social constructions emerge. It will study the interaction between the non-social action of nature's actants and the non-social action of human agents oriented to the movements of those actants, as well as the social action directed to the behaviour of other humans triggered by nature's actants. Part Two will investigate how human agents dance with moves of these non-human actants of primal nature to form socially constructed action after each extreme prompt. It will describe the transition from unperceived risk to perceived, acknowledged risk when hazard struck, as well as the many socially constructed moves that emerged out of this dynamic biophysical context. The description in this part provides a basis for learning from the adverse consequences of disastrous weather events to take into account the vulnerabilities of modern recombinant nature when faced with extreme forces of primal nature.

PART TWO

The Dance of Humans with Nature's Movements

CHAPTER 3

Vulnerability to Nature's Hazards

Citizens who live in northeastern North America take for granted that technology and, behind it, science and reason make their lives largely exempt from the rigours of the cold, dark winter environment in which they reside. Modern heating, lighting, transportation, and appliances enable them to go from the climate-controlled micro-environment of their homes in the comfort of their heated automobiles to a well-equipped office and often to enjoy the amenities of a warm shopping centre on the way home. Farmers depend upon mechanical slaves energized by electricity in their barns to feed the urban population freed from the land and manual labour. Contact with nature has for the majority of the population been reduced to leisure activity: skiing in Vermont and in the Laurentian Mountains outside Montreal, visiting the Olympic site at Lake Placid, New York, snowmobiling in Maine, skating on the world's longest rink in Ottawa, and so on.

A political scientist expressed the achievements of modern society in dealing with nature thus: "although it is a cold late winter's day, I am able to watch the wonder of breaking ice floes as they rush down the river from the warmth and comfort of my brick home; I can sip a fine cup of coffee although coffee will not grow within many hundreds of miles of my home; and my wife and I can spend many wonderful moments together, attending plays, going to the cinema, or discussing the books we are reading, because modern technology has released us from many of life's burdens. We can be confident we will not contract smallpox, nor suffer from malnutrition."[1] He then pointed to the clinching argument for modern society: a much higher life expectancy.

In northeastern North America, hailstorms and flooding have been small-scale, earthquakes mild, and hurricanes rare, and tornadoes have left only localized damage. Freezing rain occurs about twelve to fifteen times a year, falls on a small locality, lasts a few hours, and then is followed by warm air that melts the accumulated ice. The electrical transmission lines and towers have been built to withstand projected amounts of ice; for example, the steel towers were made with a special capacity to hold 45 millimetres of ice-loading instead of the usual 15 millimetres. It would take an extremely unlikely knockout blow from nature to topple these lines and pylons. The transmission and distribution of electricity were rationally planned to avoid interruption of service as much as possible.

At the time of the ice storm of 1998, the Montreal region was served by not just one or even several high-voltage transmission lines but more securely by a "ring of power" around it. This ring stepped down high-voltage electricity from distant dams in northern Quebec to substations across the region, and then the web of distribution lines carried low-voltage power to users. If one line was broken, electricity could be redirected around the fallen line. Hydro-Québec assured its clients that the network, so important for meeting their most basic needs, was reliable. Whatever damage occurred would be rapidly taken care of by well-equipped and coordinated repair crews of the electric company and the emergency measures organizations of government. Similarly, in Ontario a web of lines was constructed to bring power into the Ottawa area, so that if one line was broken, electricity could be redirected around the fallen line. The backbone of the web is 500 kilovolts of power, which supplies 40 per cent of the needs of the Ottawa region; it comes from the Pickering nuclear station and the Lennox fossil-fuel generating station. As a special security measure, two separate sets of steel towers holding parallel lines were constructed to transmit the 500 kilovolts of power, which is normally shared between the two. Lightning strikes were also anticipated; so steel cable was attached from the top of one steel tower to the next. This safety device, called a "skywire," does not conduct electricity but carries the charge from a lightning strike through the steel tower into the ground, thereby protecting the high-voltage carrier wires. The utility also developed a research branch, Ontario Hydro Technologies, to foresee, simulate, test, and overcome potential problems before they occur.

In a cold, dark climate the modern, rational mind develops the means to deal with such matters. It is only unprepared southerners who have major problems when snow and ice strike. A large-scale, intense ice storm was, in any case, only hypothetical and highly improbable, as judged by extrapolations and time series from the available data. That the forces of nature could overwhelm modern technical and social constructions was almost unthinkable. Technology, science, and reason have mastered plastic nature and replaced its climate by controlled micro-environments in which people live and work. Unlike animals, who must submit to the forces of nature, the exceptional capacities of humans render us largely exempt from the determinisms of nature. We have constructed an open relationship with nature that enables us to manipulate it to our advantage when needed. Such was the thinking in northeastern North America on the fourth of January 1998.

THE EVE OF DISASTER

The winter of 1997–98 had been one of the mildest in memory. The vast area of northeastern North America had been experiencing unusually warm weather in the first days of January 1998. To save money, the city of Montreal decided against trucking away snow from city streets, hoping the sun would do the work. Streets were therefore full of ice, since the snow would melt in the daytime and then freeze at night. Temperatures had averaged six degrees Fahrenheit warmer than usual in upstate New York.[2]

The ice storm began on different days for various communities along its trajectory. Environment Canada includes the meteorological agency for the federal government. Its offices in both Ottawa and Montreal mentioned freezing precipitation in their afternoon forecasts on Saturday, 3 January, for the next day. At 4 a.m. on Sunday, 4 January, the Montreal office issued its first warning of freezing rain. At 3:38 p.m. the Ottawa office issued a similar warning for eastern Ontario. At that time, spotty freezing rain was reported to the west at Kingston.[3] Meteorologists observed the beginning of what they called the first "episode" of precipitation on Sunday evening in southeastern Ontario and several hours later in southwestern Quebec.[4] The precipitation first occurred in the form of snow in the Ottawa and Montreal areas, but it changed to ice pellets and then to freezing rain.[5] The National Weather Service in the United States also began sending out advisories for freezing precipitation on Sunday.[6] Environment Canada issued a weather warning of freezing rain to begin in Ottawa in the early hours of Monday, 5 January. Moving in from the west, it would hit Montreal and areas east somewhat later.

The amazing capacity of modern meteorology to monitor and explain weather patterns, matched only by its unamazing incapacity to do anything about them, was confirmed. Meteorologists saw on their radar screens a conveyor belt of warm rain from the Gulf of Mexico moving slowly northeast along the west side of the Appalachian Mountains. The depression was centred over Texas. It brought warm temperatures and rain to Kentucky, Indiana, Michigan, and southern Ontario. London, Ontario, and Buffalo, New York, had begun January on an unusually warm note. But then the warm, moist air ran into a mass of cold, dry air, affectionately known among local meteorologists as the "Siberian Express," that had drifted in from northern Quebec and stalled in the valleys formed where the Ottawa River meets the St Lawrence River.[7] The two air masses collided over a vast area centred on these valleys and extending further east, but the warm air was not strong or fast enough to push the cold air mass out. Instead, it slid up and over the cold air mass. The warm, moist southern air produced rain that chilled as it passed through the lower, cold northern air and froze on any frigid surface it struck near ground level, resulting in a buildup of particularly hard, cold ice on all surfaces. People who give names to such things refer to this as "glazed icing."[8] The thermal profile of this area was ideal for the production of freezing rain, whereas to the south (central New York State) and southwest (Toronto), where there was no cold air mass, warm rain fell, and to the north, where there was no warm, moist air (northwest Quebec), it snowed.[9]

Normally, warm air comes up the East Coast of North America over water. But from 3 January on an "egg-beater effect" was created through the interplay of high and low pressure zones that pumped warm, moist air northward from the Atlantic and Carribean and eastward. Rain changed to freezing rain when it collided with the wedge of cold air from the north that had crept southward. These were complex weather systems for meteorologists to predict. Forecasting was especially difficult before 3 January because of uncertainty concerning how the systems would align themselves and because a few miles' difference in distance or a variation of a few feet in ground elevation was crucially important in determining whether precipitation would fall as rain or freezing rain.[10] The major climatological study of this event drew the following conclusion: "During the evening of Sunday, January 4, 1998, after a weekend featuring average temperatures slightly above zero, freezing precipitation was observed over southern Ontario. In the hours that followed, it moved rapidly into southwestern Quebec. At this stage, no one could foresee the shattering and unforgettable experience that was in store for the eastern part of the country, and particularly southern Quebec, as the new year began."[11] The study's conclusion about predicting conditions that favoured ice storms based on global phenomena was hardly more optimistic: "based on our current knowledge of El Niño, we cannot predict the occurrence of situations conducive to major ice storms in Quebec."[12] The Quebec commission of inquiry into the event entitled the principal volume of its report "Confronting the Unforeseeable."[13] Experts on the American side of the border came to the same conclusion: "The magnitude and duration of this storm was not anticipated."[14] There is consensus among meteorologists that using the scientific knowledge and instruments existing at the time, the intensity, duration, and scope of this ice storm were unforeseeable even as it began.[15]

MONDAY, 5 JANUARY

Ontario

Residents of Ottawa proudly call the Rideau Canal through the centre of Canada's capital the longest ice-skating rink in the world. The unusually mild temperatures during the previous weekend had made skating on the canal a slushy slog.[16] But a weather disturbance was about to cause a dramatic change. Moving in from the west and beginning just after midnight, 24 millimetres of freezing rain was dumped on the National Capital Region. This was 8 millimetres more than forecast.[17] All surfaces – branches, wires, transmission towers, roads – were coated with ice. People left for work or school in the morning to find that sidewalks and streets had been transformed into skating rinks. Many North Americans take their cars, and so ice had to be chipped from car doors to allow them to be opened and from windshields to provide any visibility. This task was becoming more and more difficult as the ice thickened. Electrical wires began falling, as did trees, bringing down more wires. Thirty thousand households in eastern Ontario were left without electricity. Emergency response crews of electrical companies were dispatched to repair the lines and manage the problem. A few schools in rural areas closed early. Traffic became more congested than usual, especially during rush hour. This seemed just a passing irritation for people accustomed to snow and ice. Automobile accidents increased, hospital emergency wards began filling up to deal with bone fractures, and airplane flights were delayed or cancelled. Environment Canada forecast that the freezing rain would end on Tuesday morning, but it issued a new warning that 10 to 25 millimetres more was expected on Wednesday. An accumulation of 10 to 15 millimetres of freezing rain was forecast to start in southern New Brunswick on Tuesday.

Quebec

Montreal, which is 200 kilometres further east than Ottawa, slowly began to experience the freezing rain during the day on Monday. It formed a thick layer of ice over the snow that had not been cleared from streets and sidewalks after a snowstorm on 30 December.[18] Christmas trees placed by the roadside to be picked up by garbage trucks were quickly coated with ice. Road crews worked all day to salt and sand streets and sidewalks, which someone described as looking like mashed potatoes. Traffic became congested, and some airplane flights were delayed or cancelled. In the afternoon the Montreal office of Environment Canada issued a new weather alert stating that the freezing rain would be weak that evening but would intensify overnight and on Tuesday. At 1 p.m. it advised the Direction générale de la sécurité et de la prévention (DGSP) that more freezing rain would begin Wednesday, and it made arrangements to provide regular briefings. By 10 p.m. traffic had become a nightmare. The Ministry of Transportation was placed on a state of alert.[19] The freezing rain brought down a power line on the South Shore, depriving 2,500 households of electricity. A line fell in the west end of Montreal Island, depriving another 2,000 households of electricity.

In cinemas across the Western world, a record number of people had been viewing a film about the collision between an iceberg and the most technologically advanced ship of its age, the *Titanic*. That icon of technological achievement had not fared well in its confrontation with a construction of nature. As electricity flickered and went dead and moviegoers filed out of the cinemas, another clash between modern constructions and an icy creation of nature, this time from the sky, had begun.

Hydro-Québec nevertheless repeated that it would be a matter of minutes or at most one or two hours until power was restored. Modern society was functioning as it should: precise weather warnings had been given, and repair crews were rectifying nature's thoughtless processes. If nature's forces are weak, modern society is robust and reacts well. Montrealers had often experienced these inconveniences of freezing rain before, so no one was particularly worried this time. A week earlier, on 29 December, meteorologists had predicted the "storm of the century," which turned out to be a normal snowfall. So this time they erred on the side of not alarming the population, which was defined as prudence, and the population itself tended to discount gloomy weather predictions in the light of the erroneous forecast a week earlier.

New York and New England

Usually, upper New York State – or the North Country, as it is known to New Yorkers – is covered in snow during the month of January. The weather started in that normal fashion in 1998. Snow began to fall Sunday night, and by Monday evening about a foot had fallen – substantial but not extraordinary. This year, however, the milder temperatures resulted in the heavy precipitation of the storm changing to freezing rain. People in the North Country driving to work at 5:45 a.m. heard weather reports that freezing rain was spreading across Ontario and Quebec. The cold front gradually worked its way to the American side of the Canada–US border. It seeped into the valley of the St Lawrence River and then was blocked by the northern hills and mountains. The area just north and west of the mountains received a great deal of freezing rain when it first penetrated the American side of the border. But the mountains were not a obstacle for long. The cold front pushed southward across most of northern New York State and then expanded by Wednesday toward the centre of the state. In this region there is great variation in height above sea level. Higher elevations experienced only the warm moist air, not the underlying cold air mass. Their temperatures stayed above freezing, and they received rain.[20]

The freezing rain started in the outlying areas of northern New York, and several schools quickly announced that classes were cancelled for the day. Fortunately, there was little wind, which would have made damage much worse. Although traffic backed up on some roads because of the layer of ice, motorists had experienced that problem before in this region. Conditions were harsh but not unusual, yet. A meteorologist headquartered in Rochester had been paid by several radio stations in the North Country to provide forecasts. During the day he telephoned to warn that his radar showed that the freezing rain in northern New York would worsen and persist at least until Friday. "If you have a disaster plan, you'd better get it going because you're going to get clobbered up there."[21] The station manager was skeptical because he had seen severe weather warnings that did not materialize. Nevertheless, he telephoned the mayor of his village of Massena. The village had an emergency plan, but it had never been used. The mayor did not wish to alarm the public or react too quickly, but he did begin to activate the plan. The freezing rain was becoming more intense. Roads were getting more slippery; trees and electrical lines weighed down with ice were falling. Road crews would clear the road and then see that more trees had fallen. Warnings against unnecessary travel were broadcast. Electrical repair crews would restore power in several areas, only to discover new outages more numerous than before. Late Monday the freezing rain began to fall in central Maine as well.[22]

TUESDAY, 6 JANUARY

The slow but prolonged fall of freezing rain left a clear, thick coating of ice on all the branches and twigs of every tree and shrub.[23] This recast them into marvellous crystalline figures. Lowly weeds were transformed by this process of nature into splendid works of art. Ice sculptures were everywhere. The most rationally planned garden became a glimmering wilderness. "A strange, unearthly beauty lurked in the storm and its aftermath. You could see it in the light of day: residents became tourists in their own neighbourhood."[24] Professional and amateur photographers spent hours in the freezing rain under umbrellas shooting the beautiful creations of a disturbance of nature.[25]

Ontario

At 7:32 a.m. a baby boy was born in the backseat of his parents' car in the middle of an ice storm because ice on the roads destroyed their plans to get to the hospital quickly.[26] The day saw another 11 millimetres of freezing rain fall on eastern Ontario, particularly from Kingston-Ottawa eastward, adding to the 24 millimetres from the day before. Each wave of freezing rain coated trees, power lines, and electrical towers, increasing their surface area and giving the next wave of freezing rain a larger surface to fall upon and cling to. This positive-feedback loop of freezing drizzle exponentially increased the weight on trees, power lines, and electrical towers. Trees fell on power lines, or lines fell because of the unbearable weight, and they in turn snapped the supporting hydro poles and towers. Electricity to 100,000 homes in the greater Ottawa area was cut. The power failure stretched to the limit the rationally planned measures that had been put in place to deal with such an occurrence. Two 115-kilovolt transmission lines to the nation's capital had been damaged. Nonetheless, hydro officials still perceived the ice storm as a major but manageable inconvenience: a twenty-four-to-forty-eight-hour power outage to a large number of customers that would be corrected by Thursday.[27] Nature was about to change this view of things.

One of Ottawa's hospitals had already reported forty fractures since the storm began. Injured pedestrians lying on the sidewalk had unusually long waits for paramedics because the very thing that caused their fracture – icy conditions – also delayed the arrival of help. Another practical problem involved garbage. Collection is based on the premise of an average number of seconds per stop. The freezing rain coated garbage bags with ice and froze them into the snow, increasing the time for collection. Trucks found driving slow on the slippery roads. They were running far behind the collection schedule. Garbage collection in rural areas was cancelled for the remainder of the week. Since salt to prevent Ottawa's roads from becoming slippery is trucked in from the St Lawrence Seaway, there were worries that the slippery roads might prevent the salt trucks from arriving, a peculiar irony.[28]

On this day the power company, Ontario Hydro, answered calls every three seconds over the twenty-four-hour period. Many more calls went unanswered: "the phone number became known as '1-800-No Response.'"[29] Even this description underestimated the need for help because the phone service was down in some areas. Power went out for 28,000 customers in a suburb of Ottawa; crews restored it for all but 1,000 of them before the end of the day. Later that day power was cut for 30,000 other customers in three different areas of the national capital; crews then restored it for 80 per cent of them that same day. It seemed as if nature was playing tag with power repair crews. Some schools closed and then reopened because of the power "blackout-restoration" flip-flop. "The way things were happening it was difficult to get a sense of impending disaster."[30]

The hydrological office for the Ontario region of the International Joint Commission began operating on a twenty-four-hour-a-day emergency basis to support the St Lawrence River water level and flow operations. The Canadian Meteorological Centre experienced the first of 107 emergency power stoppages and restarts over the next eight days. About 35 millometres (almost an inch and a half) of freezing rain had fallen in a twenty-four-hour period.[31] The freezing rain stopped in the late afternoon in eastern Ontario and soon after in southwestern Quebec, and the freezing-rain warning was lifted. Then later in the evening forecasts called for more freezing rain on Wednesday, possibly changing to rain that afternoon.[32] The two-day meteorological forecast was laden with uncertainty and called for almighty help. "A series of disturbances approaching from the south will bring us episodes of freezing rain one after the other. By Thursday morning, 15 to 20 millimetres of freezing rain is expected to have fallen. Thursday night into Friday remains to be seen. Pray for plain rain."[33] Forecasters admitted that Friday's weather was a tough call.

The Mohawk territory called Akwesasne, along the St Lawrence River, spans parts of the Canadian provinces of Ontario and Quebec and the American state of New York. It, too, experienced freezing rain, broken branches, fallen trees, and downed power lines. After sporadic outages, electricity there was knocked out at 1:55 p.m.[34]

Quebec

From 1 a.m. on Tuesday the situation in Quebec became much worse. Nine millimetres of freezing rain fell on Monday, and 16 millimetres more would fall on Tuesday in the greater Montreal area. The weight of the ice on hydro lines, towers, and trees added 70,000 hydro subscribers (households, businesses, and government departments) an hour to the list of those without electricity. Electrical current had stopped in over one-quarter of the 840 distribution lines[35] that make up the power web on Montreal Island. A major electrical blackout had hit Montreal. The situation was even worse on the South Shore, where the collapse of medium-voltage transmission lines left 124,000 households powerless by 2:30 a.m. This urban area south of Montreal was deprived of electricity the longest. It would come to be called the "triangle of darkness."

Hydro-Québec's telephone hotline to report power outages was clogged. By 8 a.m. 400,000 subscribers were without electricity. Between 12 a.m. and 8 p.m. the number almost doubled to 700,000. It has been estimated that a subscriber buys electricity for an average of 2.5 people;[36] so close to 2 million people were without electricity in Quebec alone. One of every four subscribers in Montreal and its surrounding area was deprived of the power that enables modern society to function. In urban areas, trees and branches were falling on lines, bringing them down. Along rural roads the accumulation of ice was toppling lines in a domino-like fashion. Many other transmission and distribution lines became unusable. Hydro-Québec switched to a preplanned alternative circuit each time a line went down. But these lines also fell, leaving no backup lines because the freezing rain was much more intense and widespread than had been anticipated. On this day in Quebec an atmospheric disturbance of nature became a technological disaster: the electrical infrastructure for the centre of its population and economy was being crushed.[37]

The Montreal office of Environment Canada increased its computer and communications support to twenty-four hours a day in an effort to ensure quick resolution of problems. It gave the first of five hundred media interviews over the next five days. At 11:30 a.m. the civil security agency of the Quebec government urgently requested a forecaster for a briefing. Two forecasters were sent to give eighteen-hour-a-day support to that agency for the duration of the storm.[38]

By 2 p.m. Hydro-Québec was warning municipal authorities and emergency measures organizations that its electrical distribution system in Montreal and southwestern Quebec was in danger of collapsing, but the public was not informed so that it would not be alarmed. Hydro officials later admitted that they were worried about a total failure of their electrical transmission system in the area, with perilous consequences for modern life in a frigid climate. For example, if heat, light, pumping facilities to fight fires, and normal hospital electrical power are down in one small area, adjacent areas can help their neighbours. But if a complete blackout occurs in a large, frigid, highly populated area, there would be nowhere to take shelter or seek relief from nature's impact. The ice storm was so vast that it destroyed both the main transmission lines in the electricity web and an enormous number of small distribution links to households and organizations.

Firefighters were busy with a rash of house fires caused by power surges when electricity was restored. The emergency 911 telephone service received 5,300 calls from Montrealers on Tuesday morning alone.[39] Authorities sought to reassure the population, even to minimize the impact of the storm. They nonetheless told people to stop calling Hydro-Québec, the 911 emergency service, and the city unless a serious question of life, health, or security was involved. It was evident that these telephone lines had become clogged.

Every school board in the Montreal area closed its schools on Tuesday, not knowing when they would reopen. The closure sent parents frantically searching for daycare for their children. At this point the storm claimed its first victim when an eighty-two-year-old man in Quebec died of carbon monoxide poisoning as he ran a gas generator in the basement of his electricity-deprived home.[40] Cities and towns started to organize recreation centres and public buildings such as schools to shelter those without heat and light.

Hydro-Québec sent all its available repair crews in the greater Montreal area, two thousand employees, into action to fix electricity lines and check its network. They were quickly overwhelmed by the scope of the problem. Hydro-Québec hesitated, however, to call in crews from more distant regions of Quebec for fear those regions, too, would be hit by the freezing rain. Instead, it counselled patience on the part of subscribers in cold, dark homes and work areas.

Suddenly a huge ball of fire shot into the night sky, the result of a 125,000-volt short-circuit caused by a branch hanging over three live wires. The fireworks would have been impressive had it not been for the knowledge that they sent a whole neighbourhood of twenty city streets into coldness and darkness. Live wires were setting trees afire on contact, a dangerous situation in which the electrically operated pumping stations would also stop functioning if their electricity was cut. Hydro transformers were blowing up with loud banging noises, sending intriguing flashes – sometimes blue-green, sometimes orange-yellow – into the sky and depriving more homes and businesses of electricity.

Exhausted hydro repair crews, working sixteen-hour shifts, would repair a line only to find that another falling tree had brought it down again. The dangerous work was carried out in the most difficult conditions imaginable. A tree falling because it carried ten times its weight of ice could easily hit a worker. The creaking and groaning of bending trees produced an unsettling sound as workers in the middle of them did their job. The work was especially difficult and dangerous when it had to be done after nightfall with the only light in a dark city being from the headlamps on the workers' hard hats. Accidents occurred. Job descriptions were ignored: electrical linemen climbed trees and wielded chainsaws when needed, rather than waiting for tree-pruning teams. It was impossible to keep dry and warm in the unrelenting freezing rain, even with raincoats on top of parkas. Between sixteen-hour shifts, these workers returned to their own homes, which also lacked heat, hot water, and warm food. No one knew how long they would have to keep up this frantic pace. Electricity linemen previously ignored by the public became heroes as they battled to restore electricity.

Meteorologists struggled to get their predictions right. They knew from experience that they would be laughed at and ridiculed if their forecasts turned out to be exaggerations. But they also sensed they would be sharply criticized later if their forecasts underestimated a serious problem – convenient scapegoats for an inadequate organizational response.

Even if trees or branches did not break, some would bend because of the weight of the ice and intermittently touch an electrical wire, thereby cutting the circuit because of the safety device that had been installed to minimize problems. In the ice storm such safety devices were now maximizing problems. Heavy road salting to melt ice and snow on a low-lying section of the principal expressway in Montreal had the perverse consequence of creating a flood-like situation, leaving automobiles axle-deep in water. Electrical linemen would repair ten lines and then find that another twenty had fallen under the weight of the ice.

On Tuesday afternoon people were shocked to learn that eight Hydro-Québec electrical towers on the South Shore had crumbled because of the weight of the ice on them. The main divided highway between Montreal and Quebec City had to be closed for four kilometres because of downed high-voltage live wires. Regional repair crews could not clean up these fallen pylons and wires, and a more specialized crew had to be called in. It took a full day before traffic was restored.

Modern, rational society was using all its organizational and technological means to combat the ice storm, and the storm was easily winning. Two of Montreal's four universities closed down. The airport at the west end of Montreal had to cancel seventy departing flights, and the others were delayed. The term used to describe the resulting confusion – "chaos" – was an understatement. Trains arrived and left late because trees fell on tracks and the ice froze outdoor equipment. Train service between Montreal and Quebec City, Montreal and Ottawa, and Montreal and Toronto was interrupted.[41] The suburban trains functioned intermittently depending on whether their electricity was cut off. Bus arrivals became random. Four thousand employees from the city of Montreal were dispatched to help automobile drivers and pedestrians in various ways. Up to this point, only the underground metro was reliable. The emergency rooms of hospitals in Montreal and especially on the South Shore became congested, and all non-urgent surgery was cancelled for fear of interruptions in electrical power. At this time 250,000 homes in Montreal were deprived of electricity and another 350,000 on the South Shore.[42]

Bridges connecting Montreal Island with the surrounding mainland are of the old, suspension type with superstructures high over the roadway. The Ministry of Transportation began de-icing highway signs and the superstructures of bridges for fear the heavy ice would make them collapse on motorists passing underneath or the weighty ice itself would fall on automobiles. This work would last almost two weeks. The freezing rain also led to slippery roads, streets blocked by fallen branches and wires, and many others threatened by trees and wires about to fall.[43]

In this region, electricity provides the means not only of enjoying modern comforts but also of meeting very basic needs. Homes without electricity first felt the loss of entertainment (televisions, sound systems, videos, computers) and then of conveniences (washers, stoves, microwave ovens, hot water); next they slowly experienced the temperature falling from 20 degrees Celsius to match the −8 degrees outdoors. People learned the hard way that oil and gas heating depends on electricity to ignite furnaces. Water pipes froze and then burst because of water expanding to ice. Lighting vanished at dusk – that is, at 4:30 p.m. – not to return until sunrise at 7:30 a.m. the next morning. Those with special needs – for example, an electrically powered compressor to relieve asthma – were thrown back to the pre-compressor days.

Emergency measures organizations in cities on the South Shore and then in Montreal began to set up shelters for those forced out of their homes by lack of electricity, heat, and light. Whatever conflicts of language and culture existed in normal times dissipated in these extraordinary moments. In Montreal two senior citizens' homes had to be evacuated and the patients moved to shelters because the lack of heat and light threatened their health. An eighty-nine-year-old man died trying to keep warm because a previously unused chimney overheated.[44] People in cold houses desperately sought any means of heat, such as barbecues and gas camping stoves, unaware that these outdoor appliances give off carbon monoxide, which is extremely dangerous indoors.

The weather predictions of the Montreal office of Environment Canada finally gave reason to hope: very little freezing rain was forecast for Tuesday evening and night. But they also contained an intimidating warning: "Another disturbance coming from the Great Lakes will give the south of Quebec more sustained freezing rain on Wednesday." As people huddled in their cold, dark shelters, the weather predictions supplied a glimpse at the angry beast outside about to strike. Meteorologists were asked to provide special forecasts to Hydro-Québec and to the office of the premier. Measurement of the amount of freezing rain became tenuous, however, because ice clogged the sensors used by meteorologists: the more freezing rain, the less accurate the measurement of its quantity.

Hundreds of banks had to close because of the power blackout. Quebec has a highly developed cooperative banking movement called *caisses populaires*. A reported 318 of its branches (25 per cent) and 400 of its automatic banking machines (20 per cent) were rendered inoperative for lack of electricity. Many retail outlets lost their debit card and credit card services, so that people were no longer able to pay for their purchases in the way they had grown accustomed.[45]

Virtual reality disappeared as nature cut off the electricity that powered computer screens, throwing their operators back to the real world. Illusions that the social constructions of modern society have tamed the forces of nature were shattered. In response to a reporter's question about when electricity would be restored, a vice-president of Hydro-Québec gave a remarkably candid and accurate response: "Tell me when the freezing rain will stop and I will tell you when electricity will be restored." Even high-ranking technocrats had to admit their powerlessness when confronted by such massive forces of nature. At this point it was estimated that 650,000 people in Quebec and Ontario were without power.[46]

New York

The metropolitan area of New York City was "footloose and overcoat free," as the *New York Times* called it, "momentarily freed from the prison of winter."[47] A warm, dry air mass coming from the Great Plains had brought a heat wave to the city on Friday, and meteorologists expected it to last until the following Friday. Golf courses, ice cream stores, and street vendors were doing a thriving business, unusual for this time of year. Spring "giddiness" was in the air. Ski shops, heating-oil companies, and outerwear stores were complaining that business was sluggish because of the balmy weather.

The context was different in the northernmost areas of New York State. Branches and trees laden with ice were falling and pulling down electrical power lines with them. Roads were blocked. Electrical transformers were blowing, and trees were catching fire. Electrical companies, fire departments, local governments, and radio stations were deluged with telephone calls about power outages and road blockages. The massive ice front began in the northernmost part of New York State and then pushed slowly southward. The amount of freezing rain and damage varied by area, and so did the closing of schools. The hospital and radio stations in the town of Massena on the St Lawrence River lost their electricity, but the hospital and one station had emergency generators. Regular power would be restored, only to be cut again. The town's emergency command centre went into operation.

People waited for the storm to break, but instead it became worse. The high school gymnasium was opened to give people a warm shelter. The one remaining radio station stopped broadcasting entertainment and replaced it with a steady stream of information: road closures, cancellations, stores where shipments of generators had arrived, and so on. In this small community the most frequent question was, Is the shopping mall open? There was some alarm when it was realized that the freezing rain would not end as quickly as other episodes before it had. People did not know what to do or where to go. But that feeling was replaced by a sense of acceptance as the hours passed. As planned entertainment requiring electricity was stripped away from people, they had to reinvent ways of spending time. Card games and board games resurfaced, even conversation.[48] The freezing rain led to slippery roads and to some accidents, particularly involving tractor-trailers, but in general it induced drivers to reduce speed. The number of accidents was not high for such driving conditions.[49]

New England

The freezing rain continued on Tuesday in Maine, placing a slippery glaze of ice over all surfaces, but it was not as awful as in states and provinces further west. Schools closed; walking and driving conditions were bad. Municipal road crews succeeded in their battle with ice and rain on the main roads, but the rural roads were left icy. By 5 p.m., however, the police were receiving calls about cars that had slipped off the principal highways as well. Temperatures were above freezing in some areas of southern Maine but below freezing in the centre of the state. Meteorologists predicted more freezing rain on Wednesday and Thursday.[50]

Terrible Beauty

A man observed a white birch weighed down by ice making an almost perfect arch from roots to tips across his driveway and said, "It looks kind of nice like that. I'd like to keep it that way when the storm is over."[51] Other sights were shocking and original, if that is to label them beautiful. "The poplars are all stripped – just toothpicks sticking up in the air."[52] Many of the trees that had seemed so beautiful when the first freezing rain glazed them with ice came toppling down when the next wave overburdened them. The incremental character of the accumulation of ice on all objects led many observers to fail to perceive that what was particularly attractive one day would be broken and destroyed the next. A rural resident who first thought the countryside was superbly beautiful later called it tragically beautiful.[53] One resident of Ottawa remarked, "Everywhere I go, a tree crashes. The storm is so destructive, but at the same time it's really beautiful. The light shining through the trees is almost magical."[54] A northern New York journalist described huge trees covered with ice along a street in his area as "a thing of terrible beauty" just before they cracked and fell, taking power lines down with them.[55] Another recounted "waking up to a crystallized wonderland" that turned out to be a "wintry nightmare."[56] Most impressed were those who came from southern climates. A counsellor at the Brazilian embassy in Ottawa observed: "Another thing that struck me about the ice storm is the irony of the thing, because it's so beautiful. In the first moments I even went out to take pictures because the trees looked as if they were made of glass. Normally catastrophes are so ugly."[57]

A Vermont journalist subtitled his article "If you think it's bad here, take a ride north" when he visited Montreal. Nonetheless, he wrote that "Montreal seemed bathed in an eerie beauty. The sun that occasionally peeked through the clouds found a city outlined in white, its ornate architecture gleaming with an unpredictable coating of ice. The ice was charming, but it carried danger."[58] Just south of Montreal the standardized, repetitive shapes of hundreds of massive electric power pylons were twisted into hundreds of original configurations as they crumbled into uselessness. *Time* magazine described the lamination of every surface with ice as a "grim fairyland ... in a glittering state of emergency."[59] The *Canadian Geographic* referred to it as "lethal beauty."[60] In some places eerie blue-yellow flashes lit the sky as transformers exploded. In Plattsburgh, New York, people saw wires burning on the ground. Then there "was a big fireball. There were sparks everywhere. It spread to about 10 feet wide and 3 feet high. It was so bright, it lit up the whole sky and the house."[61] At the beginning of the ice storm and for those at the periphery who did not suffer much, a little risk heightened the beauty. But for people who lacked heat and light for close to thirty days in frigid weather, the inconveniences, suffering, anxiety, and danger they experienced masked the beauty. For most people, though, an ambiguous beauty remained while the ice coated all surfaces. The ice palace was also an ice prison, denying people their normal lives.

Even in its destructiveness, the ice storm was a symphony of original sounds. Ice crystals shattering on the streets created the effect of breaking glass. The snapping of branches and the crackling of ice added to the unusual noises. "It sounds like a shotgun going off every time one of those fir trees snaps." The noise of creaking branches warned people that next would come the thunderous boom of ice-covered tree limbs crashing to the ground. "All I've heard all day is 'snap, crackle' as the branches are breaking. I woke up around 4 o'clock and heard this pretty loud noise. When I looked out, I saw this tree on the ground. The whole tree fell right from the base."[62] Unusual "poof" sounds marked the explosion of transformers. If it had not been so threatening, the ice storm would have been a delight for the ear as well as for the eye.

This was an unusually photogenic disaster. Newspapers in the affected areas produced not only stories but also photos, and photos not just of the destructiveness but especially of the beauty, emphasizing the contradiction between the two. Many newspapers published a supplement dedicated solely to the ice storm, with statistics, human-interest stories, and especially photos. Communities issued books on the consequences of this weather event in their locality, with the centrepiece invariably being photos. A hardcover book of photos became a bestseller in Canada.[63]

WEDNESDAY, 7 JANUARY

Ontario

On Wednesday morning there was a lull in the freezing rain. Much more ominous was the massive storm front coming from the Gulf of Mexico that meteorologists saw next on their radar screens. At 4:30 a.m. Environment Canada sent out an updated weather warning forecasting another 28 millimetres of freezing rain. It would later be found that this prediction underestimated the freezing rain in eastern Ontario and southwestern Quebec by about 10 millimetres. The second ice storm arrived late Wednesday morning, covering a much larger area: from Chicago and Wisconsin on the west across to Georgian Bay, Ottawa, Montreal, the South Shore in Quebec, northern New York, and Maine to New Brunswick and Nova Scotia on the east.[64] The shape, size, and location of the freezing rain footprint had shifted, but the Montreal-Ottawa region still received the most freezing rain, with the low-lying frigid air mass producing a very clear, hard coating of ice. By 3:30 p.m. Environment Canada had revised its estimate to 40 millimetres along the St Lawrence Valley. The conveyor belt bringing the warm, moist air from Texas that was the source of the freezing rain just kept conveying. This wave of freezing rain, which started Wednesday morning, lasted forty hours, with occasional breaks. Since the temperature was below freezing, the ice did not melt. The new layer of ice piled up on the previous one, the third on the second, and so on, producing an unprecedented buildup.[65]

A farmer heard a loud bang and saw live electricity lines on the ground. They were touching his barn, so electricity was running through the metal in it, and thirteen of his eighty cattle were electrocuted when they drank water. He shut off the electrical switch to his barn but did not perceive that the metal in the building was still conducting current from the broken but live line, and he narrowly escaped electrocution himself.[66]

Rockliffe Park is the wealthy area of Ottawa, where the residences of the governor general of Canada, the prime minister, foreign ambassadors, and the wealthy business class are located. It is heavily treed, with the allure of an inhabited forest in the heart of a city. The ice storm damaged 85 per cent of its trees, almost half being gravely harmed. Rockliffe was one of the first areas of Ottawa to lose power as a result of falling trees. Even the governor general and his family went two days without electricity. During a blackout, the editor of Ottawa's principal newspaper and his family were warming themselves by the fireplace in one room of their stately old Rockliffe home. The electricity came back on. They had the impression that the fireplace was giving off an amazing amount of heat. But it was not the fireplace. The restoration of power had resulted in a surge of electricity through the old wiring that was burning down the rest of their house as they sat in one part of it.[67]

Ottawa Hydro called the region's Emergency Measures Unit, admitting it did not know how quickly electricity could be restored. So it might be wise to open shelters. The director of emergency planning was reluctant to declare a state of emergency because a shutdown would harm the economy. But the electricity situation was worsening, and the weather was not improving.[68]

The high-speed highway through Ottawa became as icy as a curling rink and was ordered closed by the Ontario Provincial Police. At 11 p.m. road crews started barricading the on-ramps from both ends of the region. It took them until 4 a.m. In the meantime Ontario Ministry of Transportation trucks had been putting salt on the highway to melt the ice, and by 4 a.m. the salting had done so. "As the last barricades were being put up, the order came to remove them."[69] The road crews managed to take all the barricades away just in time for rush hour the next morning. The highway was closed again later for two hours because live electric lines had fallen across it.

Hydro transformers were blowing up with loud banging noises, sending blue-green flashes into the sky and depriving more homes and businesses of electricity. Exhausted repair crews, working sixteen-hour shifts, would repair a line only to find that another falling tree had brought it down again. Often they would sleep only four or five hours a night. One night a foreman mistakenly called a lineman back in to work after only an hour and a half. The lineman explained the situation, and the foreman excused himself and told the lineman to go back to sleep. He tried, but sleep would not come, and so he went in to work. He later admitted that the ice storm "has taken a psychological toll. He has lost his sense of time in the endless workdays."[70]

A special meeting of directors of the Regional Municipality of Ottawa-Carleton debated declaring a state of emergency, but any decision was deferred to the next day. The chair of the region informed the media that he might consider such a declaration.[71]

At 9 a.m. the Mohawk Council of Akwesasne Grand Chiefs declared a state of emergency. This was the first Ontario community to do so, and it alerted Ontario's Emergency Management Organization, headquartered in Toronto, to the gravity of the developing situation. That organization sent a liaison field officer to Akwesasne from its office in the nearby city of Kingston, but he did not get there until evening. By then the roads were all but impassable. He made it in, but the driving conditions were so atrocious he could not get out.[72] At 10:50 a.m. power had been temporarily restored to some areas. But by 10:35 p.m., following sporadic interruptions, power went completely out on the American side of the Mohawk territory.

Quebec

More freezing rain and stronger winds early Wednesday brought down a major hydro substation on the South Shore, doubling the number of customers (businesses and households) without electricity. Anxiety grew. People began to pour into shelters and hotels. Restaurants and hotels with electricity did a booming business. Roasted chicken deliveries by the Quebec chain St-Hubert more than tripled. Even the premier of Quebec, Lucien Bouchard, after spending the night without power in his home in a wealthy area of Montreal, moved his family into a hotel. By 5 a.m. the eastern half of Montreal Island, which had not been as seriously affected as the western half, went black. The emergency preparedness of municipalities was being tested, and many were failing. The Quebec government decided to give financial assistance to municipalities to pay for emergency expenses. Schools, universities, and businesses were closed.

Overall, though, not much freezing rain fell during the daytime on Wednesday. By 11:17 a.m. the weather office had changed its forecast to little freezing rain during the day but important additional quantities on

Wednesday night and Thursday because of a new weather system. Precise predictions were difficult to obtain because of the enormous physical forces involved. The number of disrupted power lines on Montreal Island decreased during the day. Although Hydro-Québec launched a massive de-icing and repair operation, calling in reinforcements from the rest of Quebec, the Maritimes, and the United States, the functioning of its electrical transmission and distribution network depended more on the coming and going of the freezing rain. The weather was relatively calm during the day. The number of disrupted power lines on Montreal Island decreased, and the number of power customers in Quebec without electricity dropped from 700,000 to 500,000.[73] At this time the feeling of many people, as expressed in a newspaper headline, was that "at least it can't get worse."

By 7 p.m., however, the situation had become much worse. In a few hours the number of hydro customers lacking electricity went back up to 700,000. More electrical towers had crumbled. The Ministry of Transportation systematically closed all the divided highways in metropolitan Montreal to de-ice its overhanging information panels for fear they would collapse on passing automobiles. Nearly all of Montreal Island was now without electrical power. Communications were also in disarray, with many telephone lines down as well. Since everyone needed the same things in the storm – generators, batteries, flashlights, candles, and the like – shortages and lineups were common. All 4,100 police officers, including off-duty ones and those on leave, were ordered to report for work. Office staff were put on street duty and replaced by officers on maternity leave. Each shelter was assigned a uniformed officer to keep order.

The accumulating ice began to halt the functioning of meteorological stations that take observations automatically. Data were lost at a number of these autostations. Anemometers (wind gauges) and rain gauges were then de-iced manually. A twenty-four-hour-a-day manned program was started at two of the more important stations to replace the automatic measuring devices in order to maintain the integrity of the data.[74]

Most depressing was the forecast of Environment Canada that another 20 to 30 millimetres of freezing rain was on its way in the following hours, along with a 40-kilometre-an-hour wind. The ever-increasing accumulation of heavy ice, already reaching 55 millimetres, on all surfaces, together with the blowing wind, promised to crack even more structures. The question of everyone's mind was the same: Will the ice storm never stop? No one had yet forecast the end.

The premier of Quebec at this point accepted the offer of the Canadian government to send the army to help. During the night the first 2,500 soldiers arrived in Montreal from a military base near Quebec City. They cleaned up streets and removed fallen trees and branches, freeing Hydro-Québec workers to do their more specialized electrical work. More than

800 line-repair and tree-trimming workers were also called in from the northeastern United States.

The transportation system collapsed even further. All trains to Montreal, Ottawa, and Kingston were cancelled. The suburban commuter rail line was suspended. Major airlines called off flights to Montreal. The intercity bus company Greyhound cancelled its runs to and from Montreal. The ice that had accumulated on the superstructures of bridges fell off in chunks, threatening motorists underneath.[75]

New York

The brunt of the storm hit 7,000 square miles of four counties in upper New York State on Wednesday night. It was centred on a line that ran across the Adirondack Mountains from Watertown in the west to Glens Falls in the east.[76] Temperatures varied not only along the usual north-south axis but also according to ground elevation, and so did the type of precipitation. A temperature inversion occurred: the higher the elevation, the warmer it was because the warm, moist air currents were riding over the cold air mass. At higher elevations in the Adirondack Mountains such as Lake Placid, as well as further south, massive amounts of snow were melting because of the exceptionally mild temperatures. The forecast was not optimistic even for areas not experiencing freezing rain. "The southern Adirondacks, Mohawk and Hudson Valleys, Central New York and the Southern Tier are under a flood watch and are close to the point of experiencing flooding. Heavy rains will move into the state during the next 24–48 hours adding to the problem."[77] When the waters from rain and melting snow flowed down the rivers to lower, colder areas, they ran into ice jams that blocked the rivers, thereby increasing the threat of flooding.[78]

Down in the cold air mass, freezing rain coated all surfaces with hard, thick, weighty layers of ice, causing extensive power outages and making roads slippery or blocking them. Trees and utility poles had fallen every few feet on side roads. Highway crews laboured just to keep the main roads open for power repair crews and emergency vehicles. A travel ban was put in place for many of the affected regions. With tree limbs falling and live electrical wires on the ground, there was some discussion about whether the travel ban included walking outdoors. Officials in power companies began to admit that electrical problems would get worse unless the weather improved. There was a rush on stores for staples. Shelves were soon bare because of the difficulty transporting new stocks and because of power outages. Surprisingly, for a storm consisting of freezing rain, one of the main deprivations was water because the power outage shut down electrically operated water pumps.

There were temperature fluctuations, the most prevalent being the usual differences between night and day. When the slightest melting occurred, ice fell in chunks as big as a foot thick and seven feet long. Ice falling from a transmission tower punched twenty holes in the roof of a radio station and destroyed the porch as two men were walking out. Danger took many forms during the ice storm.

Many radio stations on both the Canadian and American sides of the St Lawrence River were blacked out by the ice storm, but one American station had an emergency generator. For thirty-six hours it became the only source of information for its broadcast area. Its phone lines rang continuously. Callers were businesslike during the day. "People who called at night were often scared, lonely, needed someone to talk to, or were often unaware of what was going on. Some of the callers were very emotional."[79]

Many businesses, government offices, and schools closed early and sent employees and students home. Some hospital employees could not come to work because of blocked roads; others had to leave early because the daycare for their children was closing early. But the hospitals had a full load of patients and even had to open some new beds. A hospital cannot just close down like other organizations and businesses. So the call went out for staff to fill in and to work double shifts. And the call was answered.[80]

New England

Up until now the light rain and drizzle had deceived residents of Maine about the upcoming impact of the storm. But on Wednesday that form of precipitation was quickly replaced by freezing rain. The sidewalks and roads became so slippery that even walking was extremely difficult, and an increasing number of people were showing up at hospitals with broken wrists and hip injuries.[81] Falling trees brought down electrical, cable, and telephone lines themselves burdened by ice. Utility poles fell one after the other like dominoes. Ten thousand homes and businesses lost their electrical power.[82] Ironically, so did the Central Maine Power Company headquarters.[83] Branches and whole trees began cracking and falling from the weight of the ice. "It sounded like a war. You'd hear the loud explosion from the tree breaking, and something like machine-gun fire as the branches hit the ground."[84] The war metaphor kept coming back in interpretations of the damage, even when it appeared exaggerated: "The destruction was worse than anything I'd seen. Worse than Vietnam."[85] The National Weather Service warned that the main brunt of the storm was moving into Maine and that there would be severe icing. Hence many businesses and schools closed early.[86] Wednesday saw the beginning of three days of heavy freezing rain in central and southern Maine.

THURSDAY, 8 JANUARY

Ontario

The heart of the storm that generated the hot, moist air moved from Texas to Kentucky. However, the line where this southern air collided with cold northern air remained over the Ottawa-Montreal region.[87]

When repair crews restored power, it did not always bring comfort. At 1 a.m. on Thursday morning another electrical surge following the restoration of power sparked a fire that tore through a home, killing a sleeping elderly woman.[88] An outbreak of chimney fires occurred as electrically deprived homeowners overused decorative fireplaces that had not been properly cleaned. The Ottawa Hydro control room showed 50,000 homes without electricity at 4 a.m., twice as many as when its current repair drive began. Modern, rational society was using all its organizational and technological means to combat the ice storm, and the storm was still winning.[89]

Emergency measures organizations operated shelters for those who had to evacuate their homes because they had no electricity, heat, or light. Most victims were not confused or in a panic; rather, they were trying to help themselves and others as best they could. The initial response was by the victims themselves: seeking alternative sources of heat and light, informing power companies and governments about the location of problems, aiding neighbours, gathering together in the rare places with heat and/or power, and so on. The problem was that the means they used for normal living, most of which depended on electricity, had been stripped away from them by the power blackout. Another major problem was that the incremental character of the freezing rain and accumulation of ice led many to fail to perceive how serious the danger was becoming and the need for protection.[90] Risk was underestimated by many individuals, by the provincial Emergency Management Organization in its headquarters in unaffected Toronto, and even initially by meteorologists. Two-thirds of the electrical transmission capacity for eastern Ontario had been destroyed by the ice and wind. The risk was high that people would touch live power wires. "One vehicle did light up as it hit a wire across a road."[91]

Early in the morning two eastern Ontario municipalities declared a state of emergency. Soon after, two more within the Regional Municipality of Ottawa-Carleton did the same. A meeting was called to discuss whether to declare a state of emergency for the whole region. It was the responsibility of the chair of the region to make the decision after listening to the debate among his experts. But he had trouble leaving his home to attend because fallen trees blocked his driveway. A police car was sent to get him. After discussion, a consensus developed that a declaration of a state of emergency was necessary to convince the public of the gravity of the situation, to get non-essential services to shut down, to reduce traffic on streets, and to get people off the sidewalks.[92] An earlier news release with a warning had not accomplished these goals. The argument for financial assistance for the region from the province would be strengthened by such a declaration, but in Ontario it was not legally guaranteed.

At 10 a.m. the chair announced that the whole Regional Municipality of Ottawa-Carleton had declared a state of emergency, the first in its twenty-nine-year history of regional government. Shortly thereafter, classes were cancelled at Ottawa's two universities, the telecommunications giant Nortel sent its employees home, and the Canadian government ceased operations in the nation's capital, except for essential services. Within a short time, sixty other Ontario municipalities declared states of emergency, more than double the number that had been declared at any one time in the past.[93] However, even this large number of declarations and shutdowns in Ontario did not lead the provincial Emergency Management Organization to perceive that a provincial emergency existed.

The electricity was cut to the homes of the prime minister and the governor general in Ottawa.[94] The Ottawa region decided to request the assistance of soldiers to clear the way for power repair crews and to clean up debris. The federal government was so willing to help that Prime Minister Jean Chrétien was on television speaking about military aid before the brigade that was to do the work formally knew its services were being requested.[95] The region had no experience in estimating the number of soldiers needed and guessed at 200. Eventually 4,500 soldiers would struggle against the ice storm in the region of Ottawa alone. The federal minister of defence announced that the prime minister had ordered soldiers into action in the affected areas of Quebec and eastern Ontario, but he was unable to explain how soldiers would fight an ice storm. How does an army fight a war against an enemy such as weather? The contingent of soldiers grew to become the largest peacetime deployment of troops in Canada's history, eventually rising to 15,000 soldiers.

The safety device of skywires was causing problems everywhere. Whereas transmission lines give off heat, which retards the formation of ice to some extent, skywires do not normally carry electricity. They give off no heat, and so ice accumulated on them more quickly, and they drooped and short-circuited the power lines below them.[96] Paradoxically, rationally planned security measures to protect against one phenomenon of nature only aggravated the danger when a different phenomenon struck with more force than expected.

At 3:25 p.m. a skywire sagged under the weight of the ice that had accumulated on it and faulted one of the two lines that formed the backbone of the Ottawa area's power supply. The complete 500-kilovolt load was transferred to the other line, but it too was in danger of faulting for the same reason. The nation's capital would have lost 40 per cent more of its electricity in one blow if a single link of that second chain had faulted.[97] Worse still, the secondary 230-kilovolt and 115-kilovolt transmission lines in the web were already damaged and could not be used as backup. At one point, 65 per cent of the transmission capacity of eastern Ontario was lost. For three days, only one teetering transmission line separated the nation's capital from total blackout.[98] Ontario Hydro officials subsequently conceded they were concerned about a complete collapse of their electrical transmission system in the Ottawa region, an admission similar to that made by Hydro-Québec in Montreal.

At 3:30 p.m. on Thursday, Environment Canada forecasted ice pellets on Friday morning, changing to more freezing rain in the afternoon.[99] Modern technology enabled meteorologists to see the blows coming but not to prevent them. They watched their radar screens helplessly as nature sent its ice bombers toward them.

Repair crews worked frenetically, but the ice knocked down lines faster than the crews could put them up. Fallen wires were now frozen between two layers of ice. Wires still connected were sagging so low they were dangerous.[100] At noon the Children's Aid Society closed until Monday because half its employees could not come to work. Unlike some police stations, fire departments had backup generators and continued to function. They did, however, run out of signs and barricades because there were so many downed power lines and branches blocking roads.[101] At 9:20 p.m. Ottawa's largest and most modern hospital announced that, although it would stay open to care for emergencies, it had no more beds.[102] Recently, the region had decided to close hospitals and beds because of financial constraints and an assumed excess capacity. The rational restructuring of hospitals that was occurring no longer seemed so rational to those who needed a hospital bed but found none available. A cell phone tower lost power, and service went down for five hours.[103] Ontario Hydro later announced that, despite sixteen-hour workdays on the part of every hydro crew it could muster, the number of households without power in eastern Ontario had risen to 102,000 by the end of the day. The Ottawa region's twenty-four-hour phone service was so busy that night that it had to add staff, and some had trouble coping with the angry callers.[104]

A desperate need for generators had developed, especially at shelters and homes for the elderly. An hour after the state of emergency was declared, the Ottawa region set up a special group to find generators. Calls were made to cities in Ontario not affected by the storm, to the federal government, and to the military. By the weekend a steady stream of generators was arriving. A priority list was established for lending them. Records were kept of the serial numbers, and they were sent back to their owners when no longer needed. Often, generators were installed under the assumption that the power blackout would only last a few hours, and they collapsed from overuse as that time was greatly exceeded.[105]

Around Ottawa lies farmland, but not more than 40 kilometres away huge tracts of uncleared wilderness begin – forests populated by bears, wolves, deer, moose, porcupines, and other wild animals. Modern information technology – cellular telephones, fax machines, computers, email, the Internet – has enabled those forests to be inhabited in recent years by another species: urban professionals. These writers, artists, consultants, and engineers enjoy the amenities of quasi-wilderness and a sense of community with like-minded souls. They telecommute on the information highway to get their work done. They are wired into a "virtual community" in which distances around the world are crossed instantaneously.[106] This "global nervous system" or "organic neural network" is largely based on the electrical system. These professionals moved to the countryside to escape things they did not like about the city, but they still desired and received a city-like level of electrical service. Theirs was definitely not the country life that existed a century ago.

The ice storm reminded them that their sophisticated community depended on vulnerable wires strung from an interminable line of old wooden poles. The electrical grid was crushed by the freezing rain and went dead.[107] Batteries quickly discharged. Generators were difficult to acquire when everyone wanted them. Transportation was atrocious, so even people with generators had difficulty getting the gas needed to power them. Cellular communication jammed from overuse. The electrical lifeline of these professionals to modern civilization was the first to be cut by the ice storm and the last to be restored because they were at the end of the line, because of the length of the line needed to reach them in the bush, and because of the sparse population there. Since there was no underground wiring to their habitat in the bush, the freezing rain taught them that the information superhighway was as robust as its weakest old pole. Their virtual communities constructed upon the essential infrastructure of old poles was crushed by ice. Bush is what they wanted, and bush is all they got for the duration of the blackout.

In the Mohawk territory other areas declared a state of emergency, as did nearby cities and counties on both the Canadian and American sides of the border. The Akwesasne Emergency Operations Centre was activated at the Tribal Police Headquarters. A shelter was established at the Mohawk Bingo Palace, mutual aid contacts were made, supplies were assembled, and all shelters were now in full operation.[108]

Quebec

At 5 a.m. on Thursday, 8 January, electricity went out at the only shelter in Hull, a city across the Ottawa River from the capital, transforming that refuge from cold, dark homes into coldness and darkness. Hydro-Québec announced that 77,000 of its customers on the Quebec side of the river (the Outaouais) were without electricity. It hired thirty tree-cutting crews from New York State to work in the Outaouais.[109]

Partisan politics and federal-provincial tensions were shunted aside. When the premier of Quebec accepted the offer of the federal government to send in the army to help, the nearby 450-soldier Immediate Reaction Unit went promptly to the "triangle of darkness" on the South Shore of Montreal to clear debris so that electrical repair workers could do their job. An additional 3,100 troops from Quebec City began their journey to Montreal. Operation Recuperation, to use the military's bilingual expression, had begun.[110]

A woman in her family farmhouse on the South Shore was jolted from sleep in the dark night by a thundering crash, then a second, a third, and so it went. All but one of the seventeen giant, 250-foot, latticed steel transmission towers spanning her farm had crashed to the ground in domino fashion, bringing down with them six vital power lines.[111] At 9:51 a.m. nineteen transmission towers between two substations on the South Shore fell like dominoes, bringing down a 735-kilovolt line that supplied electricity to both the South Shore and downtown Montreal. The burden consisted of not only the weight of the ice on the tower itself but also the massive ice load that had accumulated on the power lines suspended from the towers. One observer estimated this weight as equivalent to automobiles hung from the wire at 100-metre intervals pulling the tower down. At 10:02 a.m. twenty more transmission towers and one conductor fell, downing the 735-kilovolt line between the first substation and the dam up north. At 12:30 p.m. the line fell between the second substation and a third, along with nine towers and three conductors, including two that fell on a highway. The "ring of power" around the Montreal region that ensured the transmission of electricity was quickly being reduced to powerlessness. The freezing rain was destroying not only the huge web of local low-voltage distribution lines but also the high-voltage transmission lines on and into the much acclaimed "ring of power." In the language of electricity specialists, not only distribution but also supply was disintegrating. Like a puppet on a string controlled by the freezing rain, Hydro-Québec changed its priority from repairing the maze of local distribution lines to reconstructing its transmission lines.

Nature's demolition of this part of the transmission network completely blacked out the urban sprawl linking three cities on the South Shore. Demolition, not damage, was the appropriate term to describe what had happened. This segment of the transmission network would have to be rebuilt, not just repaired, requiring a huge amount of work, a major financial investment, and a long period of time in the cold, dark Canadian winter. Morale sank with each of these blows of the ice storm.

Almost 3 million people were now deprived of electricity – that is, deprived of heat, light, hot water, and other amenities – in the province of Quebec alone, accounting for more than one-third of its population, and this number was climbing steeply.[112] Temperatures were below the freezing point. One power line remained to supply the city of Montreal. It was monitored closely all night by Hydro-Québec workers. If it had gone down – and Hydro-Québec bosses were bracing themselves for such a total blackout of Montreal – a catastrophe would have occurred. "Really, it was a very, very close call," the Quebec premier admitted later.[113] In the greater Montreal metropolitan area a hundred emergency shelters had been set up, with a capacity of almost 6,000 beds. The needs of people without power, however, began to exceed the available resources. The call went out for donations of food, cots, and other essential supplies. This proved inadequate, and the government had to make massive purchases of food and other goods. Improvisation had become the norm.[114]

A weather radio antenna stopped transmitting information as a result of the coating of ice and a power failure. The Gatineau Receiving Station broke down, resulting in the loss of RADARSAT meteorological data. Meteorologists nevertheless began daily briefings to an emergency centre consisting of the Montreal Civil Security Centre, Hydro-Québec, the military, and the Quebec and Montreal police.[115] The area of freezing rain changed its shape and size and at times fragmented as it slowly moved eastward, but its centre remained near Montreal.[116] At 11:33 a.m. on Thursday meteorologists in Montreal sent out a weather alert that stretched from Montreal west up the Ottawa River and east along the whole south shore of the St Lawrence River. Ice pellets and freezing rain were predicted for the whole day in the Montreal region. Once again this forecast demonstrated the limits of modern attempts to predict the actions of the beast of nature when it gets angry, not to mention the utter human incapacity to control them. At times the beast leaves the expected target in peace and unleashes its attack on others. This Thursday such a serendipitous occurrence happened to Montreal Island, which received only 2.8 millimetres of freezing rain over a twenty-four-hour period, compared to 19.6 millimetres in Ottawa and 22.7 millimetres along the South Shore. Unfortunately, much of the electricity for Montreal Island came from the South Shore.

Human error added to the problems caused by the freezing rain. At 1:03 p.m. the cable news service announced that Hydro-Québec would shut off electricity to all of Montreal at 3 p.m. in an attempt to balance its power network. This news of a total blackout of Montreal Island was retransmitted by Montreal radio stations. Hundreds of major companies immediately ceased operations and sent their workers home, creating an early afternoon traffic nightmare never before seen. At 1:39 p.m. the embarrassed news service admitted its announcement had been a mistake and apologized profoundly. One employee had stated that if a blackout occurred on all of Montreal Island, radio would replace television. Another colleague heard all of the sentence except the "if" and broadcast it. Thus a false rumour snowballed in the ice storm.[117]

In North America, urban sprawl around cities has resulted in numerous houses built on unserviced land. The homeowners dig their own wells and use electrical pumps to bring water to their taps. They also employ sump pumps to ensure that water does not enter and remain in their basement. None of these pumps worked when the electrical power was cut. So these homeowners watched their basements flood.[118] The headquarters for civil security in Quebec City was still convinced that the electricity would be restored in a few hours, mainly because it had not yet received accurate information concerning the scope of the disaster from Hydro-Québec.[119]

New York

The extraordinary heat wave continued in New York City, where the high for the day hit a record 65 degrees Fahrenheit, eclipsing the old mark for 8 January that had stood for sixty-one years.[120] The weather was not so pleasant in northern New York State, where an additional 100,000 customers had lost their electrical power as a result of the freezing rain. In some cases tree limbs crashed through roofs of houses and invaded bedrooms.[121] Tree surgeons, as they are called, were doing a booming business. People were advised to remain indoors.[122] The few stores that remained open dealt only in cash because their electrically operated cash registers no longer functioned. Banks and their automated teller machines had to close down without electricity, so people were running out of cash. In one upper New York State county, 1,000 employees were unable to pick up their paycheques today, Thursday.

Fire departments played an important role protecting against fires caused by fallen live wires and seldom-used fireplaces and chimneys and against carbon monoxide poisoning. Many fire stations were used as emergency shelters. In some cases, however, their telephone connections were cut by the ice storm. "Our biggest problem was that the Whallonsburg Fire Department couldn't receive calls. That really worried us; they were without phone service for almost four days."[123] Where communications were cut, amateur radio operators (hams) attempted to fill the breach. Since it took time for the severity of the storm to be noticed beyond the local region and for aid to arrive, villages were on their own for the harshest days. In other villages, such as New Russia, flooding forced the evacuation of residents until the waters receded.

With trees cracking around their mobile home, one family could not sleep, so they sat huddled in their living room with a candle for light. At noon they were evacuated to the Red Cross emergency shelter at the Quality of Life Building at the Franklin Correctional Facility in northern New York, where they felt much safer.[124] These prisons had backup generators. "You can't have the lights go off in these places," said the deputy superintendent of a correctional facility.[125]

A power repair crew in a central New York State community got the call at 6 a.m. to drive to a city in the north end of the state. It was rainy and 50 degrees Fahrenheit when the workers left, but as they drove north, ice began to glaze their bucket trucks. They worked through the day on Wednesday repairing lines while being drenched by freezing rain and ducking falling branches and trees. "Late at night around 2:30 and 3 in the morning, it gets rough. Once it's light, we get our second wind and go at it again."[126] On Thursday morning they were having their first warm meal since Tuesday.

The problem had become so bad that the governor of New York State declared a state of emergency in the five afflicted counties on Thursday night. Only emergency vehicles were allowed on its roads. Automobiles from Canada seeking entry were turned back by American customs officials.[127] The Division of Military and Naval Affairs was sending 500 personnel, eighty trucks, twenty ambulances, and twenty water trailers to the area.[128]

The plummeting house temperatures that were too cold for the human body were, nonetheless, too warm for frozen food. The longer power was out, the more food in freezers and refrigerators was rotting. People put some food outside or brought ice inside, using plastic bags to surround their food. Without electricity and heat there was little to do for those who were home alone, and houses grew colder and colder. "Soon it was necessary to wear a coat, then a second coat, gloves and boots, just to stay warm. He [the author's father] crawled into a sleeping bag, settled into a comfortable chair and began to read, using a battery-operated fluorescent lamp when daylight was no longer available. He stopped when it seemed like 9:00 at night, but his sense for time had betrayed him. His watch indicated it was only 6:00 p.m."[129] At night even the bed was cold despite multiple layers of blankets, so a good night's rest was almost impossible. The pattering of the sleet and freezing rain and the sounds of trees crashing to the ground provided more reason for insomnia. Blue flashes in the sky and "poof" sounds were the signatures of exploding electrical transformers. "We kept seeing flashes and explosions all over town. The sky was lit up like a war zone."[130]

When it became evident that electricity would be out for a long period, family and friends were invited to stay in any house with heat. "There were fourteen other family members camped there, mostly huddled in the one room where the wood stove had been installed. Parts of the house had been temporarily sealed off to contain the heat only to the areas deemed necessary for living. The diminishing quantity of food and other supplies was becoming a concern."[131] In normal weather towns are alive at night with street, house, and store lights and other illumination, but now even downtown was dark and empty. Everything was pitch black. Those who had to drive were confronted by electrical wires lying across the road and had to decide case by case whether the lines were dead or alive, insulated or stripped, and whether driving over them was safe or an act of self-electrocution.[132] Fearing fires, several towns asked residents to conserve water because the power outage had shut down their water pumps. The emergency 911 call centre for one of the affected counties typically received 40 calls a day at this time of year. On this day it received 769 calls. More freezing rain was forecast for overnight and Friday.

A shelter was created at the State University of New York in Potsdam. Over 800 people had arrived, more were on their way, and even more were expected the following day. But there was no one to oversee its operation or organize the volunteer crews. One person declined, citing lack of experience and training, but accepted when it was pointed out that no one else had those qualifications either and he was well-known and liked. The challenge was enormous for a shelter: one phone line for 1,300 people, 1,500 meals three times a day, no cots and not enough blankets, power going off and on because the generator had insufficient capacity for such a large operation. A bigger generator, cots, and other supplies were promised but would take several days to arrive. So people were shuttled home by truck to bring back mattresses and blankets. Some helped on the food line and cleaned up after meals.[133]

The US Army sent surveillance helicopters over the North Country to report on the damage to power lines in rugged, isolated areas. More than 500 power crews were now at work. Tree-trimming and road crews were also brought in from the rest of the state. The Department of Correctional Services sent crews of ten to twelve people to clear debris. "Inmate crews from the Moriah Shock Incarceration Facility were to assist in cleanup work in the county."[134]

Nine months earlier two fifteen-year-olds had passed their tests and received their licences as ham radio operators to enjoy a new hobby. During the ice storm the everyday communications system (telephone, fax, Internet) was in difficulty. If the emergency communication system failed as well, another means of backup communication would be needed. On Thursday morning authorities asked for someone to operate communications, and the two fifteen-year-olds were the only hams around. So they were sitting in a corner of a local fire hall relaying vital information and anchoring communications during a major disaster.[135]

New England

New England had not only weather forecasts but also the advantage of seeing what had hit regions to the northwest of it earlier in Ontario, Quebec, and upper New York State. Following the weather forecast, seventy-five trucks began salting and sanding the Maine Turnpike at 7 a.m. By dawn the freezing rain had transformed the state of Maine into an immense ice sculpture, and many Mainers make good use of their cameras. Falling trees were bringing power lines down with them. Electrical transformers shorted out, lighting up the sky, and 205,000 customers lost their electricity. Twenty-five electrical repair crews were brought to Maine from the neighbouring states of Rhode Island and Massachusetts. Walking was treacherous. Hospitals reported a growing number of fractured wrists and hip injuries. Debris from branches, trees, and power and telephone lines was everywhere. Highway workers used chainsaws to keep roads open for emergency vehicles, but they were fighting a losing battle. Authorities had to close sections of at least fifty state highways. Conditions became so dangerous that even some emergency workers were ordered to stay indoors. Sporting events and flights were cancelled: employees were sent home early; schools and a shopping mall closed their doors. The governor declared a state of emergency and requested federal disaster assistance. Forty shelters opened throughout the state of Maine.

WHEN A DANCE RESEMBLES A WRESTLING MATCH

When weather is variable within typical norms, there is a fluid coordination of the planning and practices of the modern societies of Canada and the United States with nature's dynamics. This leads nature's movements to be taken for granted and assumed to be in tune with society's needs. Safety seems ensured because nature is usually an accommodating dance partner. However, at times it reasserts its autonomy. On these hitherto rare occasions the dance partner becomes a wrestling opponent. When extreme weather struck, the previous orderly dance was replaced by the frenetic reactions of organizations and individuals to each change in precipitation, temperature, light, and wind. These responses were partly choreographed by organizational planning in advance, but much was improvised as the population and leaders tried to adjust to the unforeseen movements of their newly aggressive partner. Assumptions of safety were replaced by an emergent sense of risk because of the experience of harsh consequences and the incapacity of modern society to ensure its normal functioning. Organizations that had been rational and efficient under the normal dynamics of nature were now having great difficulty coping with its extreme movements. The severe weather exposed modern society as fragile. Previously, nature had seemed reduced to benign recreation, but now it appeared threatening and filled with danger.

This disaster was caused by freezing rain that resulted from unusually warm, moist air coming from the El Niño phenomenon in the Pacific Ocean. This air mass crossed the North American continent and collided with the usual cold air mass in northeastern North America in winter. It seems that the damage and disruption must have been a natural disaster. Or was it?

CHAPTER 4

The Natural Disaster Ends, but the Technological Disaster Continues

FRIDAY, 9 JANUARY

At 6 a.m. on Friday freezing rain started to fall on Kingston, and by 10 a.m. it was coming down all over eastern Ontario, southeastern Quebec, and southern New Brunswick. Many automatic weather stations failed, and modem communications between some meteorological centres were lost.[1] Ten more millimetres of freezing rain fell on Friday. It was now reaching a much larger area: from northern Lake Huron to southern Newfoundland. More than 20,000 homes were without electricity in the Canadian Maritime provinces as well. Despite the changing shape of the area targeted by the freezing rain, it was still centred over the Montreal-Ottawa-Kingston triangle, where it was most intense.

Ontario

Ottawa's principal newspaper started speaking of "the storm of the millennium" and wondering if it would ever end.[2] These comments were likely more influenced by all the commotion over the year 2000 than by meteorological evidence, since the newspaper did not have any records for the last one thousand years. But they did capture a prevailing sentiment of interminability among the storm's victims, which is difficult to imagine years after it ended. At 8:30 a.m. about 2,000 soldiers started to arrive in Ottawa in waves. They brought chainsaws, portable heaters, dump trucks, bulldozers, even armoured personnel carriers, and included some engineers, electricians, and medical personnel.[3] Travelling over icy roads from its base west of the city, the military convoy stopped at a professional hockey arena and used its vast parking lot as a staging area for personnel and equipment. The soldiers then moved to the most affected areas in the surrounding region on both the Ontario and Quebec sides of the Ottawa River. Lacking the electrical expertise of linemen, they were deployed in a multitude of ways: cleaning up fallen trees and branches, chopping wood as fuel for wood stoves, delivering generators, checking on homeowners without heat and light, especially in remote rural areas, and moving them to shelters. About 50 soldiers had telecommunications skills and helped telephone crews hook up individual customers. The army improvised to provide aid in unusual ways. For example, a dairy farmer had been unable to activate his machines to milk his cows because of the electrical blackout. Two soldiers raised on farms were found, and they milked the herd by hand. The military defined its role as supporting local authorities, not replacing them. "Implicit in all this was the psychological impact of a military response. The visible presence of soldiers suggests that everything that can be done is being done."[4]

Exhausted electrical crews also received reinforcements from southern Ontario. At 11:30 a.m. thousands of AT&T's Canadian customers lost their long-distance telephone service. Around 1 p.m., the power utility Ontario Hydro stated that more than 80,000 customers were without electricity in eastern Ontario. The National Arts Centre Orchestra in Ottawa cancelled its scheduled concert, the first time in its history it had done so because of the weather.[5] At 3 p.m. on Friday the state of emergency in the Ottawa region, which was to expire at noon on Saturday, was extended indefinitely. By 4 p.m. both Ontario Hydro and Hydro-Québec reported that the number of their customers without electrical power had increased despite their massive effort. Electrical lines fell as fast as others could be repaired. Ontario Hydro stated that sixty of its towers had been damaged. Eastern Ontario reached its high point of households without electricity: 122,300. People moved into shelters. These were not equipped for plants and pets, some of which died when left behind in unheated houses. Whether because of the bad roads or the declaration of a state of emergency, traffic dropped to one-third that of the previous day.[6]

Most people are losers in a disaster, but some benefit. Hardware stores were doing a booming business. One that normally received four delivery trucks of merchandise a week was obtaining four a day from outside the affected area during the ice storm to restock its shelves, especially with batteries, flashlights, lanterns, propane cylinders, camping stoves, and fuel. Locating stock became a full-time job.[7] One company specialized in supplying oxygen for people at home with respiratory problems and had 1,200 clients in eastern Ontario and western Quebec. A power failure can be deadly for people who rely on electrically powered machines to supply oxygen. Even though the company's drivers worried about live electrical wires hitting oxygen tanks, they provided round-the-clock home delivery even in remote areas.

Newspapers in this region are no longer delivered on foot after the school day by boys and girls carrying bags, but before dawn by adults in automobiles. Despite the atrocious driving conditions, the main newspaper in Ottawa reported that 98 per cent of its papers were delivered in urban areas and 80 per cent in rural areas, with a reduced number of complaints and an increase in praise.[8]

Modern technology increases agricultural efficiency in that automation enables farmers to have large herds despite using little human labour and to benefit from economies of scale. Productivity defined as production per person is thereby greatly increased. But this benefit depends on the availability of other forms of energy – in particular, electricity. Machines now milk cows, pump water, feed cattle, remove manure, clean barns, and process milk. Electricity provides the energy that empowers each farmer to be highly productive. It has replaced human manual energy to such a degree that farm labourers with the necessary skills no longer exist on the scale required. During the ice storm, about 2,000 of Ontario's 7,000 dairy farmers lacked electricity to activate the machines to care for their herds. Without electricity, water pumps from wells did not function, and without water, cows would not eat. In desperation some farmers tried to move snow into their barns by wheelbarrow to water their cows, a Herculean task.[9] The fact that the automated manure-removal system was inoperative did not convince cattle to stop producing the stuff. Unmilked dairy cattle get infections and become sick.

No one knew how long power would be out, but everyone knew that farmers in sparsely populated areas would be the last to have it restored. The consequence was enormous stress on farmers and a desperate need for large generators to power farm machines. Such generators are expensive to purchase and operate, but losing one's herd is much more costly. Farmers, particularly dairy farmers, suffered a great deal from the ice storm because it threatened to destroy the livelihood they had worked to build. They could only hope that nature would stop knocking down electrical lines faster than power crews could repair them. The Ontario Ministry of Agriculture tried to find large generators for farmers, especially dairy farmers. The fact that the minister of agriculture's own farm in eastern Ontario was without power reinforced his concern over the issue.[10]

At 5:40 p.m. the freezing rain stopped suddenly in the Ottawa area, and Environment Canada ended its warnings. Gentle, fluffy, unthreatening snow took its place. Virtually all observers agree that this change from freezing rain to snow, rather than any human action, marked the turning point of the 1998 ice storm disaster. No more ice was added to the trees, hydro lines, or pylons.[11] In this sense the natural disaster ceased abruptly at that point after sixty-three hours and forty-two minutes of freezing rain had fallen between 4 and 9 January. More than 69 millimetres of freezing rain had struck the Ottawa area in those six days, with no melting. In the forty years that records had been kept, the greatest amount of freezing rain for a six-day period had previously been 20 millimetres.[12]

The dependence on an electrical power grid resulted, however, in a continuing disaster, namely, a technological one. Serious problems remained because citizens, businesses, and government rely on electricity for the necessities of life in a dark, frigid environment: heat, light, communication, production, administration, and other activities. In eastern Ontario alone, Ontario Hydro announced that 2,800 kilometres of power wire had to be restrung, 10,750 hydro poles had snapped and had to be replaced, 84,000 insulators and 1,800 transformers had to be replaced, 300 steel transmission towers had been destroyed or damaged, and 40 per cent of its electrical distribution system had been wrecked. A three-man crew can hook up eight to ten poles a day, a smaller number if the poles have to be replaced.[13] Huge numbers of crews were needed, and still progress was slow. They dared not take more than a half-hour for lunch or they would lose the energy to get out of their chairs. Then back up they went in their open buckets suspended in the air, affectionately called "cherry pickers" in English and *giraffes* in French. When they laboured near homes, residents applauded their work and, on their return to ground level, hugged them and gave them cookies and coffee.

As the emergency continued, some workers acquired a great deal of knowledge about how to proceed. They worked extremely long hours and felt irreplaceable. "Because no one expected such a lengthy disaster, there were no plans in place for rotation."[14] The families of emergency workers were also affected. Those workers were needed on the job, but they were also needed at home. Even if their homes had heat, who would mind the children when schools were closed? Those whose homes had no electricity had much work to do at home and decisions to make to ensure the safety of their families. The ice storm brought out conflicting demands on them from work and family. They did not abandon their emergency roles, but they did suffer the anxiety of role conflict. Role abandonment and role conflict are two very different concepts. These competing demands did lead to burnout in some emergency workers.

Regular telephone service was adversely affected, but much less so than regular electricity. The telephone system was much more decentralized: there were batteries in the remotes that serve each small area and generators at the switching stations. The batteries ran down after five hours, but the phone company was more or less successful in getting generators to the remotes.[15]

A part of the Akwesasne territory lost water pressure. Six power crews were sent from New England and the Carolinas to repair electrical lines. Things were going well at the Mohawk Nation Longhouse shelter, with wood and food supplies arriving. But the generator at the Tribal Police Station failed, and the Akwesasne Emergency Operations Centre had to be moved to the Head Start Building.[16]

Quebec

Overworked Hydro-Québec crews in the west of the province were reinforced by sixty repair crews from the United States and twenty-five from other regions of Quebec.[17] Despite the superhuman efforts of the crews, the number of Quebec customers without electricity kept rising as long as the freezing rain fell. On Friday, 9 January, it increased to 1.4 million subscribers, that is, 3.5 million people, or one-half the population of Quebec.

On this day all five high-voltage lines to Montreal save one were broken by the ice storm. What little power remained relied on just one line. Montreal was fortunate that that line did not collapse too. To avoid overloading that precarious line, Hydro-Québec intentionally resorted to "load shedding": cutting off the supply of electricity and blacking out other large sections of the downtown core and western suburbs.[18] The premier of Quebec did not tell Montrealers they might lose electricity throughout the city. Later he said he had withheld that information for fear of causing a panic.[19]

Montreal has two oil refineries, operated by Shell and Petro Canada, both requiring electricity to function. On Friday they lost power and were no longer able to operate. Major distribution terminals owned by Esso, Ultramar, and Petro Canada also ceased operation when their pumps lost electrical power. Only one major distribution terminal owned by Shell had a backup generator and continued operating, but since it had only a two-day reserve and a non-functioning refinery, soon it too would have no more refined oil to distribute.[20] Service stations and gas bars that were lucky enough to have electricity were put out of business by these inoperative refineries and distribution centres. Many others lacked electricity themselves, and hence their own pumps no longer functioned. The power blackout broke many links in the oil refining and distribution network.

High-tension cables carrying electricity across the St Lawrence River fell on it, so the St Lawrence Seaway, used by ocean-going vessels to carry goods to the interior of the North American continent, had to be closed downstream from Montreal. The city has a good subway system, and underground public transportation should not be affected by freezing rain in the atmosphere. Since the system operates on electricity, however, two major power blackouts paralyzed a large section of the network for two hours on Friday. Driving was difficult because of icy roads: visibility through ice-coated automobile windows was poor, power lines and trees had fallen on streets, and others threatened to fall. Power outages blacked out many traffic lights and street lights, halted the operation of pumps in tunnels and underpasses, and disrupted work in traffic-coordination centres. Three bridges to the South Shore were closed because of ice falling from their superstructures and in order to allow workers to de-ice them. Those bridges would remain closed for four days. Only one bridge and a tunnel continued open in that direction.[21]

Montreal is an island city, an island upon which almost 2 million people were quickly becoming trapped. In normal weather large gasoline transport-trailers are not allowed to use the tunnel. But communities on the South Shore, especially those in the triangle of darkness, where the power blackout was the most severe, were threatened by the possibility of running out of gasoline to power emergency generators for their essential services (hospitals, shelters, homes for seniors.). As an exceptional measure, the transportation of dangerous materials was allowed in the tunnel under the supervision of the Ministry of Transportation and the police.[22] A disaster forces authorities to take extraordinary risks.

In all, less freezing rain fell on the city of Montreal (41 millimetres) than on Ottawa (70 millimetres). Montreal suffered more, however, because it depended on the South Shore for much of its electricity. There, 100 millimetres of freezing rain fell, crushing the transmission lines and depriving the city of most of its power.[23] Damage and suffering were determined not only by the freezing rain but also by vulnerability resulting from the specific way lines in the electrical grid had been placed. In the afternoon of 9 January Quebec's civil security agency mobilized all its forces even though it had not yet officially received information from Hydro-Québec about the state of the power network. The scope of the disaster had become self-evident. Only later that evening did Hydro-Québec officially inform Quebec's civil security agency about the precarious state of its electrical grid.[24]

The temperature rose slightly by the end of the day, causing huge sheets of ice, some weighing fifty pounds, to begin breaking off the tops of buildings and bridges. Authorities reacted by cordoning off a large part of the downtown area and closing bridges. Even on the remaining streets some pedestrians wore bicycle helmets.[25] Blocks of ice broke off the roof ledge of a hospital in a small Quebec city and fell, breaking windows above waiting rooms on the first floor. Patients there received a shock to go with their illnesses. The waiting rooms had to be closed. As they melted, some ice blocks were big enough to cause water damage on the level underneath.[26] Walking was treacherous on the slippery streets. Forty-six regional highways were also closed. One commentator described the mood as one of "fear: fear of the weather, fear of chaos, fear that the systems and networks that hold our society together were simply falling apart."[27] The government of Quebec requested more troops on this, the "worst day of the crisis." The storm hit the provinces of New Brunswick and Nova Scotia on Friday, but not nearly as badly as Ontario and especially Quebec.[28]

In Quebec, as in Ontario, the freezing rain ended in the evening. The peak in the numbers of customers without electricity had been reached. Since lines no longer fell faster than others were restored, progress could finally be made in the restoration of electricity. But that would take over two weeks more, leaving victims without heat and light in a frigid, dark climate.

An elderly couple died when their house caught fire, as did a ninety-year-old woman who insisted on staying in her unheated home. A woman in a Quebec town was killed by a falling chunk of ice.[29] House fires were becoming much more frequent than normal because people were using unusual means to keep warm. Hundreds more residents left their homes to seek shelter at schools and community centres with generators, but most people stayed in their homes and tried to keep warm with portable heaters and fireplaces.[30] Where shelters were full and their homes were cold, some families tried to sleep in their cars with the motor running,[31] a risky practice indeed.

Although shelters were warm in both the physical and the emotional sense, they were boring for many people. Time passed slowly. Television was its usual distraction, but it seemed even more unreal than usual. How many soap operas could you watch or games of bingo and cribbage could you play before ennui set in? Insomnia, already a problem because of the stress, was aggravated by the snores, moans, complaints, giggles, and prayers. "In one shelter, an old woman shouted 'Lorraine! Lorraine!' all night long. But no Lorraine appeared. A friend of mine got up in the night to visit the toilet, and almost crushed a sleeping child who had rolled off his mat onto the bare floor."[32] Even shelters were vulnerable to power outages. When a windowless gym used as a shelter lost electricity, it was transformed into pitch-black darkness that terrified its inhabitants. They were also vulnerable to social problems. Montreal's downtown convention centre was used as a shelter and suffered an outbreak of thefts and drug abuse. People in a shelter located in a big high school in the "triangle of darkness"' were terrorized by a group of teenagers until the latter were expelled. Conditions were particularly difficult for the 2,000 displaced people in that shelter because it was situated in the large area where electricity was completely cut off and there was no heat whatsoever: only cold mattresses to sleep on, cold water to wash with, and cold sandwiches for food. Shelters also became breeding grounds for flu. But for some lonely, elderly people who lived alone in normal weather, the experience provided an exciting opportunity to have company.[33]

On Friday the Quebec police, with the help of the Royal Canadian Mounted Police and the Canadian military, began visits to all residences in zones without power to verify whether people were in danger. On Montreal Island 237,000 residences were contacted, as were another 500,000 in the remaining electrically deprived regions of Quebec. Thousands of people were relocated, including several hundred who had to be urged to leave their homes. Another 900 residences potentially at great risk were visited daily. The police estimated that these inspections saved at least ninety lives on Montreal Island alone, particularly elderly people.[34]

Since everyone needed the same things that are little used in normal weather, from generators to candles and batteries, all these essentials of extreme weather were in short supply. The largest circulation newspaper in Montreal described the response of the population with the front-page headline "It's hell."[35] Ice had replaced fire in this new meaning of hell. A Hydro-Québec official attempted to reassure the population by claiming that the ice storm was a one-in-ten-thousand-year event. Some people believed him because they had never seen anything so severe or long in their lifetime of fifty-odd years, but others were skeptical since the official produced no evidence for the remaining 9,950 years.[36]

New York

On this Friday, 9 January, the wind joined with the freezing rain to bring down trees and electrical lines, close roads and airports, and leave 130,000 people without power, forcing thousands into shelters. The severity of the problem led the governor of New York to mobilize 1,500 rescue workers and 1,500 National Guard troops from all parts of the state to travel upstate bringing generators, food, and other desperately needed supplies. The *New York Times* described it as "one of the state's largest emergency responses ever."[37] Forty military ambulances, 60 water trailers, and 120 trucks were sent. Utility companies from across the state brought hundreds of specialized crews and vehicles to repair power and telephone lines. The Office of Mental Health dispatched a tractor-trailer containing 8,000 pounds of food. The State University of New York set up emergency shelters at its Potsdam and Canton campuses. The Department of Correctional Services organized fourteen crews of 120 inmates to clear debris and snow from the roads. The governor visited the area himself to assess the damage. By late Friday emergency workers had set up ninety emergency shelters throughout the five affected counties. At least 4,000 people had been forced out of their homes by the storm at this time. Some things cannot be put on hold by an ice storm. The largest hospital in the hardest-hit community, Watertown, delivered twelve babies on Thursday and Friday using a backup generator after its normal power line went down. Typical was a father who cleared fallen branches from his driveway to drive his wife to the hospital and assist at the birth, and then drove to a distant city to purchase a generator so that their house would have some light and heat when mother and baby returned home.[38]

The National Guard delivered food to people who remained in their homes and took others to shelters. One senior citizen, typical of many, tried to ride out the storm covered in blankets in her home, but she became fearful as her house got colder and colder. In the morning she cried out for help from someone walking down the frozen street and was taken to a shelter. Mayors of towns established curfews so that electrical repair crews could work without fear that their bucket trucks would be sideswiped by cars sliding on the slippery streets.[39]

Those fortunate enough to retain telephone service nonetheless had a problem: there were fewer and fewer people to phone. Some had lost their service because nature had disconnected their lines. Others had evacuated their homes. Businesses had closed. In the shelters the rumour mill increased its production: stories multiplied about dams bursting.[40] As the storm dragged on, the novelty of life in a shelter wore off. Games, singing, and other such activities lost their capacity to distract from the storm. Many people, especially children, became more scared and worried with time. "I've got an eleven-year-old here who's getting really panicked. He's scared that something bad is going to happen to all of us."[41]

Lack of electricity resulted in a shutdown of sewage treatment facilities and breakdowns in pumping capacity for the water supply. This problem led the Department of Health to issue a boil-water order throughout the region. However, not everyone had a source of heat to boil water. The acute-care capacity for all fourteen hospitals in the affected area was 526, yet they were now caring for 698 people because of storm-related injuries, and more emergency cases were expected to arrive. Long-term care capacity was 917, but 968 patients were now in care. The filling of acute-care hospitals and long-term care facilities in the region beyond capacity meant that doctors, nurses, and other specialists were approaching exhaustion. Two truckloads of 2,000 meals prepared by Prison Industries were to arrive shortly with a police escort. One correctional facility lost both its power and its telephone service. The state police were called in to help the Department of Correctional Services ensure that order was maintained.[42]

The ice storm was so vast that any affected county was surrounded by other counties, states, or provinces that were also affected. Neighbouring regions were unable to help one another. As supplies and help came from afar, they were often redirected to aid communities along the way. The rerouting of promised generators led communities at the end of the supply line, who saw it as "commandeering," to propose that the equipment be escorted by the National Guard.[43]

Thousands of utility workers from as far as Louisiana to the south and Michigan to the west came to help northern New York in its crisis, bringing with them utility poles, generators, and other equipment. The job required specialized skills, precision, and brawn, but the working conditions were atrocious. Workers were hoisted in their plastic buckets to the tops of electrical poles. There they spent hours cutting branches that threatened wires, smashing ice to free wires, thawing frozen support lines, and laboriously reattaching power lines. All this was done with thirty-mile-per-hour howling wind gusts that plunged the wind chill factor to 30 degrees below zero Fahrenheit. Most discouraging was that the freezing rain kept wrecking lines they had just restored. Temperatures varied from day to night and from hour to hour. The workers' clothing became soaked when it rained and then froze solid when temperatures dropped. Every few hours they took refuge in the cramped cabs of their trucks with the engines running, drying their clothes on the steering wheel, and eating cold food.

Working with electrical wires is dangerous at the best of times, and homeowners were inadvertently adding to the perils faced by repair crews. By attaching gasoline-powered generators to the wiring of their houses, thousands of homeowners were feeding power into downed lines, increasing the risk of electrocution for repair workers who assumed that the lines were dead because the main power source had been cut off. After an eighteen-hour workday, workers would sleep only four or five hours at night in unheated motels.[44] Then the next workday would begin. It was a remarkable experience, especially for workers unprepared for frigid temperatures, such as those who flew up from balmy New Orleans. Workers often pushed themselves to the limits of endurance. In one case three people on life-support systems were being evacuated to another hospital, but the street was blocked by fallen power lines. A utility crew preparing to take a break after an eighteen-hour shift was spotted and the situation explained. The crew then worked for several more hours to clear the road.[45]

The storm did not show any sign of relenting. There were many indications of the exceptional severity of the storm. Friday, 9 January, was the first time the *Watertown Times* newspaper had failed to publish since it was founded in 1861. The CEO of the power company that serves most of upstate New York stated that this was by far the greatest power failure faced by the company in its hundred-year history.[46] The governor toured the area by car and announced that 127 shelters were housing 7,300 people. He released a letter he had sent to President Bill Clinton asking for federal aid to rebuild the area. A seventy-eight-year-old man was found semi-conscious and taken to hospital by emergency crews, where he later died of hypothermia. Authorities had tried to convince him to go to a shelter on Wednesday, but he refused.[47]

Most people think of New York in terms of a world metropolis, but it is also a state that ranks third in milk production in the United States. There were 1,400 milk producers in northern New York State. Two of their four major dairy processors were knocked out of service by power outages, and seven of their nine major cheese processors.[48] Farmers without power were particularly hard pressed because electricity has become the energy source to operate the modern farm. Without power, cows are not milked, which severely distresses them and could affect future milk production. Even farmers fortunate enough to have generators had problems because the generators produced sufficient power to milk a large herd only sporadically. So farmers were stuck. "If it was just me and my family, we could pack up and go. But you just don't move 300 head of livestock."[49] Large factory farms had emergency generators that kicked in when electricity was cut, but smaller farmers could not afford them.

On a typical 250-acre farm in the area, which is not a large one, the farmer does the milking himself with the help of electrically powered machines. But when the ice storm crushed the power supply, all family members were called upon to replace electrical power with human muscle. Cows drink ten gallons of water daily. When the electric pump no longer had electricity to function, water was obtained by melting ice on a kerosene heater. Ice was bagged to cool the 575 gallons of milk in the tank. The whole family rose at 5 a.m. to work in the barn. Their twenty-year-old daughter "Becky concedes that she does not much care for dairy farming, and as she shoveled manure an hour before breakfast this morning she conceded that she was 'not a happy camper.' She cannot wait to return to her [non-farm] job." With limited water and heat, dirty laundry piles up and the house takes on the smell of the barn. The seventeen-year-old daughter described the effects of the ice storm this way: "You sleep less. And you stink."[50]

Unfortunately, there was still no electricity to pump the milk into the hauling truck when it arrived on Friday. Some was given to neighbours in containers, but the remainder had to be poured out. "When Tom had to dump the first milk, he just kept saying that there was no hope, that we're all done for. He was sick at heart." The storm was very stressful for hard-hit farmers who rely on electricity. The crisis nonetheless drew the family together, and they laughed when they heard city people complaining on the radio about nothing to do without electricity. There were signs of local solidarity. After two days without electricity and fruitless attempts to find an emergency generator, this family was loaned one by a local Mennonite who brought in several from other Mennonite communities. Another neighbour loaned them a gasoline-operated milking machine. At 3:45 p.m. the freezing rain stopped in one county and shortly thereafter in the other counties as well.

New York City was still basking in unseasonably warm weather. The *New York Times* quoted a grocery-store owner as typical of its citizens: "Winter is non-existent, except when I look up at the trees and see their leafless branches through the fog. It's almost surreal. I took a walk on the beach the other night – the warm, humid air, the fog, the mist – incredible."[51] The high of 63 degrees Fahrenheit in Central Park was within one degree of the record for 9 January, set in 1937. Since the new year began, four days had registered highs of 60 degrees or more, and the other four were between 50 and 60. Metropolitan New York's weather was caused by a slowly evolving pattern in the jet stream that rendered it an island of contentment surrounded by a massive ice storm to the northeast and unrelenting rain to the southwest. "There's a warmth that envelops me, puts me in a great mood."[52] Nonetheless, other New Yorkers found the warmth and peculiar fog at the beginning of January rather unsettling: there was "something weird and ethereal about the fog. It takes away any sense of where we are in time, where we are on the calendar."[53] Most New York City residents found this a small price to pay for the warmth and comfort brought by nature.

New England

On this Friday Central Maine Power Company reported 275,000 customers without power, Bangor Hydro Electric had tens of thousands in the same situation. New Hampshire was not as hard hit as Maine, but power outages there still peaked at 70,000 customers without electricity on Friday morning,[54] and the governor of New Hampshire declared a state of emergency.[55] Normal power was cut even to Central Maine Power headquarters in the capital city, and it had to depend on backup generators. That utility answered 200,000 calls for help, but it could not estimate when repair crews would arrive at the callers' homes. There were so many live power lines down in one area that road crews were told not to start work until dawn, when there would be enough light to see the danger. Maine Public Television went off the air for eight days. It had been intended as the backbone of the emergency broadcast system, but that failed as well.[56] So did Public Radio for a long period. Many private radio stations in south-central Maine lost power for transmitters and went dead. Four Portland television stations located just outside the area hit by the ice storm organized a telethon to raise funds for the American Red Cross. All functioning stations and publications attempted to link people offering services and goods with those in need. Over 3,000 Mainers moved into 130 shelters, and tens of thousands relocated to the houses of relatives or friends fortunate enough to have power, but most toughed it out at home.

There were, nevertheless, winners. Taylor's Trustworthy Hardware, which opened its store at 7:30 a.m., lost power an hour later, but it kept functioning using a propane lamp and flashlights and did a booming business in batteries, propane, kerosene heaters, and other emergency supplies.[57] The rare gas bars that had operational pumps were mobbed.[58] Tree-trimming businesses thrived as a result of the damage to trees and the need to cut trees around power lines. A supplement about the ice storm printed later by Maine newspapers carried large advertisements by Lucas Tree Experts, the Asplundh Tree Expert Company, and the like, which hoped to profit from the newly acquired fear of trees.

In Vermont the freezing rain ended late Friday. Temperatures rose above freezing and remained there for a day, stripping ice from trees and power lines. Nature itself was removing the danger it had created.[59]

SATURDAY, 10 JANUARY, AND SUNDAY, 11 JANUARY

Ontario

Saturday began with a strong southwest wind that by noon had blown the warm moist upper layer of air out of the region. It was sunny, with the temperature rising to slightly above freezing.[60] Chunks of ice began falling from buildings and trees onto sidewalks and streets, presenting a new peril for pedestrians and motorists. A man was hit on the head by falling ice while walking in front of the American embassy in Ottawa. Officials warned pedestrians to keep their heads up.[61] But that was not the only problem on the horizon. Meteorologists saw an Arctic air mass approaching and forecast falling temperatures,[62] hardly welcome news for the millions of people still without heat. The dangers were driven home when a family of three outside Ottawa were rushed to hospital by ambulance after breathing carbon monoxide from an improperly ventilated generator.[63]

At 11 a.m. on Saturday the Dairy Farmers Association of Ontario ordered farmers to dump their milk because the processing plants had no electricity.[64] Three cases of carbon monoxide poisoning occurred on Saturday night because the fumes from a generator operating in a garage had penetrated the attached house. The people survived because they were alerted to the odourless, colourless gas by a detector.[65] The telephone company installed extra phones in shelters and provided free local calls, but it then had to set up a control system when some people called overseas.[66]

Prime Minister Chrétien and Mike Harris, the premier of Ontario, visited Ottawa's Emergency Response command post as well as hard-hit rural areas. Because of the emergency, they and the Quebec premier postponed their departure on a trade mission – dubbed Team Canada – to Latin America. Another fifty generators were donated to eastern Ontario farmers by their counterparts in the western part of the province, and a hundred more were on their way from various areas of the United States. Since many telephone lines went down, the Ontario government sent satellite telephones to municipalities for emergency communication. Nonetheless, life went on. Two huge bridal shows – the 27th Annual Bridal Fair and the Wedding Palace Bridal Show – displaying the latest in wedding dresses, were sold out, attracting many people who had no electricity at home.[67]

The novelist Jane Urquhart lived in an area of southwestern Ontario the storm missed, but her twenty-year-old daughter was studying at Queen's University in Kingston, 150 miles away. Kingston's metropolitan population was 122,000, and it was the westernmost city struck by the freezing rain. All week the Urquharts saw on television the fallen trees, wires, and cables, as well as the twisted electrical towers. They received daily telephone calls from their daughter reporting how she and four other electrically deprived friends had moved into one room hoping their body heat would increase the now unheated room's temperature. The cat was passed from lap to lap for warmth, whether it be warmth for the cat or for the laps. All week long the Urquharts discussed what to do, and early Saturday morning they decided to act. "A state of emergency had been declared; trains had been canceled. The authorities were asking people to stay off the roads. But I wanted my child out of there."

The Urquharts joined a convoy of farmers in pickup trucks driving generators to other farmers in Ontario and Quebec. Normally, Highway 401 is one of the busiest in North America, transporting people and freight between Toronto and Montreal, but this day it was deserted except for long convoys of military vehicles. Approaching Kingston from the west, Jane Urquhart described how she "entered a brittle, icy world in which everything had collapsed. Electrical towers had crumpled into oddly human postures. The few trees that had not snapped or split were bent toward the earth as if in despair. Even the outcroppings of limestone were altered, their roughness obliterated by inches of beautiful ice. Entering the town, I drove through an obstacle course of fallen poles and downed wires, avoiding the streets where the police had put up the most yellow tape." The Urquharts found their daughter and her friends huddled in sleeping bags. "As we drove west, leaving the world of ice behind, the evacuees argued about which had been worse, the dark or the cold. We talked about the frail network on which we stake our survival."[68] A Canadian military convoy and a fleet of Detroit Electrical Company trucks heading east to help brought loud cheers from their car.

Although most of the work done by the soldiers was hard manual labour in frigid conditions, they were happy to come to the aid of the population in this civil emergency. "Finally now we're doing something helpful, as opposed to just training all the time. I mean, today all we were going to do is throw grenades. And we always throw grenades, and that gets boring."[69]Military helicopters flew over the affected area attempting to locate downed lines and bring patients to hospitals.

More than a thousand people waited in line at 7 a.m. on Sunday morning at a Home Depot store because they had heard that a supply of generators had arrived, each costing $798. The *Sunday New York Times* appeared in newsstands with the headline on the cover of its travel section reading, "Ottawa laughs at winter."[70] The title and the article promoting Ottawa for a winter holiday might have rung true on other occasions. But this was very bad timing. In the middle of this ice storm, a more appropriate headline would have been, "Winter laughs at Ottawa and at modern society." At 3:30 p.m. on Sunday, Environment Canada forecasted a mix of sun and clouds and a high of −5 degrees Celsius, which is around normal. Electricity in the city of Ottawa was restored on this day, but the blackout was still very serious in rural eastern Ontario, where tens of thousands of people heard they would have to wait another ten days for heat and light. In one rural area 90 per cent of the population of 34,000 were without power, and officials warned them that three to four weeks would be needed to restore it. Two men died from heart attacks while shovelling heavy ice.

Only now, almost a week after the freezing rain began, did the Ontario Emergency Management Organization realize the total impact of the ice storm.[71] Ontario Hydro produces electricity for almost all customers in the province, but it sells it directly to only a portion of them. Many municipalities have their own electrical utilities, which purchase electricity from Ontario Hydro and then sell it to clients. When Ontario Hydro announced the number of households without electricity in eastern Ontario, it only gave the number of its own customers. The Emergency Management Organization for Ontario had taken a whole week to realize that these figures did not include electrically deprived customers normally served by the forty-five other utilities in eastern Ontario. This misunderstanding made the conditions seem much less serious than they were and created confusion. In all, 700,000 people in eastern Ontario lost power, some for three weeks.[72]

Quebec

The area that suffered the most damage from the ice storm was the region on the south shore of the St Lawrence River just southeast of Montreal. It consists of a 45-by-45-by-20-mile triangle that is partly rural and partly made up of city commuters; together it is home to a million people. The electrical system was crushed by the ice: utility poles, distribution lines, transformers, transmission towers, and transmission lines. It would take up to four weeks to rebuild. The impact of the freezing rain led this area to be named the "triangle of darkness," the "black triangle," and the "triangle of ice." Huge, high-intensity steel transmission towers lay twisted, and long lines of transmission wires were splayed out across frozen fields.

European settlers had learned from the North American Native peoples how to tap sugar maple trees to obtain sap and reduce it to a sweet syrup or sugar by boiling. The six-week sap run in early spring was named the "maple month" or "sugar month" by the Ojibwa. An Iroquois legend tells of using the "sweet water" obtained by piercing a maple tree to cook venison, the beginning of maple-cured meats. Maple syrup is one of only two pure, natural liquid sweeteners, the other being honey. The world production of maple sugar is limited to the hardwood maple forests running from the midwestern United States through Ontario, Quebec, and the New England states, to the Canadian Maritime provinces. This area roughly corresponds to that hit by the January 1998 ice storm when it was at its most extensive. There are an estimated 20,000 maple-syrup producers in North America, generating 22.7 million kilograms of syrup in a normal year, more than half of it being from Quebec.[73] Maple syrup and sugar are consumed principally in North America, but small amounts are exported to France, Germany, and other countries. Many rural people earn their living by tapping the sugar maple trees on their farms and boiling the sap to make syrup. Buckets have been replaced by a vacuum-tubing collection system. The heavy coating of ice crushed the sugar bush, causing much stress for owners. "My father spent a couple of nights of hell lying in bed listening to the trees breaking. Every 20 seconds or so there was a big sound like thunder, and just by the sound you could know exactly what was happening with the trees. There's no insurance for a sugar bush."[74] Some of his destroyed trees were over one hundred years old. It takes twenty-five years for a maple tree to start producing marketable syrup and fifty years before it is producing well, so the recovery period will be long.

The Canadian prime minister and two of his cabinet ministers held a televised news conference this weekend with the premier of Quebec[75] to reassure the population that everything possible was being done and that the two levels of government were cooperating in this emergency, even if they do not do so in normal weather. The Quebec premier praised the impeccable coordination between the Canadian army and Hydro-Québec workers.[76] The military brought in supplies and personnel from all over Canada to an airport outside Montreal, using nine Hercules planes, a chartered 727, an Airbus, and two rented Antanov transport aircraft. The US military also helped to fly in supplies from western Canada. Fourteen flights arrived on Sunday, 11 January, alone.

Meteorologists officially declared this ice storm finished on Saturday, 10 January, around 7 p.m.,[77] but that did not end the misery. The freezing rain had stopped, but power was not restored. The number of inoperative automatic banking machines and closed bank branches reached its maximum on that day.[78] The breakdown of the debit card and credit card system had got worse because of the lack of electricity. In many municipalities it became impossible to purchase essential goods because people did not have access to their own money. Telecommunications companies also suffered heavy damage to their infrastructures, delaying or preventing communications.

On Saturday 3 million Quebecers, almost half the population, were still without power. Many cables and conductors fell on roads and railway lines, blocking them.[79] Traffic lights did not function. A concert in Montreal's Olympic Stadium by the Rolling Stones had to be cancelled. So was a hockey game featuring the city's "Glorieux," the Montreal Canadiens.[80] Crime decreased by 57 per cent. Hydro-Québec stated that it would take two more weeks to restore electricity fully, especially in the hardest-hit area south of Montreal. By Sunday 11,000 soldiers had arrived in the affected area.[81] Hotels and motels with electricity were fully booked by local people. Since the supply of electricity was so precarious, the premier of Quebec asked businesses and other enterprises in the centre of Montreal to remain closed. Most big ones did – head offices, department stores, banks, universities. But some small ones, such as a sex shop, marched to a different drummer and opened for business.[82] A huge shopping mall with a curious name that draws on religion to promote materialism, Les Promenades de la Cathédrale, left the decision to the discretion of each store manager. The result was an open boutique with rock music blaring from a giant speaker and a hundred spotlights blazing on rare shoppers outnumbered by staff.[83]

Worry about their homes was an important source of stress for those who had had to abandon them. They often made daily visits on nearly impassible roads to check, an additional cause of fatigue. In other cases there was electrical power at work but not at home, so it became difficult to reconcile work responsibilities with those of the family, especially for emergency workers. Although this disturbance of nature hit all classes of society, it had particular impact on those who were most vulnerable. The elderly and one-parent families with young children found it especially difficult to cope without the essential infrastructures of modern living. People with few financial resources, those without family or friends, and recent immigrants unfamiliar with the language, resources, and ways of life were particularly defenceless when nature's disturbance struck.[84]

Life went on in spite of the ice storm, and so did death. A man in Montreal died during the storm of causes unrelated to it. The poor man was beyond the need for heat and light at his wake and funeral, but even these social occasions depend on electricity. The embalming of his body required it, as did the mourners. The ceremony of death was disrupted by ice.

Outside Montreal a sixty-year-old woman was hit on the head by falling ice and died. Police closed off sidewalks beside skyscrapers because no helmet could protect against ice falling from such heights. City authorities closed every bridge leading into the city from the south and east because of ice hanging from the girders. Motorists were forced to take the only road left connecting the island city of Montreal with the South Shore. It went through a three-lane tunnel that quickly became clogged with a mile-long traffic jam. Parts of the city seemed like a ghost town without street and traffic lights. Some shelters had become full. A 1,500-person shelter was a tumultuous place filled with crying babies, running children, and card-playing seniors. The ice storm did have its advantages for those fortunate enough to have a fireplace in their homes. Many were glad to have intimate family time sitting around the fire talking, without the distractions of television and video games.[85] On this Sunday the government of Quebec issued a decree creating a ministerial coordinating committee composed of fourteen cabinet ministers presided over by the deputy premier, who was also the minister of finance. This committee was given sweeping powers to directly decide on necessary measures.[86]

During a rest period after raising poles in a small Quebec town near the American border, seven telephone linemen brought in from Newfoundland by Bell Canada bought $3 worth of lottery tickets. They won

$1.89 million.[87] No information is available on the number of linemen who bought lottery tickets and lost. The profitability of lotteries suggests that losers form a massive and very quiet majority. In this sense, the introduction of lotteries has transformed the vast majority into a society of losers based on the principle of luck determining destiny. There are similarities with extreme weather, which from the individual's point of view brings losses to most and riches to a few and is determined by chance.

New York

Just under 400,000 people live in the five sparsely populated but vast counties of upstate New York, and the majority were now deprived of electricity. The freezing rain had crippled the area's power system and created floods that closed bridges. Tens of thousands were in isolated rural areas without power or telephone service and were prevented from moving to shelters by blocked roads. Plans were made to send rescue teams to check on them in their homes. Ice two to three inches thick weighed down and toppled trees, power lines, utility poles, and transmission towers. In some areas, most of the utility poles along streets had fallen. Entire cities, albeit small ones, were without power. A state trooper described the situation with a military analogy: "It looks like a B-52 made a practice run. Things are down all over the place."[88] A mother attempting to survive the storm in her electrically deprived home with her two-year-old daughter and eight-month-old son was persuaded to move to an emergency shelter in the high school when a tree limb crashed into her garage and a live electrical wire hit the roof of her porch. She found that many of the people trying to sleep in cots in the hallways of the school were elderly. A woman who had moved to the area from Iowa noted: "I grew up in Minnesota and I've never seen anything this bizarre."[89] More than 5,000 state employees from sixteen agencies had been dispatched to the disaster area by Saturday, 10 January. The Federal Emergency Management Agency sent 1,625 power generators, 25,000 cots, 50,000 blankets, 750,000 meals, 500,000 quarts of bottled water, and other supplies.[90] Since airports were closed, all these were being trucked in along treacherous roads with an escort of state troopers. Trees and power lines falling without warning along roads made travel hazardous. Many roads were closed, and in some areas non-emergency vehicles were banned.

A nursing home lost power, but at least it had gas stoves for cooking. Some of the elderly residents enjoyed the chaos brought by the freezing rain and found it exciting. One eighty-four-year-old woman was thrilled one morning when staff used a candle to bring her French toast: "It's the first time I ever had a candlelight breakfast."[91] In another area, neighbours found a seventy-one-year-old woman, who lived alone, dead in her living room of carbon monoxide poisoning.[92] In one New York county, 911 operators handled more than a thousand emergency medical calls. Electrical well pumps that supply rural homes with water did not operate without power, so the National Guard shipped in thousands of gallons of water. On Saturday the sun came out for the first time since the beginning of the storm, but this improvement in the weather led to more traffic on the roads, which proved to be dangerous for power repair crews.[93]

On Saturday President Clinton declared the five upstate New York counties a federal disaster area, making them eligible for 75 per cent of the cost of emergency services and debris removal, for projects that reduce future disaster risks, and for grants and loans.[94] This assistance went to local and state governments, public and non-profit organizations, and Native tribes harmed by the ice storm.[95]

Niagara Mohawk Power Corporation, whose pretty name has nothing to do with the Mohawk nation, had 1,500 employees working to restore electricity to its 100,000 customers (80 per cent) without power in the 7,000-square-mile region served by the company. Officials predicted it would take weeks to restore normal service since six electric transmission towers had collapsed as a result of the ice and high winds, the whole distribution system was in shambles, and working conditions in the freezing rain and wind were awful. Repair work was advancing slowly because many collapsed lines were located in remote or heavily forested areas.[96] Freezing rain and ice jams were also causing some rivers to flood, and bridges were closed. Another woman died from carbon monoxide poisoning on this Saturday.[97] A forty-eight-year-old man, who was a corrections officer and popular soccer coach/referee, started a generator in his basement. He, too, was overcome by carbon monoxide and died.[98] It was not just the elderly and the infirm who failed to perceive risk. The supervisor of a town without electricity for two weeks stated that the town's devastation was "across the board. In some areas, it actually looked as if we had been bombed."[99] The military metaphor returned again and again.

The change from unaffected area to heavily damaged one was sudden in places. A New York State senator driving up to the North Country to assess the damage remarked that "it was almost as if someone had dropped a curtain. It was not a gradual change. It was just all of a sudden, things were covered with heavy ice ... trees down ... lines down."[100] He noted the contradiction between the destructiveness and the beauty of the ice storm. "Everything was covered in a heavy white crystalized ice. Some of it almost looked like a very fine sort of cut glass or blown-glass sculpture." Then he, too, added the military metaphor: "The forest areas were the most striking. It almost looked like someone had come in by air and bombed." The senator was in the region to meet the state governor. Shaking hands every few feet at the shelter, the governor promised the visit was not a photo opportunity and that all the state's resources would be available to meet the needs of people.[101]

On Saturday a baby boy was born at a hospital functioning on emergency power. His parents commemorated the birth during a disaster with a middle name for the baby: Devin Storm Piercey.[102] A forty-five-year-old contract worker was killed in northern New York on this day as a result of the ice storm. He was trimming trees in a bucket forty feet in the air when the truck holding him slid down an icy road hurtling him to the ground.[103]

A wedding had been planned for Saturday at a church on Military Turnpike. A local woman was to marry a soldier from the nearby town of Peru, who was stationed at an air force base in Arizona. The couple had not seen each other for four months, but a great deal of time had been spent making the arrangements. The ice storm struck as the soldier was flying home on Wednesday, and the flight was diverted further south to the New York State capital of Albany. The bride drove down to get him. The return trip in the storm took six and a half hours, quadruple the usual time. "But at least she had the groom," the bride's mother said with relief. When the ice storm had got worse by Friday, the caterer backed out. The reception was to be held in the American Legion, but it had no power. The limo service and the photographer cancelled. Only the disk jockey confirmed his presence. The soldier's leave was short, and the couple had to go to Arizona immediately after the wedding, so it could not be postponed for a week.

The bride's mother would not let her only daughter's wedding collapse, and she started making other arrangements. The Howard Johnson Hotel could hold the reception but could not provide food. Perky's Flowers had flowers but could not deliver them. So friends picked them up. Keeseville's justice of the peace would be able to perform the ceremony. The owner of the Wedding Belle opened his store, which had no power, just for the family. The bride and her mother went in with a flashlight and got the wedding gown and tuxedos. But at 6 a.m. on Saturday there was still no food for the wedding. Finally, a store was found that had chips, salsa, and other snacks – not quite the stuffed chicken breast that was planned but food nonetheless. The line at the cashier was very long, and the wedding hour was approaching. "That's when I almost lost it," the bride's mother admitted. She pleaded with other shoppers to let her through because her daughter had to get married, and they did. The wedding was particularly emotional, as all the guests in jeans munching their chips at the reception could confirm. This happy outcome lifted people's spirits in the town of Plattsburgh during the ice storm. The local newspaper wrote that "Mother Nature's iciest efforts couldn't prevail against the warmth of another mother's love."[104] Perhaps not, but the first mother sure can cause havoc for the second.

The temperature rose on Saturday into the 40s, but then the weather roller coaster headed in the opposite direction. Most depressing were the forecasts from the National Weather Service, with meteorologists predicting temperatures falling to single digits Fahrenheit on the next two nights and winds gusting to 30 miles per hour.[105] This combination could bring down hundreds more trees and utility poles, burst water pipes, and endanger families waiting out the storm in unheated houses. The plummeting temperatures would mean that the layers of ice on all surfaces would not melt, and the weighing down of trees and the electrical infrastructure would persist. Thus the technological catastrophe of a crushed electrical system would remain and perhaps even worsen, despite the fact that the freezing rain had stopped. Because trees and power lines were falling on streets and highways without warning, authorities urged residents to stay indoors and then declared a curfew from 5 p.m. to 7 a.m.

On Sunday snowflakes fell instead of freezing rain – four inches instead of the forecast twelve – and temperatures remained in the 20 degrees Fahrenheit range, rather than the single digits predicted by meteorologists. The improved weather enabled 3,000 electrical workers to accelerate their restoration of lines and repair of transformers. Hope grew that normal living would return at last. But still there were problems. The electric company tried to send six helicopters to assess damage accurately, but the flights were grounded because of the snow and grey skies. The snow created new worries that people would inadvertently touch live wires hidden on the ground. About 300,000 people remained without power in upstate New York.

A couple ran a gas-powered generator in a desperate attempt to warm their small house and were found dead from carbon monoxide poisoning. The woman was discovered with covers up to her neck, only her face and wool cap showing. A neighbouring farmer had previously come to their door but had driven off when there was no answer, busy as he was with his own problems. Two pharmacists in Watertown filled prescriptions by the light of a battery-powered lantern. Forty nuns at the Sisters of St Joseph Mother House were blessed with a generator and passed the time praying for the less fortunate and playing cards and board games. They extended an invitation to anyone who was cold to join them.[106] At 7:20 on Sunday morning the emergency generator at Watertown's largest hospital collapsed from the strain of continuous operation over a two-day period, leaving the hospital without power. Nurses began to manually pump air for patients on ventilators in the intensive care unit. A backup generator restored some power after seven minutes, and the electric company ran a special line that was operational by 3:30 p.m., but in the meantime it was "a real tense situation."[107] Also on Sunday another couple was found dead in their home of carbon monoxide poisoning.[108]

New social problems developed after the freezing rain ended. People became impatient for supplies, and some communities lifted the travel ban. But workers suspended in buckets high in the air had already had bad experiences with motorists violating the ban, driving too fast for the icy conditions, and almost striking the repair trucks. Often utility trucks had to park several feet onto the road for stability because brakes did not function well on ice. In order to protect their workers, the utility companies threatened to take their trucks off the road where the travel ban was not enforced. A compromise was reached whereby school buses were used for a four-hour period on Saturday and Sunday to shuttle people to stores to purchase needed supplies.[109]

In northern New York State several different utility companies supply power in various areas. People began to make comparisons about which one had the electrical circuits repaired fastest. Equitable comparisons were difficult, however, because the freezing rain began sooner and was more intense in some areas than in others. The spokesperson for Niagara Mohawk Power Company put it this way: "As the days went on, the problems became more significant till we finally saw the [electrical] system actually start to disintegrate."[110] Several days into the freezing rain, the power companies, assisted by meteorologists, saw what was coming, but they could only repair and rebuild at that point because their material proved to be not sufficiently resistant. The Division of Military and Naval Affairs mobilized 1,900 troops to work in northern New York and sent sixty ambulances, fifty-four water trailers, fourteen tractor trailers, thirty-eight dump trucks, and other vehicles.[111] This was indeed a mechanized, military-like fight against an ice storm.

Meanwhile, the unseasonably warm weather around the New York metropolitan area was causing a different set of problems. Orchard owners were watching their peach, pear, apple, and other fruit trees swell up with sap and their plants bud prematurely. The trees could bloom and then be damaged by frost when the colder weather returns. Fruit farmers are always vulnerable to nature's disturbances and prefer regular cycles.[112]

New England

Maine is a geographic curiosity. Other than Alaska, it is the only state in the United States that lies north of the main population centres in Canada.

The freezing rain pelted the whole centre of the state, whereas the colder northern part received snow, and rain fell on the southern tip. But almost all of the state was affected because of the width of the band of freezing rain and its effect on the electrical system. On Saturday some 115-kilovolt transmission lines collapsed, including one that was the principal source of electricity for the capital, Augusta. At a residence for seniors in Pumpkinville, a seventy-one-year-old woman said: "I couldn't sleep for the trees. They kept me up all night – breaking off, you know, you could hear 'em."[113] When the home's electricity was knocked out early Saturday morning, the gas-heating system no longer fired up and exhausted properly, so the residents converged around a wood stove in the community room. One remained in her cold apartment bundled in three layers of bathrobes because the wood smoke irritated her asthma. In another community, officials noted the decrease in water pressure and assumed people were running water to prevent pipes from freezing, but they then found a break in the water main under a pile of fallen trees. At one emergency shelter the generator functioned for three hours and then broke down because it could not handle the load.[114] On this weekend Maine reached its peak of powerlessness – 500,000 people without electricity, or half its population.[115] The governor, whose home was without electricity for a week, declared the ice storm to be "the biggest disaster of this kind that has ever hit this state."[116] Hospital emergency rooms became full of sick and anxious people. "Because of the medical needs of many of the people who showed up in our emergency rooms, the hospitals became shelters, though that was not our intent."[117] About 9,000 people moved into emergency shelters, but even more spaces were available. Most people chose to remain in their homes despite the bone-chilling, pipe-bursting cold.

As Mainers tried to solve the problem of cold homes by using generators, space heaters, camp stoves, and charcoal grills, they often created a more serious, imperceptible one. One man ran a generator in his basement to keep the house warm. It generated not only heat but also colourless, odourless carbon monoxide, which seeped upstairs where he and his wife were sleeping. He died, and his wife was at last report in critical condition. Another man was found dead in his basement beside his generator. A hospital in the area hit by the ice storm had to treat twelve cases of carbon monoxide poisoning within twenty-four hours, the same number of cases they normally had in two years. Four residents of a nursing home using a generator were admitted to hospital for carbon monoxide poisoning. Barbecuing food indoors, faulty kerosene heaters, and even gas generators in garages were causing similar problems. Maine experienced the largest incidence of carbon monoxide poisonings in its history. Yet people were reluctant to leave generators outside for fear they would be stolen. Officials became afraid that they would face an epidemic of poisoning.[118] Some homes and institutions were equipped with electrically operated carbon monoxide detectors, but without electricity these failed to reveal the presence of the dangerous gas.[119]

Down Easters, as they call themselves in Maine, had created a catchy slogan for their state: "Maine – Life as it should be." Fair enough under normal weather, but it did not seem quite right when the weather became extreme. Motorists entering the state from the south would see that sign and then a few miles further a new one warning: "Fuel unavailable on turnpike north of Portland," because gas stations without electricity could not pump gas.[120] On Sunday 229,000 homes and businesses in Maine lacked electricity, 41,000 more in New Hampshire, and 36,000 in Vermont.[121] One hundred and thirty emergency shelters were opened in Maine. In several communities the water-supply and sewage-treatment systems stopped functioning for lack of electricity, creating fire hazards and contamination problems.[122] Since neighbouring states were also affected, power companies in Vermont brought in repair crews from as far away as Arizona and Hawaii. National Guard troops clad in camouflage used chainsaws to clear roads of debris from trees. As one guardsman said, "When we do things like this, the Guard is very rewarding. Desert Storm was a good experience, but doing this is helping the people here."[123]

The experience of life without electrical power and the sight of utility crews repairing other neighbourhoods led to frustration and resentment. Complaints came pouring into electrical companies. For example, on Saturday Burlington Electric received a complaint that the University of Vermont and its surrounding area had power, whereas the caller's nearby street did not. "I want to know why all the lights are out here, where the working people live, when they've got all the lights on for the progressives and communists," the caller asked.[124] The power company spokesperson was trained to respond that no neighbourhood had priority over any other and that the company was trying to assign crews to every area. Power was first restored in the triple lines that feed hospitals, sewage plants, and industrial and commercial centres and then in the single lines that supply individual homes in neighbourhoods. This process reflected predetermined priorities for the restoration of power: first to protect public safety, next to ensure electricity in hospitals, downtowns, and population centres, and lastly to supply individual homes. It is not known whether the complainer was convinced by this response or still believed that the communist-ridden university had got its electricity restored first.

DEPENDENCE ON AN ELECTRICAL GRID AND DISASTER

The two modern societies of Canada and the United States had developed an efficient economy, modern comforts, and conveniences that depended on energy from a centralized electrical grid. Farmers as well as city dwellers relied on this technological advance and on specialists who knew how to make it function. Even satisfying needs as basic as heat and light in a cold, dark winter now depends on electricity. Previously, people had wood stoves or coal furnaces and kerosene lamps; so they would have at least had heat and light and hence would not have faced an emergency. Freezing rain would have had little effect on their homes. This disaster occurred because people were deprived of an electrical infrastructure that had been made essential. Even when the freezing rain ceased, the emergency continued because so did electrical deprivation. Technological progress, in the form of an electrical grid upon which people were now dependent, brought the harsh consequences of freezing rain into their dwellings and communities. Organization and technology so rational in normal weather internalized nature's extreme dynamics into society. Claims of the "end of nature" are a misleading conception assumed during the average dynamics of nature. They need to be replaced by the understanding that modernization internalizes nature's dynamics into society, and in cases like this, it manufactured a natural disaster.

CHAPTER 5

The Arduous Return to Normality

Although normal winter weather for January had returned to this vast region, the destructive effects of the unexpectedly long, intense freezing rain were felt for some time. How long it took to be rewired to normal living varied from one local area to another, as lines were hooked up or transformer towers rebuilt. The return of this most essential infrastructure, upon which others depend, varied from 12 January to 6 February. Restoration was difficult and slow because at this time of year normal weather meant below-freezing temperatures, cold winds, and only nine hours of daylight.

Ontario

On Monday, 12 January, Canada's capital remained in a state of emergency for a fifth day. At 3 p.m. Prime Minister Jean Chrétien spoke on radio to those without electricity, whose only means of receiving information were battery-powered radios. "On behalf of the people of Canada, I want to let you know that you are not alone – and you will not be alone as long as you are in need."[1] Premier Mike Harris of Ontario visited a farm and promised residents that they would be compensated for their expenses, a promise widely reported in the media.[2]

The Ottawa Fire Department's communication centre was forced to use emergency power for four days, and four of its fourteen fire stations had to resort to generators. Traffic lights at eighty intersections in the region lost power, and there were not enough generators to get them working again. This situation forced planners to make decisions about which intersections were more important and to place temporary stop signs at the remainder. Some traffic lights that had their power restored lost it moments later. The Social Services Department was severely overworked. The homeless and other dependent people need aid whether there was a storm or not, but now there were many additional people in need because of the emergency. The department had its own problems with electricity and had to shut down one office for a day and a half when it was too cold indoors to work.[3]

Despite being deprived of the normal tools of everyday life they had grown accustomed to, people organized themselves as best they could. A small town called Alfred, some 70 kilometres east of Ottawa, lost all electricity. The situation was especially bleak for the eighty elderly people without heat or light in a residence for seniors. Then some members of the community remembered an enormous generator lying in a field that was used for summer carnivals in rural areas. They dug it out of the frozen snow and used two backhoes to bring it to a local hall. The generator then powered a shelter for the seniors and a hundred other temporarily homeless people, with hot meals cooked there by the town's caterer.[4] The first responders in a disaster are the victims themselves, providing for themselves and helping others long before professional help arrives. Life became simpler but more arduous. People reverted to being gatherers: lining up for candles at one store and then joining a three-hour lineup for gasoline at a rare station that had operating pumps.[5]

Schools had been closed in Ottawa since Thursday. A man died after falling from a ladder while trimming branches. All "non-essential" government workers were told they did not have to report to work, the insult of being declared non-essential easier to endure because of the opportunity to stay home with family. On Tuesday the federal government returned to normal operations in the national capital region. At 3:30 p.m. the regional government of Ottawa-Carleton announced that the state of emergency had been lifted for the urban areas. It continued in rural areas, which began to complain that power companies were slow to restore their electricity.[6] Repair crews still worked under exceedingly difficult conditions; for example, on 15 January in a wind chill of minus 40 degrees (Fahrenheit or Celsius does not matter, for that is just below the point where the two scales meet). It would take almost another two weeks, until 24 January, to restore power to everyone in eastern Ontario.

Shelters were used not only as temporary homes but also as drop-in centres. Even people who chose to stay in their cold, dark houses came to shelters to warm up, take a shower, have a hot meal, obtain supplies, see a nurse, and meet neighbours. For example, in Ottawa "about 3,500 people dropped into a shelter and about 100 people stayed the night."[7] On

Monday, 19 January, 50,000 people in Ontario remained without power. Restoration of electricity was proceeding day by day. The provincial government began to distribute relief cheques to farmers: $1,000 for every week without power.[8]

Ontario Hydro was criticized for a slow response, poor routine maintenance, and lack of planning for severe weather. It was denounced for inept communications concerning the overall power problems and misleading answers to questions about when power would return, especially in rural areas. When pressured into giving the date when power would be restored, it was not cautious in its estimate; so target dates were repeatedly missed. Furthermore, asked when power would be restored in the village of, for example, Embrun, the spokesperson would answer that power would be restored to the Embrun substation on 19 January. Days after that date, people in Embrun were annoyed that their power had not yet been restored, only to be told that the substation had been reconnected on 19 January and the spokesperson had never said that their homes would be reconnected. In some cases the substation was not even the one that served the village with the same name. Ontario Hydro was slow to appreciate that this bureaucratically correct miscommunication only enraged the population shivering in heat-deprived houses. "Persons can cope with the truth no matter how alarming; they find it difficult to accept statements that turn out to be misleading."[9] Often, not only the main electrical line on a street was broken but also the connection from the street line to individual homes. Ontario Hydro crews were not, however, allowed by their superiors to make repairs on private property. People freezing in their homes would see the power crew repairing the main line on their road, and their hopes would rise. Then they would see the crew drive away and not return, leaving families still freezing in their homes without power. Their sense of betrayal led to numerous complaints. As a result, the rules were changed.[10]

When the chair of the Regional Municipality of Ottawa-Carleton telephoned the premier of Ontario to express his displeasure, the chair of the board of Ontario Hydro and some of its other officials showed up the next day in Ottawa to answer questions.[11] Communication through normal channels from the bottom up produced meagre results, whereas the top-down approach worked miracles. After the chair of the Ottawa regional government said that he had difficulty getting answers from Ontario Hydro, the chair of the utility told a news conference that even he had trouble getting replies from his own staff. Following this "press conference from hell," as it was called by Ontario Hydro's public relations department, it was open season to criticize the utility.[12]

In Ontario there was the additional problem of an array of local electrical utilities existing alongside the provincial utility. They had not got along well before the ice storm, and tension continued during the response to it. Cooperation and coordination of action during the blackout were poor. "Instead of getting together, sharing information and reaching agreement on what resources should be called and where they should be sent, the local utilities and Ontario Hydro acted independently."[13] There were some turf wars between utilities. In some cases Ontario Hydro rejected help from local crews. In others it insisted on retaining control to ensure that a line would not be energized by a crew from another utility. These arrangements may have been justified, but they took up valuable time. In still other cases Ontario Hydro sent its crews elsewhere so that there would be no interference between the crews of the two utilities. These last-minute changes delayed the restoration of power. "That not only made Ontario Hydro's forecasts of when power would be restored appear inept, it gave the impression power was restored only when Ontario Hydro stepped aside."[14]

The Emergency Management Organization assumed that the utilities shared information with one another, but they did not. So the EMO failed to add the blacked-out clients of local utilities to the Ontario Hydro totals and grossly underestimated the extent of the disaster. To correct these public relations disasters, Ontario Hydro implemented what it called a "hug a customer" program during the last few days before power was restored to the remaining customers. There is no data to determine whether customers were warmed sufficiently by the hug to compensate for the cold caused by delays in the restoration of heat.

There has been long-standing tension between the city of Ottawa, the capital of Canada, and Toronto, the larger city and the capital of the province of Ontario, where Ottawa is located. Many Ottawa residents refer to Toronto as Hogtown, and many Toronto locals call Ottawa a "fat-cat city." When the city of Ottawa first contacted the provincial Emergency Management Organization in Toronto to request generators, it was told none could be supplied; so Ottawa looked elsewhere. Days later the Toronto-based organization wanted to know what generators had become available and where they were being used. Irritated by the initial refusal, the Ottawa team ignored this attempt to take control and continued to conduct the search for generators themselves.[15] The EMO insisted that requests for outside supplies be sent to the agency in Toronto, and it would forward the supplies to Ottawa. This policy resulted in delay, with some supplies arriving at their ultimate destination when they were no longer needed, and so it was opposed by the region of Ottawa. In ways such as this, the ice storm disaster exacerbated tension between these two capital cities.

It has been estimated that 75 per cent of Ottawa's trees on municipal property were damaged by the storm: 10 per cent would have to be cut down and the remaining 65 per cent would require surgery.[16] These trees were not clustered sufficiently for natural regeneration to occur. The cost for municipal authorities to bring the urban landscape back to normal would be enormous. Buying and planting one tree costs the city $350; this figure would be multiplied by the six thousand to be replaced and increased by the forty thousand trees that would require surgery. Regeneration of widely dispersed trees by authorities to bring life to the concrete and asphalt setting of the city would be slow, awkward, and expensive, contrasting sharply with nature's rapid, efficient, and cheap process of regeneration in a forest.[17]

Quebec

Hydro-Québec reported that 590,000 residences in Montreal and outlying areas were still without electricity on Monday, 12 January, as temperatures fell from 0 to −18 degrees Celcius. The combination of home-heating systems rendered inoperative by a lack of electricity and a cold spell created dangerous conditions for those attempting to tough out the situation in their homes. Police were now going door to door to warn residents, particularly the elderly, that they should move to community shelters. More than 10,000 people were already living in 120 temporary shelters in Montreal.

One seventy-seven-year-old woman in a shelter for six days, playing cards and worrying about her home, summed up the feelings of many elderly people: "I thought it would be one day or two days, so I didn't bring anything. This isn't normal. It's like the end of the world."[18] A twenty-one-year-old McGill University student moved into a city shelter after electricity was cut to his apartment. He commented, "You basically almost get to realize what a homeless person lives like."[19] Some people were pushed from place to place by power blackouts. One family lost electricity in their suburban home on the South Shore of Montreal and went to stay with friends two hundred kilometres away, but then power was cut there as well. The family moved back to a different part of Montreal to stay with relatives, but that home lost its electricity too. They finally ended up with the wife's mother in a community outside Montreal. Their homelessness consisted of moving between four different lodgings in less than a week.

Improvisation and resourcefulness were called for. A car wash in a town near Montreal had power, but many homes in the community did not. People were not washing their cars during and following the freezing rain; in fact, without electricity to heat water, they had trouble washing themselves. So the owner re-engineered his car wash into community showers: ten for women, ten for men.[20]

After the freezing rain ceased, repair crews working through the nights were finally making progress in restoring electricity, some areas having been without power for six days or more. Electricity was restored to the downtown business core of Montreal, but the power grid was still fragile. For example, between 9 and 19 January, service in the subway system was interrupted for a total of fifty hours. Authorities were worried that if the fragile electrical system was overtaxed by businesses coming on line, this could lead to a collapse of the restored circuits, and the whole city would be plunged back into darkness.[21] So they asked businesses, colleges, and universities to remain closed. Even the subway system was forced to reduce its electrical consumption by 50 per cent, which it did for ten days by closing one line, decreasing the number, acceleration, and rhythm of trains on other lines, and shutting off escalators.[22] This reduction enabled Hydro-Québec to repair and rebuild lines on the city's south shore that fed Montreal without worrying that electrical demands from downtown would provoke a blackout once again.

Saint-Hyacinthe, a city of 41,000 inhabitants, held the unfortunate distinction of being the hardest-hit community. Almost 10,000 of its residents fled the city to stay with friends or relatives or in motels, another 5,000 were living in shelters, and the remainder were trying to tough it out at home. The regional hospital functioned on an overworked generator, the one temporary electrical line to the city being too fragile. A resident at a shelter commented: "Everybody's anxious. Everybody's tired. Everybody wants to believe this will end on the 25th, as Hydro-Québec says ... But it's hard to live with 600 or 700 other people in a shelter."[23] The shelter residents got tired of their new routine. "The days are long and tedious, and I'm fed up. We've been here since day one. Everybody just wants to go home."[24]

A spokesperson for the Montreal Health Department stated that there had been 162 reported cases of carbon monoxide poisoning, including 3 deaths, 13 reported cases of hypothermia, and many calls reporting stomach problems.[25] People trying desperately to keep their homes from freezing would use outdoor gas-heating devices indoors and be poisoned by carbon monoxide without perceiving it. Apparent solutions to perceived risks created new unperceived risks. As food thawed out in refrigerators without electricity, people would eat it, not realizing it had already gone bad. Perceptions can deviate significantly from material reality.

A hotline was established to connect people with power willing to give shelter to people without electricity. Calls offering help outnumbered requests for help by a 20-to-1 margin.[26] Volunteers had been rendering exceptional service of all sorts since the ice storm began, especially in shelters. By the second week, however, they became exhausted. The human body had reached its limit of what it could accomplish with minimal sleep.[27]

The return to normality was a roller-coaster ride. The freezing rain followed by partial melting and then by an Arctic chill left huge mounds of hard ice that wrecked 30 per cent of Montreal's snow-removal equipment.[28] On Saturday, 17 January, Hydro-Québec announced it had replaced 460 transmission towers and that it expected to restore power completely within ten days. On Monday a 735-kilovolt line was restored, re-establishing a major link in the ring of power supplying Montreal. Nevertheless, there were still a half million people without electricity in Quebec two weeks after the freezing rain began.

By this time claims were pouring in to insurance companies. High school and university students in Montreal returned to classes. The electrical system that had been repaired hurriedly was still fragile, as 110,000 customers in Montreal who had had power restored found out on Tuesday, 20 January, when they were blacked out for another four hours. The next day another 12,000 customers lost power after regaining it only days earlier. Hydro-Québec was forced to admit that it would not be able to meet its target of restoring power to everyone by 25 January. Three weeks after the freezing rain began, 150,000 people in Quebec still were deprived of electricity. Some residents who had been told they would have electricity restored by 25 January now heard they might have to wait until 15 February. Hydro-Québec's repair crews were exhausted from the long hours of work under difficult conditions over the last three weeks. On 27 January 220 linemen arrived from Manitoba and British Columbia , as much as 5,000 kilometres away, to help rebuild the power grid in Quebec. Hydro-Québec announced that in three weeks it had used up a five-year supply of materials. Power was restored to everyone on 6 February, a month after the beginning of the ice storm.[29] Compassion came from unexpected sources: Quebec's two biggest breweries issued coupons to replace beer that had frozen and spoiled during the power outage.[30] The public security minister announced he would introduce legislation to oblige municipalities to have plans for an emergency and to keep them up to date.[31]

On 21 January there were only 75 homes left to connect in Montreal, but 166,200 disconnected homes remained on the south shore of the St Lawrence River in the triangle of darkness.[32] Their story, however tragic, had become old news, and so the media moved on. As normality returned to the cities, the media in Canada as well as the United States changed to other sensational topics, such as the affair between President Clinton and Monica Lewinsky. This new focus annoyed the electrically dispossessed, who felt their plight was being overlooked. Conflict began to develop, not along linguistic or class lines, but between those with electricity and those still deprived of it, which usually meant tension between city and rural folk. In Quebec many victims of the ice storm outside Montreal felt neglected and excluded by Hydro-Québec, by the provincial government, and even by the media, all of which focused their attention on Montreal, especially its downtown core. These victims were convinced that interest had diminished greatly once power was restored to downtown Montreal.[33] Months later mistrust of institutions was still present. "Municipal politicians attacked the civil protection agency; the public security department criticized the local politicians; Hydro-Québec came under frequent fire."[34] The ice storm evoked not only solidarity but also conflict.

A branch fell on an electrical line connecting a house to the main line on the street. Since it was on private property not under the responsibility of the power company, the woman who owned the house contacted an electrical entrepreneur, who repaired the line and sent her a bill for $2,273. It included charges of $100 per hour for each of the three workers, including one who was tracking down materials and did not come on her property. The bill also included an hour's cost for each worker to travel to the woman's home, which was located very close to the contractor's office. Furious, the woman denounced the contractor to a Quebec television program that specialized in complaints. The program told the woman's story and publicly criticized the contractor as exploitative in its broadcast of 13 January, when hundreds of thousands of people were still without power and heat. The program also talked about the story in five subsequent broadcasts.

The contractor later sued the television network and the two journalists responsible for the program for $3.4 million, claiming that he and his family were harassed and threatened because of the television report and he had lost 75 per cent of his clientele.[35] When the journalists heard of the lawsuit, they suggested to the woman that she complain to the Quebec electricians' association, which she did. The contractor stated that he had warned her that the hourly rate would be double for work done on the weekend, which is common practice, and suggested that she wait until Monday, but she wanted her electricity and heat restored as soon as possible. It was also the norm to bill for an hour of travel time no matter how far away the client lived. The contractor billed for parts at the price suggested by the least expensive manufacturer. He argued that he gave the woman a prior estimate ($2,000 to 2,500) for the work, which the woman and the insurance company accepted. The contractor was acquitted and judged to have respected the norms. As the other side of the story came out, criticism mounted concerning the way the journalists had handled the case. For this and other reasons, one of the journalists named in the lawsuit committed suicide.

In the affected regions of Quebec, two-thirds of business establishments had their activities disrupted as a result of the electrical blackout caused by the freezing rain. Half a million employees lost 2,300,000 days of work. On the other hand, 55,000 employees were paid for 1.7 million hours of overtime. While some workers were exhausted from overwork, others had nothing to do at home and in shelters. The ice storm thrust decisions on employers about whether to pay employees in a situation where no work could be done. About 90 per cent of businesses had no policy for such a circumstance, and they responded very differently: 62 per cent of employees were paid completely and unconditionally, even though there was no work; 8 per cent received partial payment or employment insurance; 14 per cent used holidays or made up time later so as not to lose money; and 16 per cent received no remuneration whatsoever.[36] Inequities were numerous. Rough justice was the order of the day.

This disaster involved the collapse of the electric grid, and so the ice storm provoked questions about Quebec's public electrical utility. Since its formation through the nationalization of many small privately owned electric companies, the government-owned utility of Hydro-Québec has been one of the most important symbols of Québécois' desire to be masters of their own territory. Its origin has been closely associated with the Quiet Revolution of the 1960s, and its rise has been an important part of the political and economic dynamics of Quebec. Former Quebec premier Robert Bourassa vigorously promoted the construction of the largest hydroelectric station in the world at the time on James Bay, part of which is now named after him. He thought it wasteful to let nature just pour the waters from rivers in northern Quebec into the ocean without using them for human purposes because, after all, oceans already had enough water. The highly profitable Hydro-Québec became the symbol of the province's economic and technological expertise. Most Quebecers, whether French or English, secessionist or federalist, have taken pride in the accomplishments of the utility. However, they have been puzzled when their assumption that hydroelectricity is an environmentally benign form of energy has been challenged by environmentalists.

Some Quebecers are, however, starting to believe that Hydro-Québec has become too big and monolithic. There has been grumbling concerning the construction of transmission towers in settled areas. But the major challenge to the credibility of Hydro-Québec has come from the ice storm.

The freezing rain was not only a physical threat but also a symbolic menace. As the extreme weather subsided, difficult questions arose. Why did four of the five lines to the island of Montreal collapse? Why did Hydro-Québec not have a computer simulation to foresee such a big power failure? Why did it not have a permanent crisis-management department? Why did it not have a contingency plan to work with municipalities? The criticism of poor organization during the storm was made: "Hydro-Québec called in workers from as far away as B.C. and Texas while 4,000 of its own employees were sitting at home on full pay."[37]

The union of electrical engineers raised questions concerning the adequacy of preparations, the cheap system of utility poles as opposed to more expensive but safer underground installations, the priority given the profitable exportation of electricity over security of supply within Quebec, and the province's uniquely high dependence on hydroelectricity. Millions of Quebecers shivering in the cold and hundreds of twisted transmission towers cast a shadow over Quebec's symbol of engineering know-how. Damage was done by nature's extreme weather not only to the electrical transmission and distribution system but also to the reputation of Hydro-Québec. The Quebec government's inquiry concluded that the extreme weather stimulated a critical re-evaluation which marked a turning point in Quebec society's relationship with its monopoly public power utility.[38] Some commentators have argued that it now "will become an ex-icon."[39]

Hydro-Québec had years earlier placed an immense, electrically lit, stylized Q atop its headquarters in downtown Montreal as its emblem. It symbolizes the company's power, efficiency, and reliability for all in the city to see. This beacon blazed night after night during the ice storm, even as businesses and residences were darkened by the loss of power. Eight days after the beginning of the freezing rain, company officials finally switched it off as a gesture of solidarity with customers still shivering in their homes and promised to keep it off until power was restored to every residence. Hydro-Québec had belatedly realized the perverse symbolism of its waste of electricity at a time when its customers were suffering from electricity deprivation and the grid was fragile.

Despite an insufficiently robust electrical grid being the source of vulnerability to this hazard of nature, Hydro-Québec was very effective during the ice storm in its public relations counteroffensive. Its president and CEO, André Caillé, joined with Quebec premier Lucien Bouchard in daily televised press conferences to inform the public about how the situation was being managed and to reassure the population that the utility's material infrastructure and its technical excellence would survive this disturbance of nature.[40] The CEO's white turtleneck sweater even became one of the emblems of the management of the crisis. A public relations spokesperson for Hydro-Québec, Steve Flanagan, was one of the most articulate television personalities during the disaster and won an award for excellence in strategic communications from the Canadian Public Relations Society.[41] Rhetoric attempted to spin weakness as strength, and it succeeded to some degree.

On 15 January two Hydro-Québec repairmen dangled from the side of a helicopter above the St Lawrence River in below-zero temperatures and brisk winds repairing a 200-megawatt line that supplied the island of Montreal. Since this exploit was planned in advance, a news network rented its own helicopter to fly perilously close to the linesmen and televise the spectacular mission live to homes that still had electricity. The linesmen's supervisor cursed the cameras from the ground but could do nothing about them. When all went well, however, his employer, Hydro-Québec, was doubly happy at restoring both its electrical line and its reputation.[42] The utility would have been condemned if it had carried out such a dangerous publicity stunt. This way it received the praise when the mission succeeded, but the television station would have been the target of criticism for grim reporting if something had gone wrong.

The government of Quebec distributed $30 million to hundreds of thousands of residents deprived of electricity. Far from bringing praise, however, this aid package was condemned for not being sufficiently inclusive. The criteria specified that $70 per week per person ($280 for a family of four) would go to people without power for at least a week starting 12 January. "Howls of outrage" were heard from powerless people who did not qualify[43] and who judged it unfair that being without power during the first week of the storm did not count. They claimed that the compensation clock should start ticking when the electric clock stopped.

Quebec has never declared a state of emergency for any disaster.[44] Instead, it has proceeded ad hoc, issuing decrees to confer central responsibility for the management of a disaster to a particular political authority. The Quebec government attempted to justify this practice by the need to make decisions rapidly in an urgent situation. It adopted fifty-seven decrees between 7 January and 12 August 1998, which led it to be accused of governing by decree. There was widespread suspicion that the government was using "urgency" as an excuse to bypass laws designed to protect the environment and to ensure local consultation. Thus eight decrees concerning the re-establishment and reinforcement of Hydro-Québec's power grid were challenged by citizens in court. In one *cause célèbre* the Quebec government authorized by decree the construction of a 735-kilovolt line, arguing that the ice storm demonstrated the necessity of reinforcing the electrical network in this way. A group of local citizens opposed the construction of this new transmission line in court. They contended that the government was using the ice storm as an excuse to expropriate their land and push through the construction of a line of transmission towers that would diminish the aesthetic appeal of their property and perhaps be dangerous. They would not receive any new or improved service. They argued that the situation was not so urgent that it required the circumvention of the law requiring public consultation.[45] The judge of the Quebec Superior Court agreed with the claimants and nullified this decree and several others on the grounds that the government had illegally exempted these projects of Hydro-Québec from the authorization procedures specified by law.[46] The government did not appeal this judgment, and construction of the proposed line was stopped.

In a public forum before the Quebec commission of inquiry, the Red Cross sought the exclusive mandate to collect funds and manage them for the voluntary side of disaster preparedness. Some other groups, notably the Assembly of Quebec Bishops, objected to such a mandate.[47] Even caring can be the subject of turf wars, and even between reputable organizations such as the Red Cross and the Assembly of Quebec Bishops.

One study documented that the private insurance industry shouldered 33 per cent of the costs of the ice storm in Quebec, the federal government 27 per cent, and the Quebec government 40 per cent, in addition to costs borne by victims themselves. Included in the provincial government costs were those carried by the monopoly public power utility, Hydro-Québec, which in many other jurisdictions would be in the private sector. Some issues erupted between the federal and Quebec governments concerning whether Hydro-Québec should be compensated by the federal government for damages incurred during the ice storm. Such payments would mean that the Quebec government would indirectly receive additional compensation over and above what it had already received under federal-provincial agreements. So Hydro-Québec did not receive such compensation.[48]

New York

In New York State the decision was made by the power companies to give priority to restoring electricity to large blocks of customers first, which increased the numbers restored but left sparsely populated areas that had been disconnected early without electricity for a particularly long period. Lake Placid had become the electrical Olympics, with long lines of trucks carrying utility crews criss-crossing the region.[49] Over seven hundred inmates from the Department of Correctional Services, supervised by seventy-two correction officers, were working on debris clearance.[50] As power was restored, people left the shelters to return to their homes. Some of the older occupants who lived alone were almost reluctant to leave because they were enjoying the companionship of others.[51] Newspapers and hospitals began speculating about a baby boom nine months after the ice storm. But the storm caused problems that might deter the necessary activity. For example, it cut the power needed to heat water for bathing. The electrically deprived had to find friends with power and travel on impassable roads if they wanted to bathe – not exactly ensuite accommodation. People riding out the storm in their unheated houses were warned about hypothermia and counselled to wear enough clothing. "Remember, 70 percent of your body heat is lost through your head."[52] Wearing a warm hat in the house was strongly recommended.

Risk did not disappear as power came back on. There was the danger that power surges would damage appliances left on or spark fires. The National Weather Service added a further prick to the balloon of growing optimism, forecasting three inches of snow, followed by winds gusting to 30 miles per hour and temperatures plummeting to –10 degrees Fahrenheit.[53] People cooped up in shelters for days began to lose patience, and quarrels broke out. In one shelter stomach flu infected a quarter of the five hundred occupants with vomiting and diarrhea. Many of the staff, weakened by fatigue, were also infected. Attempts were made to use the old method of quarantine to isolate the sick from the not-yet-sick.[54] Fortunately, power was restored in the area, and people left the shelter for their homes before a crisis occurred.

A forty-eight-year-old guard at a correctional facility was found dead in his home on Monday, 19 January, of carbon monoxide poisoning. He had moved his generator to the basement after hearing stories that generators were being stolen. "He never knew what hit him," said the local mayor, who had previously "thought we had been doing a pretty good job warning people about generators."[55] Despite being without power, a father and his eighteen-year-old son remained in their home with the wood stove going to keep the water pipes from freezing. The son noticed a growing red spot at the point where the pipes met the ceiling. The two climbed into the attic and saw flames covering the wall above the wood stove. They desperately tried to douse the flames using a five-gallon bucket of water, but the effort was futile without electricity for the water pump in this rural area.[56]

The harshest conditions occurred in isolated, rural areas. Rescuers found some people disoriented from the long-term cold. Elderly folk were trapped in their houses because the doors and windows had frozen shut.[57] National Guardsmen went door to door in remote regions to check on people and provide assistance. One elderly resident with respiratory problems had to be pulled out on a sled. Another ran out of heart medication, and a volunteer firefighter walked in the medicine. Live power lines contacted barbed wire fences, which then became electrified unbeknownst to people who might touch them.[58]

As noted earlier, farmers today use advanced automated systems that run on electricity. Supercows have been bred to produce much more milk.[59] Efficient technology under normal conditions results, however, in pain for these factory farm animals when conditions are not normal. When the electricity went out and milking was impossible, the cows got skittish, bellowed, and even ripped off nearby hinges.[60] Supercows suffer intensely when the machine that milks them is inoperative. They get infections, and as a result of the ice storm, some died of mastitis. A dairy farmer described the experience of her cows during the power blackout this way: "it is not uncommon for good cows to produce over a hundred pounds of milk in a twenty-four hour period. One can imagine the stress placed on this animal when it is not milked ... The stables weren't cleaned, the milk was getting warm in the tank, the barn was filing up with steam, seventy cows hadn't been milked and every one of them was bawling at the top of her lungs. We could not hear each other's voices over the noise that those cows were making. Even from the house we could hear them, and we were totally helpless to do anything."[61]

The cows also need to drink a great deal of water. When the electrical pumps did not work and farmers tried to bring pails of water to their thirsty cows, they would fight to get to the scarce water. Combativeness is not a usual trait of cows. Manure accumulates rapidly in a barn with a large herd when the electrically operated cleaning system lacks electricity. "If the gutters are not cleaned, they may overflow onto the cow stands and alleys. Soon, inevitable, the entire barn is full of you-know-what, and along with that comes the smell and buildup of ammonia gas, which can cause animals to get pneumonia. Appalling conditions for man and beast."[62] About 1,200 of 20,500 cows in two counties of New York State died as a result of not being milked, watered, or cleaned because of the lack of power for farm machines during the ice storm, and of the remainder, over 6,000 suffered from mastitis.[63]

For people without electricity or telephone, the battery-operated radio became the sole means of receiving communication from others. It did not, however, enable people to initiate communication. A 911 emergency telephone service presumes that all telephones will work, which was not always the case. How were aid workers to know which isolated houses with a downed phone line had an urgent need for help? Messaging systems were improvised. Radio broadcasts went out telling people to put a bag over their mailbox if they were in distress or needed assistance. People in really remote areas were instructed to place bright-coloured objects in the form of a H that could be spotted by a helicopter if they needed help urgently. Several hundred North Country residents were rescued or provided with emergency aid as a result of these improvised signalling procedures.[64] The Salvation Army coordinated a railcar catering service: two stationary box cars, three dining cars with kitchen, and two coach cars serving 3,000 hot meals daily in a chosen site.[65] No matter how bad things were, reports suggested that the situation was worse elsewhere. A newspaper in a northern New York community ran a story that residents of a Quebec town fifty miles farther north were ripping apart porches and decks for wood to burn in a desperate attempt to keep warm.[66] In response, a caravan of trucks full of firewood was sent to the town.

The ice storm inspired about four thousand people in northern New York State alone to volunteer their services clearing roads, checking people in unheated houses, feeding people at shelters, keeping generators running, and the like. In addition, six hundred American Red Cross workers were sent from all over the United States. The Red Cross designated the ice storm in the United States as a level 4 disaster on its five-point scale, the 1997 Red River floods in North Dakota being level 5 and Hurricane Andrew a catastrophic level 5. The designation determines how much money the Red Cross will spend. The government released funds for 450 general-labourer jobs paying up to $12 per hour that could last for as long as six months. These jobs, such as clearing property of debris, went to those who lost their employment because of the ice storm and to the chronically unemployed.[67]

The disaster led the government to relax certain conditions for obtaining welfare. "The State also has secured approval from the USDA to ease requirements to allow people who are doubling up with relatives or neighbors because of lack of heat or power to still be eligible for food stamps … and waive certain eligibility requirements for employment and training so that childless, able-bodied adult food stamp applicants and recipients restricted by the storm will not be penalized for failure to comply with work requirements."[68]

As the freezing rain vanished, social frictions began to appear. A visit by the governor provoked mixed reaction among farmers. One said that "there's a pretty good feeling to see he's up here. It's a big deal to me." But another farming wife who had just spent $10,000 to purchase three generators to keep the family's 300-head dairy farm operating complained: "Honest to God, I can't tell you how disturbed I am when I hear the governor on the radio tooting his own horn saying he's getting help for us. Where the hell are all the generators they're sending."[69] Farmers were feeling frustrated by what they saw as a lack of help from state and federal officials. When the Housing and Urban Development secretary held a news conference via satellite to announce $12 million in aid for the North Country struck by the ice storm, a county legislator commented: "This is a nice gesture, but $12 million is nothing."[70] He noted that not a penny would go to farmers to pay for lost milk, dead or sick cows, or fuel for generators. A dairy farmer himself, he insisted that he would "hound every federal and state agency known to mankind in search of help for farmers."[71] The governor's appearance was perceived by some as merely a ceremonial visit that usurped credit from the real heroes who had done all the work, namely, the local firefighters and volunteers. Hence the visit was not well received.[72]

The New York attorney general received complaints about price gouging by some companies and business people as a result of the spike in demand for specific commodities during the ice storm. Most complaints concerned inflated prices for generators, but some were about increased prices for bread and milk. Investigations were launched, charges were filed, and promises were made that the New York State Legislature would introduce new legislation to streamline procedures and stiffen penalties against the practice.[73] There was, nevertheless, progress to calm troubled waters. For the vast majority, electricity was restored in time to watch the Super Bowl on 25 January, the leading attraction in American life. Normality was indeed returning.

The northern New York counties that were struck by the ice storm chronically suffer the highest unemployment in the state. The result was "that the vast majority of home construction in the declared counties is inexpensive with plastic water and sewer pipes that break more easily when they freeze."[74] Water in pipes froze and expanded; as a result, water and sewer pipes ruptured, and water poured into the basements. The poor with weak defences were harmed more than the better protected. "The storm affected all backgrounds, creeds, economic strata. It exempted no one, but impacted everyone differently."[75] Social vulnerability resulting from the way society operates in normal weather led to the biophysical vulnerability of some classes more than others to this extreme hazard of nature.

New England

January 1998 put much of New England on a weather roller coaster, but not a pleasant one. The freezing rain in early January was followed by colder temperatures until 12 January. Next came mixed precipitation and then strong and gusty winds with single digit (Fahrenheit) temperatures. This was followed by six to ten inches of snow over several days and then sunny and cold weather for two days. On 23 January warmer air moved in, changing snow to sleet and then to freezing rain in the south-eastern regions. This freezing precipitation continued until 25 January and brought a significant accumulation of ice.[76] On Monday, 12 January, there were still 400,000 people in Maine without power. In homes without heat, indoor temperatures fell to 34 degrees Fahrenheit, and fish in aquariums froze and died.[77] The number of places in shelters had been increased to 15,000, but there were only 3,000 people in them, sleeping on anything from cots and mats to bubble wrap. The forced togetherness in shelters, in extended families, and among friends, as well as the mutual aid, brought pleasure to many. As one volunteer stated, "When all is said and done, everyone will feel really good about coming together."[78] The publisher and chair of the group of newspapers that published a supplement on the ice storm concluded "that while we can never completely depend on technology, here in Maine we can always depend on each other."[79]

During the ice storm, electrical repair crews worked from 7 a.m. to 10 p.m., the most dangerous time being in the last five hours of their shift after dark. "We're fatigued and visibility is severely limited. You not only have to worry about the lines, but traffic."[80] Workers were concerned that a car could slide on the slippery roads and crash into the trucks that held them in their little buckets up in the air near live wires. Their worries were not unfounded: one worker died this way. Electrical repair crews were brought to Maine from as far away as Baltimore to restore the lines. Hundreds of National Guard troops worked to check homes, bring in supplies, and clear roads. Maine's Republican senator requested that the Navy bring a power-supply ship in as an emergency source of electricity for areas along the coast.[81] On Tuesday, 13 January, President Clinton declared the state a disaster area: 75 per cent of the cleanup costs would therefore be refunded by the federal government for fifteen of its sixteen counties.

The southern coast of Maine was, fortunately, unaffected. But a little more than two weeks later, on 24 January, as workers were reconnecting the last of the homes in the centre of the state, southern Mainers were hit by a different ice storm, which cut electricity to 165,000 more people.[82] The second storm surprised meteorologists, who "had predicted that warmer temperatures would quickly turn today's freezing rain into rain, preventing icing and severe power outages. That warming trend never materialized in the state's southern coastal areas."[83]

A homeowner with an artesian well on his property in China, Maine, had 150 to 200 people coming each day to fill containers with water, some waiting in line for two hours. One motel stayed open without power, but a client tipped over a candle, starting a fire that destroyed three rooms because the sprinkler system could not operate without electricity.[84] The only heat one couple had for days was provided by the four gas burners on their kitchen stove. Many people living in severe conditions found solace in their belief that they were fortunate in their misfortune because it could have been even more dire: "so many people had it worse than we did."[85] Some people with electricity felt guilty that they had been spared.

Relations were not all characterized by solidarity. Maine's largest telephone company had between twelve and fifteen generators stolen from switching stations, which resulted in temporary outages. Most were stolen, not out of desperation for heat, but rather to sell for money.[86] Storm-related stress triggered several fist fights and domestic assaults. On Wednesday, 14 January, with 100,000 customers (representing one-quarter of a million people) still deprived of electricity and heat and with numbing cold and strong winds slowing repair work, angry people threatened Central Maine Power repair crews.[87] One man, frustrated by the cold and darkness, claimed he would take a repair crew hostage until they reconnected his electricity. The company had to hire off-duty police officers to guard its offices and to escort repair crews into places where threats had been made.[88]

Carbon monoxide is particularly perverse because it results in poor judgment and confusion. "I've had more than one family come in and tell me, 'Gee, we were all feeling pretty crummy – even the dog had the flu.' That's a tip-off because dogs don't get the flu."[89] In Maine two men died of carbon monoxide poisoning from their generators, and hundreds of other people went to emergency wards with nausea and headaches from exposure to the gas. Mainers poured into hospital emergency departments complaining of nausea, light-headedness, and confusion. Doctors who normally see two or three cases a year saw eleven to fifteen daily during the ice storm. At higher levels of carbon monoxide, victims pass out in minutes. Residents were trying to keep warm using risky methods because nothing else was available, and meteorologists forecast that 14 January could be the coldest day of the winter. Many people were using generators indoors because of reports of outdoor generators being stolen. A perceived risk led them to practices that were much more risky.

Central Maine Power (CMP) had 20,000 miles of lines serving 520,000 businesses and homes. Information and problems concerning transmission lines and power substations were provided by computer and displayed at headquarters on an 11-foot-high, 66-foot-wide electronic map. Flashing lights indicated a break in electrical transmission. Technology could not, however, do everything. For information needed to flash the lights, the company depended on complaints, which were routed to an automated answering service in the midwestern United States and then back to CMP headquarters. Assessors were then sent out in the field to check the damage. Next the mountain of confirmed reports of power cuts were analyzed at headquarters according to a pre-established priority list to determine which lines would be repaired first. Essential services (hospitals, fire and police departments, nursing homes) had top priority on the formal list. Major circuits servicing large numbers of people were second. Branch lines connecting neighbourhoods were next. Lowest priority went to the ends of power lines serving small numbers of people. Line workers, tree cutters, and other employees worked in the field following the decisions of their superiors to immediately fix or not fix particular lines.

These objective criteria for the restoration of power did not eliminate tensions. When commercial and high-density areas regained light and heat and their residents began to speak of the ice storm in the past tense, the numerous unheated felt all the more forgotten and frustrated. The fact that people with electricity problems could not speak to a live person at Central Maine Power and could only leave a message on its automated system intensified the sense of helplessness. The owner of a tiny data-consulting business in Poland, Maine, was enraged because he depended on electricity and believed that the continuing electrical outage in his small town on 21 January resulted from favouritism elsewhere. He believed the electrical company "reacted to the powerful in Maine most quickly."[90]

The ice storm hit people who were already suffering for other reasons particularly hard. A woman who had recently had major surgery was caring for her fifty-four-year-old husband who had been left speechless by a stroke. Then the electricity went out in their home in Belgrade, Maine. Because they had a propane heater, she took in her eighty-four-year-old mother, who had no heat.[91]

Washington County is one of the poorest areas of Maine, where people eek out a week-to-week living from the woods, the blueberry barrens, and the ocean with as many jobs as they can find as fishers, farmers, woodcutters, and worm-diggers. About 35,000 people live in its 2,500 square miles, in houses often a mile apart. When the ice storm downed the major electrical power line, the county was thrown into coldness and darkness, and many workers lost employment, wages, and also food in their freezers. Most families chose to weather the storm at home. Many already had wood stoves, kerosene heaters, and oil lamps. Others bought generators and kerosene heaters with credit they could not afford. "When you're already living on the hard edge of civilization, you do what it takes to get by – and worry about it later."[92] They toughed it out in trailers or old farmhouses, covering windows with cardboard and stuffing towels in doorways, but the wind whistled through the old wooden walls all the same. The small wood stoves could not produce heat as fast as it was being lost in the poorly insulated dwellings.

The front door of the home of an eighty-year-old woman froze shut, so she took the casing out of the window and crawled through to replenish her supplies. Volunteer firefighters shuttled rare generators from house to house to give each a few hours of warmth. When wood stoves and gas heaters had to work full blast and candles were everywhere, some homes caught fire and were destroyed, leaving the families homeless. The fire chief expressed his powerlessness: "The only help I'd like to see now is from the Lord. I'd like to see some warm weather."[93]

Residents tried to save food from their freezers by covering it with ice outdoors in a shady spot and then worried as the temperature fluctuated that the house would get too cold or the food too warm. Volunteers delivered the meals on the icy roads to those who refused to leave their homes. As one local volunteer put it, "people are very independent here. But they watch out for each other, disaster or no disaster. They're very proud people and they can't seem to accept the help; that's why we deliver. It isn't whether you're rich or poor ... if you don't have means to heat things, you need help. We're all in dire straits."[94] But more of the poor were in dire straits than the rich.

WHAT NEEDED TO BE RESTORED?

This extreme weather proved to be the most costly disaster in Canadian history, as well as the most costly in the history of northern New York State and the state of Maine and expensive in the states of Vermont and New Hampshire. The storm resulted in more than 840,000 insurance claims from policyholders in the two countries – 20 per cent more than the previous largest claim-producing disaster in the United States, Hurricane Andrew in Miami in 1992.[95] The major problem with this disaster was its scale. As one insurance executive put it, the difficulty was not the size of the claims but their volume. One claimant represents about three people who have suffered physical damage (for example, the average family), and therefore 3 million people suffered insurable damage as a result of this one weather event. The uninsured damage and other economic losses brought the total destruction to well over US$5 billion.[96] The normal routines of business, government, and daily life were unplugged from their source of energy. In the words of the insurance industry, "business, commerce and manufacturing were brought to their knees."[97] More people were directly affected by this extreme weather event than by any other in North America.

When telephone lines went down, people turned to cellular phones, that system jammed, and they had no modern means of communication. In some places, emergency 911 telephone service and lines to fire stations did not work. Some hospitals had to close for lack of electricity. Emergency wards of those that had electrical generators were filled with people suffering from broken bones or from carbon monoxide poisoning. Thirty deaths were directly attributable to the ice storm in Quebec[98] and Ontario, and seventeen in the United States. The overall death rates in the affected areas increased in January and February compared to those months in previous years, suggesting an indirect effect as well.[99] An entire region was brought to a standstill by the storm. In Ontario "the portrait was not of a single incident but of dozens of disasters over a wide area, any one of which would normally require a major provincial response."[100] The confidence that modern society can deal with nature gave way to feelings of vulnerability and helplessness as the freezing rain kept falling and damage kept growing.[101]

The destruction of the electrical system was enormous and extremely costly for the electrical companies. Hydro-Québec lost a thousand transmission towers and 24,000 poles.[102] Price was no guide to security of supply: Mainers pay three times as much for their electricity as Quebecers do, yet their electrical system went down as well.[103] Electrical distribution lines have fallen before, and transformers exploded, but this time nature destroyed, damaged, or made vulnerable to future storms the whole transmission system in a large populated area. The massive freezing rain broke so many links in the rationally designed electricity web that the web itself stopped functioning. Many of those who lost their electricity were effectively thrown back a century in time and became, like their ancestors, hewers of wood and drawers of water. Just how much modern society depends on electricity becomes evident when that society is deprived of its usual electrical current.

A massive and costly effort was undertaken to deal with the problem. Emergency measures organizations at all levels of government went into action. Firefighters and police were mobilized. The Canadian army sent 15,000 soldiers into combat, the largest peacetime deployment in the country's history. In the United States the National Guard engaged 1,800 troops in the fight in New York State and 1,000 more in Maine. All electrical power crews worked double shifts in Quebec, Ontario, New York, and New England. Reinforcements were brought in from as far away as British Columbia and California.

As a result of this extreme weather, the president of the United States declared a state of emergency in five counties in New York, six in Vermont, and all the counties of Maine and New Hampshire except the coast.[104] The disaster prompted him to activate the Long-Term Recovery Task Force, consisting of cabinet members for agriculture, housing, energy, commerce, and other departments and chaired by the director of the Federal Emergency Management Agency. It concluded that much work remained to be done in order to protect against future utility losses, and it proposed "the goal of creating communities that are resistant to the disasters of tomorrow."[105] Nice words, but was anything learned from this disaster that will help attain that goal? This was quite a dance of these two modern societies with the forces of nature, but will a better choreography be developed to harmonize modernity's movements with nature's dynamics in the future?

CHAPTER 6

Learning from Disaster

For some, the answer to the question of what was learned from the disaster is "not very much." An American commentator wrote that "within weeks services were restored, and to the vast majority of residents the 1998 ice storm became just one more memorable weather event."[1] A Canadian observer arrived at a similar conclusion: "Now, months later, it's tempting to take electricity for granted once again. We use credit cards and debit cards routinely, forgetting the harsh days when bank machines went dead but cash spoke loud and clear."[2] A specialist on disasters concluded that "what usually happened after a disaster is that for a time everyone is ready and willing to make adjustments to get ready for the next incident. However, before long, there are too many day to day demands to allow much time for emergency planning."[3] This chapter and, indeed, the remainder of the book will examine what can be learned from the disaster. It will focus on broad issues because disaster post-mortems have already published reports with sections dealing with specific technical or administrative "lessons learned."

INTERPRETATIONS OF DISASTER

The Dependence and Fragility of Modern Societies

The most general lesson learned from the disaster by some people was our dependence on essential infrastructures and our vulnerability to nature's disturbances. Despite the power and sophistication of technology, the ice storm demonstrated how much our modern lives depend on wires coming into our homes. "They were surprisingly fragile, surprisingly mortal. In each strand of metal, perhaps, we should still be able to glimpse the northern rainfall and the swirling river, the turbines and transformers of our lives."[4] That comment is well expressed from the perspective of Quebec, where power is generated by giant dams in the far north. The people of Ontario and the United States should learn to glimpse the not-so-distant nuclear reaction or coal combustion that powers their light bulbs, television screens, stoves, washers and dryers, furnaces, and computers. We are immersed in nature's dynamics, whether they are redeployed to produce electricity in nuclear reactors and turbines operating from the gravitational pull on water or in their natural state, such as the disturbance of the ice storm. An employee at Emergency Preparedness Canada concluded, "If there's one thing I've learned over the years, it's that all the biggest problems are caused by Mother Nature."[5] He was mistaken, since humans can cause big ones too, including disturbances unleashed by human activities that nature would have otherwise kept under control. His intuition that nature's forces can cause major problems is nonetheless fundamentally sound.

On 14 January, just after the freezing rain had ended but before power was completely restored, an editorial in the *New York Times* summed up the situation as follows: "Deep cold, undiluted darkness, unsolicited silence – that is what remains when the power goes down in winter and stays down."[6] A temporary loss of electricity in the summer, by contrast, is much less consequential. The electrical grid is embedded in seasonal cycles and powerful dynamics of nature. The editorial then drew the following conclusion: "A storm like this reveals the shallowness of technological civilization – how swiftly the grid collapses. But it also reveals its depth – into how many reaches of ordinary life electricity has penetrated and how high above the fundamental concerns of existence it allows us to float. This storm demonstrates the slow, brutal strength of the cold. The wonder is not that cold is so powerful, but that we are so seldom aware of its power, of its capacity to fracture a community, to isolate it, and, as events have proved, to draw it together as well."[7] Similar conclusions have been drawn by other authors: "The storm and its aftermath reminded us of some immutable truths – of our fragility, of nature's strength and of our remarkable ability to pull together in times of need."[8] Disaster makes visible the vulnerability of even modern, advanced societies in their confrontation with nature's disturbances, if we are willing to see it.

The author of a book on the ice storm in northern New York claimed that "disaster taught us just how well we could respond under extreme adversity, especially as a community, yet we saw how very fragile and dependent our existence can be ... However, born from this adversity was a society better prepared for future. Local Red Cross chapters and other emergency response agencies have since devised more detailed and comprehensive Disaster Preparedness Plans."[9] Because communities were unprepared for such a long, intense ice storm, problems had to be dealt with on the fly, and solutions improvised. Shelters had to be set up spontaneously as a result of the ice storm.[10] The storm drew attention to the fragility of even American society in the face of these massive forces of nature. This experience led to more preparations on both the individual and the organizational level: post-storm assessments and planning. There is, however, something paradoxical about all this. The failure of planning and forecasts when human beings are confronted by overwhelming disturbances of nature leads to even more detailed, comprehensive planning and forecasting. If the fragility and shallowness of technological civilization when it is confronted by the strength of nature's dynamics are to be taken seriously, then the possibility that nature will once again surprise our most detailed and comprehensive plans must be taken seriously as well.

As Maine struggled to return to normality, the city manager of its capital stated that "the biggest hit from this storm may be to our complacency. It'll be awhile before we take things for granted again."[11] An upstate New York author concluded, "Maybe we lost our sense of security, our innocence or naivety."[12] The "maybe" is important. More detailed planning and new technology based on the experience of this ice storm could well lead once again to an overblown sense of security and naïveté concerning the human capacity to forecast and master the dynamics of nature. The lessons of the crushed electrical system upon which modern society has become dependent may fade as time passes after a disaster. Whether complacency will return and whether benign, malleable nature will be taken for granted remains an open question.

Scapegoating Nature and Blaming God

Because the intense, persistent freezing rain was a surprise, the easiest response is to claim that this disaster was an act of God or a phenomenon of nature. But things are not that simple. Most of the fatalities, suffering, and damage occurred because of human decisions: to become dependent on a centralized system of electrical energy for heating, lighting, and other essential amenities by replacing the decentralized sources of energy of the past, to construct electrical distribution lines above ground suspended from poles instead of underground, to build transmission towers and lines with a specific but inadequate resistance to ice loads, to save money by eliminating redundancy and do away with stockpiles of supplies such as generators, and so on. If these decisions had been different, the suffering and damage would have differed as well. Scapegoating nature or blaming God to absolve humans of their share of responsibility is misleading. It is the interaction between nature's hazards and a community's vulnerability that results in a disaster, not one or the other alone. That vulnerability was the product of decisions and priorities. Nature's disturbance of freezing rain became a costly human disaster because of the construction of vulnerable communities in its path.

The Construction of Redemption Using Freezing Rain

The Canadian military has a valued reputation both within Canada and outside the country, particularly as a result of its contributions to peacekeeping around the world. Several years before the 1998 ice storm, however, that reputation was tarnished by an ugly incident in Somalia when several Canadian peacekeepers tortured and killed a Somalian. Subsequently, neo-Nazi initiation rites were discovered at the Camp Petawawa military base, northwest of Ottawa. There were also accusations that some Canadian soldiers had not behaved properly in their peacekeeping role in Bosnia.

When the military was dispatched to help contain the massive 1997 floods of the Red River in Winnipeg, Manitoba, that action helped to rehabilitate its reputation with the Canadian population. The 1998 ice storm provided another opportunity for redemption. Soldiers and defence analysts were well aware of this potential "public-relations windfall"[13] from the freezing rainfall. Prime Minister Chrétien enthusiastically seized this opportunity, offering military aid on television even before informing the military. The government took advantage of the ice storm to frame a more positive meaning of the military: it was to be recast from torturers in the deserts of Somalia to saviours of Canadians during an ice storm. Many of the troops sent to help in the Ottawa region came from the previously criticized military base. The population greatly appreciated the presence and aid of the military in these uncertain conditions. Not only power lines were being restored, but also the reputation of the Canadian military. Even a disturbance of nature can be used as raw material for socio-political constructions.

Some military analysts worried that these opportunities were also threats. If the population views the military in terms of disaster assistance rather than military goals, then it may support the purchase of generators but be unwilling to spend large sums on warplanes and attack submarines.[14]

LEARNING FROM THE EXPERIENCE OF DISASTER

Democracy Temporarily Enhanced by Disaster

The experience of this intense, persistent freezing rain and power blackout created uncertainty concerning the recurrence of extreme weather, the vulnerability of the electrical infrastructure essential for a modern society, and, more generally, the fragility of modern technologically based society. Meteorological data sufficient to make accurate forecasts of recurrence do not exist, yet decisions have to be made about whether costly standards of resistance should be built into the electrical grid and whether expensive emergency preparedness measures should be developed. With all the uncertainties concerning extreme weather events, how was learning from this disaster accomplished?

After the storm ended and power was restored, the worst-hit region, Quebec, created a commission of inquiry into the disaster. It was made independent of government and of the power utility. Its members comprised the president of the order of engineers, a former dean of a university faculty of science and engineering, a high-ranking public servant, a lieutenant-general in the military, the mayor of a city, the vice-president of the human rights commission of Quebec, and a sociology professor specializing in disasters and risk management. It had a substantial budget to order scientific studies of the disaster and to hold public meetings to allow input from concerned groups. Collaboration between technical experts, political representatives, and concerned lay people was fostered to determine what was known and what should be investigated. This combination of the best available scientific research and highly visible public democracy was the self-reflexive means the state used to attempt to demonstrate that the government and the public power utility were being held accountable to the people, that lessons would be learned, and that risk would be minimized in the future.

The extensive collaboration between specialists and lay people[15] to discuss the boundary between the technical and the social was effective in clarifying the issues and in reassuring the population in the context of uncertainties engendered by nature's construction of extreme weather in an electrically dependent and vulnerable society.[16] The disturbance of nature, in which normal weather was temporarily replaced by an extreme event, inspired a social event in which routine democratic practices were temporarily replaced by enhanced practices. The creation of such a commission was not unique to Quebec. Its construction has typically been prompted by the experience or forecast of disaster.

A State of Dependence

Many people demonstrated a great deal of ingenuity and resourcefulness by improvising to try to meet their needs and those of relatives and neighbours during the power blackout.[17] However, the complexity of the means used to heat and light homes largely disqualified people from helping themselves when the electrical grid collapsed. Decentralized means of heating and lighting had long been abandoned and replaced by technologies dependent on centralized sources of electrical power. Residents no longer had wood or coal furnaces they could ignite to get heat. New technologies of electricity rendered the individual more dependent on public and private organizations when nature's disturbance struck. The specialization of knowledge and division of labour resulted in the inability of people to solve their own problems of lack of heat and light. They could not simply splice and reconnect their power distribution line. Only trained and experienced linemen could work on the electrical lines in freezing rain without getting electrocuted. "As the system disintegrated, as tree limbs came down and across live conductors, there were flashes that occurred. Some of the limbs would burn. This was a very dangerous situation for anyone close to those facilities."[18] The population had become dependent on specialized power crews to repair the electrical lines, on even more specialized crews to rebuild the transmission towers, on emergency measures organizations to establish shelters, on power companies to construct ice-resistant lines and pylons, on suppliers to bring in emergency generators to power modern equipment, and on governments to organize and set regulations to ensure that all this is done.

The individual was helpless and dependent on public and private organizations when nature's disturbance struck. Most of the population were reduced to playing board games to pass the time while specialists struggled frenetically to restore this essential infrastructure of modern society. Reliance on complicated centralized technology reduces the self-reliance and autonomy of the population in times when that complicated technology fails. Lack of self-reliance is not always an individual trait. In cases like this, it is instead a social-structural characteristic embedded in complex technologies that most people do not understand. This cause of a lack of self-reliance is one lesson that needs to be learned and incorporated into planning for future emergencies.

The Federal Emergency Management Agency (FEMA) in the United States noted that many people suffered extreme hardships because they were not prepared to be self-sufficient, so it recommended a campaign to promote the understanding that communities and individuals should be prepared to live independently for three days until utilities are restored.[19] However valid that suggestion, it is paradoxical that temporary self-sufficiency was proposed after the change from permanent self-sufficiency to permanent dependence.

Homelessness

The ice storm destroyed a few houses by bringing trees down on them. It also resulted in the fiery destruction of several others because of the overuse of fireplaces and the misuse of candles and through power surges when electricity came back on line. But by far the most frequent way that the ice storm created homelessness was by rendering houses unlivable. More than 4.7 million people were deprived of electricity in Canada and an additional 1.5 million in the United States; that meant deprived of heat, light, often water, and other essentials in a dark, frigid winter environment. Over 700,000 people were still without power three weeks after the storm began. The home was transformed from an abode of rest and leisure to an unheated, unlit shell for days in freezing winter weather. Many abandoned their unlivable houses to stay with relatives or friends or in hotels or shelters. Many more were forced into mobile living, searching for a hot shower and warm meal before returning to their cold, dark homes. Houses sheltered inhabitants from the freezing rain and the wind but not from the numbing cold. Dwellers could see their breath as easily indoors as outdoors. Houses became dangerous places because of the lack of heat, light, and supplies, because of roofs weighed down by ice, because of trees and live electrical wires crashing down all around, because the desperate search for heat resulted in perils of fire and carbon monoxide poisoning, and because the thawing and refreezing of foods in incapacitated freezers created the risk of food poisoning. Under these frigid, frightening conditions, insomnia became widespread.

Instead of the home being a warm, comfortable shelter protective of its residents, an inversion occurred. Many residents stayed to protect their cold houses from bursting water pipes and other dangers. Everyone assumed the new form of homelessness would be a temporary state, but no one knew when it would end. It was not only the poor who experienced this form of homelessness. In some cases the wealthy in their wooded neighbourhoods were more affected than the poor living in bare areas because falling trees brought down above-ground power and telephone lines. The prosperous had, nevertheless, more resources and could flee to hotels outside the devastated zone rather than to shelters, although that was not always possible because of the atrocious travel conditions. Shelters tended to receive isolates who were not integrated into social networks or whose friends also resided in houses rendered powerless by the storm.

Disturbances of nature, such as this persistent, intense freezing rain, create homelessness. Hurricanes, tornadoes, earthquakes, and floods similarly destroy homes or force people from their homes. The scope of this type of homelessness was vast, affecting millions of people during this extreme weather event, but its duration was relatively short and its effect less intense than the homelessness of the poor, the addicted, and the mentally ill. It was different in kind from the homelessness experienced by those groups during normal dynamics of nature.

Crime during a Disaster

When the power was blacked out, many burglar alarm systems were inoperative. Crime rates nevertheless decreased. For example, the Quebec police found that crimes declined by 31 per cent for the first twenty-five days of January 1998 compared to the previous year: break-ins were down 29 per cent, stolen vehicles 62 per cent, and assaults 58 per cent.[20] In part, this pattern was the result of the atrocious weather conditions. Why steal a car for a joy ride when it would take time to crack ice off the windshield to see where one was going and when the roads were blocked? Breaking into a house becomes a special challenge when the windows are frozen shut and laminated with multiple layers of hard ice. There were no potential victims on the streets to assault. There were, nevertheless, many opportunities for crime, but it never materialized. This reduction in crime rates during the 1998 disaster confirms the findings of other studies showing the same tendency.

People did not, however, believe that the rate of crime was unusually low, partly because the rare crimes that did occur were reported in sensationalist fashion in the media and partly because residents felt vulnerable when the normal infrastructure of society collapsed. This was particularly true of people who had to evacuate their homes. But police patrols and military patrols helped to persuade them that their homes would not be broken into. The patrols provided reassurance that convinced some to go to a heated shelter instead of remaining in their dangerous houses just to protect them.

Shelters or Drop-in Centres?

In all provinces and states there was a relatively low use of shelters as places to stay overnight. In Quebec, for example, it has been estimated that only 4 per cent of people without electricity stayed in a shelter overnight, even though 25 to 56 per cent (depending on the region) left their unheated houses.[21] Even in municipalities without power the longest, the proportion of victims who stayed at least one night in shelters never surpassed 10 per cent. People preferred to remain at home even if they had no heat and light, stay with relatives or friends, or use their own personal resources to rent a room. There was, on the other hand, heavy use of shelters as drop-in centres to get warm, eat a meal, and have a shower. Suggestions were subsequently made to reduce the size but increase the number of shelters in order to make them more attractive and diminish social tensions and the spread of infectious diseases.

The Cockatoo Quandary

An expensive hotel called the Château Laurier in Ottawa offered free rooms during the disaster for those without heat and electricity. One couple brought a cockatoo with them. The hotel at first said it could not allow a bird in the room, but it finally relented. This "cockatoo quandary" led Ottawa Social Services to suspect that some people might be placing themselves in danger by remaining in unheated houses to protect their pets. In Quebec, too, it was found that leaving domestic animals alone in an unheated house where water froze caused stress for people who went to shelters.[22] Thinking about their beloved pet without food or water in a frigid house was a worry for owners in shelters. But bringing those animals to a shelter caused havoc: the cat belonging to one person did not get along with the dog brought by another, and both frightened the bird introduced by a third. Pets adored by owners bothered other sheltered people. Shelters had trouble enough coping with the demands of humans and had no resources for dealing with pets. People have affectionate relationships with their domestic animals, which may be housed separately in normal weather and rarely cause problems. But when a disturbance of nature forces people to live together, it creates the issue of what will be done with incompatible pets. Caring for people without thought of their pets leaves the care of people incomplete. Ottawa Social Services improvised an agreement with the Ottawa Humane Society to care for pets when people vacated their homes for shelters,[23] but not all humane societies have resources for such disaster pet preparedness.

Competition or Cooperation in a Disaster?

In a widespread disaster such as this one, all affected municipalities were competing for the same scarce resources, such as generators. Municipalities tend to address their own needs before they give resources to an adjacent municipality, even if the latter is experiencing greater trouble.[24] A study of this disaster in eastern Ontario concluded that the current provincial policy based on competition for resources is inappropriate.[25] Cooperation, rather than competition, is called for in times of crisis. The failure of the Ontario government to coordinate the actions of the electrical utilities occurred at the time of a chill against government participation in the economy. In Quebec this issue did not present a problem because there was only one electrical utility: the publicly owned Hydro-Québec. Paradoxically, the more a public utility supplying an essential service is broken up, privatized, and deregulated, the more government planning and regulations are needed in an emergency to ensure that actions are coordinated.

Technical Solutions

Can we rely on technical fixes to make the present system less vulnerable to nature's forces? For example, should we build transmission lines that could carry even more ice, install "physical fuses" that would break under a specified load so as not to buckle the transmission tower or snap the hydro pole at either end, or put the wires underground? Such solutions are technically possible, but they add expense, and safety devices that protect against one hazard of nature may provoke additional dangers when a different hazard occurs, such as freezing rain short-circuiting drooping skywires installed to protect against lightning strikes. The technical fix chosen depends on risk assessments, which are often underestimations. For example, although major ice storms occur every twenty years in this area, Ontario Hydro's $66-million-a-year research branch, employing 450 scientists, engineers, and technicians to anticipate and solve problems in advance, admitted that it had never imagined or simulated freezing rain this severe, lasting this long, or covering such a wide area.[26]

The ice storm resulted in gaps in the technical communication network, which in turn created problems of communication even between government officials. After the storm, the Interagency Hazard Mitigation Team in the United States recommended attempts to identify potential trouble spots and to locate cell phone towers better to alleviate this problem. Realizing the low chance of success of this strategy when confronted by surprises of nature, the team also recommended the compilation of a list of ham radio operators who would volunteer their communications expertise in an emergency.[27] Extreme disturbances of nature throw modern society back to a reliance on primitive but flexible technologies. Technical solutions are desirable where possible, but they cannot be relied upon as a cure-all.

NATURE

The Turning Point

One commentator in Montreal wrote: "There was no outstanding moment, no dramatic symbol of the city's return to health; just a gradual reawakening, a slow shaking-off of the trauma, one street, one block, one sidewalk at a time."[28] Although this observation was true for the human part of reconstructing normality, it is inaccurate in a broader sense. During the freezing rain, nature succeeded in knocking down more transmission lines, poles, and towers than the combined forces of Hydro-Québec, Ontario Hydro, the Canadian army, American hydro workers, and the National Guards could repair. As human, rationally planned action took one step forward, nature pushed them two steps back. The turning point occurred when the freezing rain changed to fluffy, unthreatening snow. "If any single moment marked a turning point in the Great Ice Storm of 1998, it was the end of the ice rain that had showered the region off and on since early Monday morning."[29] Nature thereby removed the treadmill that had until that point thrown power repair crews back no matter how laboriously they tried to move forward. From that moment on, progress could be made in the restoration of normality on all fronts. Although many problems remained and much work had to be done, the abrupt end of the freezing rain was the dramatic event provided by nature which determined that normal living could begin returning. Thus it was not human action, with all the resources modern rationality could muster, that determined the beginning of the return to normality; it was the end of the freezing rain. If the freezing rain had continued, so would the state of emergency. The dependence of modern society on a return to normal weather was confirmed.

In Quebec the freezing rain stopped late Friday, 9 January. Despite Herculean efforts by repair crews and the electrical utility, a graph of its customers without power is a replica of the accumulating ice load on the lines: rising on Monday, 5 January, when the storm began, decreasing Tuesday as the freezing rain abated, rising steeply as it returned in force on Wednesday and Thursday, and reaching its peak on Friday, then decreasing sharply from the time the freezing rain stopped.[30] It was as if repair crews did all the work but the presence, absence, and intensity of the freezing rain determined whether that work would have any effect.

A Niagara Mohawk Power Company spokesperson in New York State put the turning point slightly later but still admitted that it consisted of a change in nature's dynamics rather than a human accomplishment. "On Saturday, when the temperatures got above freezing and the ice started to melt, that was certainly the turning point as far as restoration was concerned. We were having extreme difficulties trying to put conductors back in the area that had three, four, five inches of ice accumulated on it. The additional weight made it virtually impossible to get those conductors back up in the air ... But when we received that thaw on Saturday, a lot of the ice left and that certainly helped us."[31] The technological disaster of a crushed electrical system began to be reversed when nature's dynamics started to cooperate with human goals.

Technology That Makes Trees Dangerous

There is no particular problem when tall trees grow in a wilderness region. But when a power line is constructed through an area with tall trees, whether urban or wilderness, it brings the risk that a disturbance of nature, such as freezing rain, a tornado, a hurricane, or an earthquake, will cause trees or their branches to fall and break the line. In northern New England, high-voltage transmission lines and communication towers sustained little damage because wide corridors were kept cut for them to pass through. Lower-voltage transmission lines and distribution lines, however, took a beating because they were located in narrower rights-of-way; hence they were vulnerable to damage from sagging, ice-laden branches and falling trees "because a) homeowners and municipalities plant trees and resist tree trimming and removal for aesthetic reasons, and b) trees are ideal collectors of ice."[32] Where there were no trees, power outages only occurred with very high ice loads on lines.

Trees are transformed into dangerous objects when the costly electrical power lines upon which society has become dependent are constructed nearby. The normal growth of trees thereby becomes a risk. The electrical line must be continuously protected against this life force of nature. To avoid damage, power companies spend millions of dollars on tree trimming. FEMA proposed the replacement of tall trees with short species near power lines, which it saw as "the single most important mitigation action that can be taken."[33] If trimming and pruning are resisted, more drastic measures are advocated: "removal of the tree to the ground would remove the problem, preventing recurrence and minimizing long-term cost."[34] The suggested solution is the cutting down "of danger trees adjacent to the right-of-ways"[35] and the monitoring of communities "to assure that the community has 'the right tree in the right place.'"[36] Herbicide applications to reduce future vegetative growth are recommended to keep the surrounding land barren when power lines are constructed through a forest. This measure is also referred to as the "replacement of inappropriate landscape with appropriate landscape."[37] Landscape is to be judged in terms of its appropriateness to power lines: bald strips through the forest hillside are appropriate landscape. If one believes that beauty is only a cultural construction in the eye of the beholder rather than an objective phenomenon, then modern societies will just have to learn to redefine these barren strips through the forests as beautiful!

Inadvertent, Perverse Consequences of Managing Nature

Ecologists concluded that the direct negative effects of this extreme weather on ecosystems were temporary and unimportant to their survival. Tree mortality was more likely to result from human actions following the freezing rain than from the long-term effects of the storm itself. In urban areas of Quebec there were many cases of savage trimming of trees that exacerbated damage initiated by the freezing rain.[38] In the northeastern United States most forests are privately owned. Forest owners "panicked and cut down trees that could have been saved. Their reaction could have a negative impact on the economy beyond the actual damage wrought by the disaster,"[39] not to mention the negative impact on the ecology. They cut trees that had become temporarily unaesthetic as a result of the storm, but many of those trees would have eventually returned to their appealing appearance if left alone. Landowners authorized loggers to enter their forests to cut injured trees but failed to implement inspection and control; hence more profitable uninjured trees were often cut as well. Logging after the storm increased the prevalence of diseases such as root rot, caused either directly by heavy equipment damaging roots or by the reduction by logging of transpiration, leaving water to accumulate in pools around roots for a long period of time. Trees undergoing what is defined as good management in normal weather were more prone to injury when this disturbance of nature hit than trees in a pristine forest. Large crowns tend to grow on trees in thinned forests, in forests planted in rows, and on urban trees. These crowns had a greater surface area to collect ice, and more of these trees toppled during the freezing rain.[40] In this sense, human management of trees increased the impact of the extreme weather.

In wilderness forests, the freezing rain had more impact on human activities than on the forest itself. Debris consisting of fallen branches and trees blocked snowmobile, cross-country skiing, and hiking trails. Broken branches hanging in the overstorey could come crashing down at any time on unsuspecting loggers and hunters. Like all predators, human hunters had more trouble finding their prey in the enhanced underbrush. "While events like the ice storm may be devastating to people, the forest itself will survive and ultimately thrive."[41] The "ultimately" is important.

The forest recovers more slowly than landowners and investors are willing to tolerate. "The storm caused great damage to pocketbooks of people with financial interests in the trees,"[42] such as sugar-bush owners and tree farmers. The time frame of forest recovery did not correspond to their profit-seeking time frame. One of the hardest hit areas was in Rougemont, Quebec. It was devastated because it was constructed as an apple orchard with short-term expectations of profit, and nature did not meet those expectations.[43]

The major threat to forests is that this extreme weather event will provoke a shift in perceptions. It may lead landowners to view tree growing as a high-risk, low-return business, to turn the land to non-forestry uses that appear less risky and more profitable, and to destroy the forest.[44] Ecologists contend that this threat could be diminished by financial and technical aid from government. Unfortunately, the policies of government agencies at times have the perverse effect of providing an incentive to kill trees rather than to save them. "Because FEMA funds can be used for tree removal but not repair, many communities removed trees that could have been saved."[45] The ice storm aggravated the belief that trees are a dangerous source of problems. Since some ice-laden trees brought down some power lines, it was assumed that all trees which could possibly grow higher than power lines would have to be eliminated.

THE ECONOMICS OF DISASTER

The ice storm resulted in a short-term loss of $1.6 billion Canadian in economic output in the country's manufacturing, transportation, communications, and retail sectors. In addition, there was a $1 billion loss of income from, for example, a short rise in unemployment. New housing starts dropped by 4 per cent in Canada because of the ice storm, and the GDP by 1.5 per cent, which are significant figures given that the freezing rain hit only one region of Canada. At the height of the power blackout, 30,000 retail outlets in Quebec had their business interrupted, 10,000 of them for more than twelve days. Production by the majority of the 116 big companies in the affected region was suspended, in some cases for a relatively long period. Overall production diminished in January. The costs of the ice storm to federal, provincial, and municipal governments, hence to taxpayers, was huge. Soldiers, police, firefighters, medical personnel, and emergency workers had to be paid, usually overtime. Infrastructures such as bridges and roads had to be repaired, and debris cleaned up. It was costly to shelter and feed thousands and provide emergency services and administration. There was also the expense of relief payments and disaster assistance to victims. All this was in addition to the destruction of the electrical grid.

Resilience and the GDP

The Conference Board of Canada estimated that the storm would inject $1.4 billion into the Canadian economy through insurance claims and monies from the two hydro companies and from the federal, provincial, and municipal governments to repair the damage. Hence the storm would stimulate economic growth for the remainder of the year, with the additional growth being greatest in areas that had suffered most damage. The storm would thus give a bigger boost to growth in Quebec than in Ontario, to Montreal than to Ottawa. It would reduce unemployment, at least in the short run. When the multiplier effect, estimated to be 1.3, of secondary spending by workers is taken into account, the economic output gained as a result of the ice storm would be greater than the output lost.[46] Even during the month of January, when the storm hit, economic growth regained steam as Hydro-Québec invested heavily in repairing and reconstructing its power system. February, March, and April saw a surge in economic growth as people and companies made repairs and storm-delayed purchases. If, and only if, nature cooperates by stopping its extreme dynamics, modern rationally planned action is impressively resilient.

The ice storm had an overall positive effect on indicators of economic growth, such as the gross domestic product (GDP), in part because they only measure economic activity and fail to take destruction of goods into account.[47] The greater the destruction, the greater the reconstruction. Hence destruction is good for the GDP even if the country is no further ahead and has less wealth after than before the disaster. The benefits are more apparent than real, in that they consist of rebuilding what was already there, rather than contributing additional constructions or infrastructures. This phenomenon has been called "the economic paradox of natural disasters," but it should more accurately be called the weakness of the gross domestic product as an economic indicator. It "disguises the real loss to society. Changes in wealth are not part of the GDP calculation. In the current context, the capital losses experienced by Hydro-Québec, Ontario Hydro and the insurance companies do not appear."[48] Nor do the uninsured losses of individuals or governments. By entering only the construction side of the ledger and ignoring the destruction of capital, the GDP gives the misleading impression that society is further ahead economically. Production after a natural or technological disaster consists, not of growth of capital, but only of rebuilding the capital that was destroyed.

The GDP tells us how fast we are running when pushed back by the treadmill of nature, but it is important to realize that we have not advanced beyond the starting point.

Absenteeism during a Disaster

This long, intense experience of freezing rain also made visible particular social issues and provoked reflection on them. If a business was closed for lack of electricity and not making money, should employees be paid? If it had electricity and was functioning, perhaps even especially busy because of the storm, should employees be paid who did not come to work because of difficult driving conditions, parental obligations as a result of closed schools or daycare, and other obstacles, or, on the contrary, should they be fired? These questions are easy to answer if one day is involved, but they become more difficult as days change into weeks. They are especially difficult issues for small and marginally profitable businesses. The Quebec Labour Board received thirty-eight complaints as a result of the ice storm: fifteen alleging forbidden practices and twenty-three claiming dismissal without cause.[49] At the time of the January 1998 ice storm there were no laws in Quebec explicitly governing dismissal or withholding pay during a disaster. As a result of the experience of the storm, the Quebec minister of labour proposed that explicit mention be made of a disaster in the labour code and that a person who temporarily takes time off to assist family or protect property from loss or serious deterioration during a disaster not be subject to dismissal.

Compassion, Vote Harvesting, and Rewarding Irresponsibility

When the president declares a disaster in an area of the United States, the declaration authorizes the Federal Emergency Management Agency to pay a large proportion of public costs. For example, the ice storm resulted in an estimated US$6.1 million damage to public property in the state of Vermont. The federal government paid $4.8 million, the state $762,000, and the local communities the rest. The federal government also provided $300,000 in mitigation grants to the state for projects to prepare for disasters in the future. Many community leaders view FEMA as having been responsive and helpful, although they do complain about the amount of paperwork. FEMA teams criss-crossed Vermont to determine whether it was eligible for federal disaster assistance and took a week to decide. When an affirmative decision was made, things improved. The emergency management coordinator for Vermont stated: "Once the presidential disaster is declared, everything takes on a whole new life. Feds bring suitcases full of money. That's the best way to help communities: money."[50] This mechanism spreads the cost of disaster over the whole nation so that the localities that experienced the disaster can be resilient and bounce back.

There is economic and political pressure on the president from the affected areas to declare a disaster, which is checked by resistance in the wider society to paying taxes. The compromise is that damage from small, frequent dynamics of nature is completely paid for by the local community and state, whereas destruction from massive, rare forces of nature is largely covered by the federal government. This approach still leaves a great deal of discretion to the president to interpret events of nature. "Between 1989 and 1994, the President of the United States declared two hundred ninety-one disasters, at a cost to the U.S. Treasury of thirty-four billion dollars. Substantial numbers of disasters continue to occur. In 1997, there were forty-three Presidential disaster declarations throughout the United States."[51] During the period 1989–94 two different presidents, representing the two opposing political parties, held office.

The largesse of politicians after disasters is not always appreciated by taxpayers. Nature's extreme weather has the capacity to fracture not only the local community in a physical sense but also the broader community in a social way. Nine months after the ice storm, John Tierney wrote a column entitled "The Courage to Vote No on Ice Aid" in the *New York Times* under the byline "The Big City." He raised the question: "Why do those upstate dairy farmers deserve our money any more than Mongolian nomads do?"[52] His focus was not aid to Mongolian nomads but, rather, opposition to a congressional bill granting over $25 million to farms and businesses hurt by the January ice storm. Dismissing an obligation to fellow New Yorkers hurt by a natural disaster beyond their control, he asked why the North Country's hardy souls, "said to be guided by traditional virtues and the timeless rhythms of nature," were so unprepared.[53] The victims deserved sympathy because of the exceptional severity of the extreme weather. They just did not deserve money. Freezing rain is a known business risk in the Snow Belt, just as market turbulence is on Wall Street, and federal aid does not flow to brokers and secretaries who lose their jobs.

Tierney claimed that it is simpler to prepare for bad weather with insurance and other precautions than to prepare for business risks. But dairy farmers refuse to buy storm insurance, Californians do not purchase earthquake insurance, Midwesterners build on flood plains, and Floridians construct expensive homes on shores frequently struck by hurricanes. Why? Because they know that politicians will throw the taxpayers' money at them when things go wrong. Tierney quoted an analyst at the Competitive Enterprise Institute who argued that disaster aid buys votes and provides photo opportunities for politicians, but it wastes billions of dollars, infantilizing Americans by giving them the wrong incentives. He pointed out that two decades ago Americans in the heartland regarded New York City residents as deadbeats and delighted in turning down the big city in its hour of need. Those residents absorbed the losses themselves and bought insurance. "So when those improvident upstaters come begging, we have the right to politely demur: CITY TO STICKS: ICE HAPPENS."[54]

The January 1998 ice storm struck the "sticks" and not the big city in New York State, but the opposite was the case in Quebec. There, too, discontent arose: some victims believed that the criteria for government aid were too strict, whereas some non-victims contended that governments were too generous giving their tax money to people who had refused to purchase insurance. The Quebec commission of inquiry concluded that the role of government had degenerated recently from a last-resort insurer for disasters not covered by private insurance to a first-resort money dispenser for people who did not wish to take responsibility for their own fate and sought to displace that responsibility onto governments. This attitude, it argued, is bad in itself in that it leads citizens to act irresponsibly and also places a huge financial burden on governments, which in turn will undermine their capacity to give universal last-resort protection in times of disaster.[55] Thus the commission tried to clarify the responsibility of governments without downgrading individual responsibility.

In the future, activities in the big city may be related to problems in the sticks. If extreme weather events become more intense or frequent because of greenhouse-gas emissions, then such events will occur in rural areas in part because of the millions of vehicles, greenhouse-gas–emitting factories and airports, methane-producing garbage dumps, and other activities in the big city. Until proof positive of the connection is in, however, big cities continue their improvident ways. Big cities in the industrialized world emit more greenhouse gases per square mile than any other part of the planet, with the big city of New York being near the top of the list. Geographers have developed the concept of "ecological footprint" to describe the extensive impact of modern societies on a wider environment. A similar concept could be developed for the atmosphere – perhaps the "atmospheric cloud" produced by a geographically small area such as the big city, which then hovers over a much vaster area affecting it too. The big city externalizes its costs by dumping its greenhouse gases and acid rain into the atmosphere, where they form a massive cloud that floats over rural regions, causing problems.

Safety versus Cost

The electrical grids could have been made sufficiently robust to resist ice-loading from this freezing rain, so their collapse and the resulting disaster were technically avoidable. However, this degree of ice-loading was not foreseen in any risk assessment, and the cost of sufficient robustness would have been enormous. "Wires, for example, are produced to the same standard across North America, which makes them affordable. Replacing thousands of kilometres worth with some customized replacement is not economical."[56] Decisions are difficult when safety devices can be installed but the cost is very high. It is tempting for electrical utilities to be influenced by cost considerations even in their assessments of recurrence and risk.

In North America the period of time people are allowed to remain in hospitals has been shortened, mainly to save money. This financial efficiency necessitates adequate home care for the patients, or else their condition will deteriorate and they will have to return to the costly hospitals. The ice storm disrupted home health-care services, particularly in rural areas. Electrically operated life-support machines failed to operate in homes without electricity. Many health-care providers were unable to travel to homes to supply services.[57] Whereas a hospital usually has a backup generator, few homes did. The roads leading to one hospital in a region can be quickly cleared, but it takes much more time to clear the roads to widely dispersed homes. The very means used to increase efficiency and reduce the cost of patient care under normal weather exacerbated problems of care and safety when extreme weather struck.

Tight Coupling and Lack of Redundancy

Tight coupling and the elimination of redundancy that was cost-effective in normal weather also aggravated vulnerability and augmented costs when extreme weather occurred. The linear character of power transmission lines and communication towers connects regions together such that a broken link at one point cuts power or communication elsewhere. By this fact, extreme weather events in one locality wreak havoc in areas where the weather is less extreme. For example, power outages on Montreal Island resulted not just from broken lines there but especially from the more intense freezing rain on the South Shore that brought down transmission towers and lines to the island. Problems were propagated by the tight coupling of technology well beyond the area affected by the most extreme weather. When regions are connected in a network technologically, their vulnerabilities to nature's dynamics are linked additively. Since the frequency of freezing rain *anywhere* in the network is greater than the frequency *at any one particular point*, "the risk of exceeding the 50-year return-period ice load increases with the horizontal extent of the transmission line."[58] There are technical and non-linear solutions that diminish vulnerabilities, but they are costly.

Different essential infrastructures have also been tightly coupled to one another. When the freezing rain brought down the electrical system, it not only stopped the heating and lighting systems from operating but also indirectly provoked a major dysfunction of telecommunications, financial services, transportation, and water supply because of their reliance on electricity.[59] With respect to transportation, lack of electricity caused grave problems for the functioning of gas pumps and refineries, underground subways, street and tunnel lighting, ventilation of tunnels, the pumping of water out of underpasses, traffic lights, and traffic coordination centres.[60]

Placing power lines alongside roads made efficient use of those corridors and reduced costs in normal weather by serving a double function, but when a power line fell during the ice storm, the result was not only an electrical blackout but also a blocked road. Even some major divided highways were blocked because power-transmission lines fell on them.[61] The decision to provide transportation and electrical transmission in the same corridor made it easier for a disturbance of nature to devastate both of these essential infrastructures.

The Federal Emergency Management Agency in the United States has concluded that its "power distribution system is under designed. The poles that carried a single phase and a phone line 30 years ago, now usually carry two additional phases, or bundled conductors, as well as additional phone cables, cable TV and fiber-optics. Heavy transformers may have been added to poles without upgrading the poles."[62] Adding more equipment to existing old poles saved money and produced apparent efficiencies under normal dynamics of nature, but it increased the devastation when nature produced disturbances. FEMA proposed that existing poles should be replaced by stronger ones to withstand ice storms and earthquakes, utilities should improve maintenance and inspection, and power "dead-ends" and "design-failure" should be built into the network so that a break in one pole would not pull down adjacent poles on the line. These technical solutions would result in a more robust network, but they are expensive and result in higher electricity bills. Even then, there is no guarantee that the redesigned, stronger structures will resist the next powerful disturbance of nature. After all, by the beginning of January 1998 Quebec had made its electrical grid more robust than at any other time in its history, yet it was crushed by unexpectedly intense, persistent freezing rain.

Before the ice storm, the suggested purchase of backup generators by the city of Montreal for its filtration plants seemed a costly and useless redundancy because Hydro-Québec assured the city that it could always provide power. When nature prevented the power company from keeping its promise, a perilous and potentially very expensive situation was created. Redundancy looked much more prudent. Safety measures that seem an unnecessary cost in good weather appear much more attractive during a disaster.

"Just-in-time" organization of production is cost-effective in normal weather because it dramatically reduces the need for expensive and redundant inventory. When a disturbance of nature blocked transportation, cut electricity, and reduced communications, however, there were no nearby stocks of essential goods to draw on to meet the needs of the population. Furthermore, production problems in the area affected by the disturbance were spread to other regions because of their dependence on it for components, especially when the area was a major centre of production such as Montreal.

These technological and organizational advances in redundancy elimination and the associated cost-benefit analyses were based on an underestimation of risk, and they rendered the essential infrastructures more vulnerable.[63] After studying what happened during the January 1998 ice storm, the Quebec commission of inquiry concluded that the interdependencies among essential infrastructures constituted one of the fragile elements of modern society. An impairment in one is quickly propagated and becomes a weakness in most of the others. The commission recommended that, where possible, these interdependencies should be limited.[64] In almost all cases, doing so would imply building in costly redundancies. A technical solution becomes an economic problem. But a disaster is enormously expensive, so that the costs of defences would be compensated for by the diminished cost when a disturbance of nature struck.

Urban Sprawl as a Source of Disaster

In the United States and Canada the fastest growing communities have been ten to twenty miles outside cities. Here land is less expensive, gas taxes are the lowest in the developed world, and toll-free expressways make commuting easy and fast. The provision of utilities and public services is, however, more difficult and more costly per capita in this urban sprawl than in a densely populated area, and so is the prevention of disaster. In high-density cities it is cost-effective as well as aesthetic to bury low-voltage electrical distribution lines, telephone lines, and other cables, making a city less vulnerable to extreme weather events such as this one. But such underground wiring becomes very expensive with low-density urban sprawl. Hence vulnerable wires are strung from pole to pole.

The difference between North America and Europe is striking in this regard. Urban sprawl in North America has resulted in electrical and communication lines hung from poles, where they are exposed to extreme weather events such as freezing rain. High-density living in Europe has made it economical to bury that infrastructure underground, where it is protected from extreme weather. European gas taxes promote high-density communities and provide revenue for public transportation and the burial of electrical and communications lines. Those taxes take into account more of the hidden costs of the automobile. In North America that cost is less visible because it is paid in indirect ways, such as the bill for the ice storm, but it has to be paid all the same.

The Insurance Industry Response

The insurance industry portrays itself, with some justification, as the avant-garde of safety in the private sector, claiming that it has been the champion of initiatives to mitigate losses of all sorts. "It was insurers who brought the first fire departments into existence, initiated the first building codes, commenced boiler, machinery and elevator inspections, established the Underwriters Laboratory (UL) and promoted the use of airbags in automobiles. Insurers have given premium credits (incentives) for the installation of sprinkler systems, smoke detectors, burglary alarms and the use of airbags."[65]

"Sustainable Development" was the title of a chapter in the industry's analysis of this ice storm. It defined sustainable development as referring "to the structures' ability to withstand the rigours of natural hazards and endure over time."[66] The implication, which is not drawn out in this report, is that the construction of the electrical grid in both Canada and the United States was not sustainable development because it was not given the capacity to withstand the rigours of this natural hazard and did not endure over time. Instead it collapsed.

The insurance industry advocates a "holistic approach to mitigation" and a "sustainable hazards mitigation" method that seek to avoid burdening future generations with risk. It argues that savings derived from stronger energy codes, materials, and practices would be particularly helpful in mitigating losses from nature's disturbances related to society's reliance on electricity. For example, better insulation would equip buildings to retain heat longer and, together with more efficient lighting, would enable generators in short supply during an electrical blackout to accomplish more. Insurers argue that photovoltaic cells and windmill farms should be thoroughly studied as alternative decentralized sources of energy which do not require long transmission lines, and that land-use issues which would diminish damage must be considered. The emphasis would be on the precautionary principle to diminish the reliance on fossil fuels, nuclear energy, and distant hydroelectric megaprojects. This approach does not put all hope in resilience, that is, bouncing back after disaster. Instead, it also emphasizes improving constructions so that they would be less vulnerable to nature's disturbances in the first place and thereby avoid disaster altogether. Prevention and robustness would have pride of place.

These principles have a nice ring to them, but as usual, both the value and the devil are in the details. "How people and structures that are in harm's way are handled is extremely important to the sustainability of an environment that is subject to natural hazard events."[67] True, but this extreme weather event was so enormous in scope and so persistent that more than 6 million people were in harm's way. It demonstrated that for some massive disturbances of nature, almost everywhere is in harm's way, and this characteristic will likely be exacerbated by climate change.

The insurance industry contends that "the holistic approach must evaluate the risks that are present and provide an indication of the potential loss frequency and severity facing inhabitants, businesses and governments that are susceptible to the vagaries of the various hazards. Also, the plan must give an indication of the cost benefits to be derived if a comprehensive course of action is pursued."[68] Some of the preconditions of insurance might not be met even if the very reasonable holistic approach is adopted. It is uncertain that state-of-the-art technologies and methodologies can provide the capacity to accurately evaluate the risks and the potential for catastrophe, recurrence cycles, loss frequency, and severity of extreme weather events such as the January 1998 ice storm. All the parties – insurers, governments, power companies, and meteorologists – using state-of-the-art technologies and methodologies failed to anticipate the severity and duration of that event. The cost-benefit calculations of comprehensive action remain laden with as much uncertainty as the assumptions upon which they are based. The insurance industry adds a rather ominous warning: "A holistic approach should not be viewed as guaranteeing the availability or affordability of insurance in the private sector."[69] Will insurers simply refuse to make insurance available or affordable?

Modern societies can now build resistant structures, but will they? Technology has provided the capacity, and economics the constraints. The temptation is great to assume, on inadequate evidence, that the 1998 extreme weather in northeastern North America will not recur for a hundred years, in order to avoid the cost of development that is sustainable rather than collapsible. New technological advances intensify the interaction of modern societies with nature, thrusting additional decisions upon leaders and the population and forcing them to choose whether the improvements and added resistance are worth the cost in the context of overwhelming but uncertain disturbances of nature.

RISK

Perceived Risk and Unperceived Risk

People ran out of supplies because the freezing rain hit unexpectedly. They had to drive elsewhere to acquire necessities because in North America many stores are of the "big box" type distant from residential areas, and an automobile is needed to go from home to store. Telephone lines were sometimes cut as well, so people had to drive to get help. Those who went to shelters or to stay with relatives or friends or at hotels drove there. Driving, however, became an extremely hazardous endeavour. The freezing rain covered all roads with a thick, slippery coating of hard ice. Trees, branches, poles, and electric wires – some dead, some live – were lying across the roads. Clearing, sanding, and salting of roads were nullified by the next wave of freezing rain and falling debris and wires. On a few roads drivers could round a curve and encounter a fallen transmission tower. They also ran the risk of being hit by trees, poles, and wires falling from above. Street lights and traffic lights were blacked out. At the start of their journey, drivers would have to hack thick ice from their windshields, only to find another layer forming as they drove. They sprayed a gallon of antifreeze on windshields for a ten-mile trip home, yet even that yielded only minimal visibility.[70]

A remarkable fact about the disaster was the almost complete absence of fatalities, injuries, or accidents on the perilous roads.[71] Some cars and trucks did slide off into ditches, but there were virtually no reports of anything serious. This absence contrasts sharply with normal weather, when vehicle accidents are one of the main causes of death and injury. In an area where people had experienced the effects of freezing rain in the past, the risks of driving under such visibly atrocious conditions were clearly perceived. A state trooper in Maine remarked that "when you have to scrape a half-inch of ice off your windshield, I think people quickly realize they need to be careful."[72] Driving was reduced to what was strictly necessary, and great precautions were taken. Municipalities and counties placed bans on non-essential travel in many areas, but enforcement was spotty.[73] Unlike speeding bans in normal weather, these restrictions during the freezing rain were obeyed by individuals, not for fear of being caught by police, but rather because of apprehension about the material consequences of recklessness. Risk was clearly perceived and acknowledged, and hence danger was successfully avoided.

Most of the deaths and injuries during the ice storm were the result of behaviour in which risks were either unperceived or unacknowledged or some combination of the two. For example, in Quebec thirty people died: five from hypothermia that crept up on them; ten from burns as a result of fires and explosions caused by the overuse of previously little-used fireplaces or unanticipated power surges; six from imperceptible carbon monoxide poisoning because they used generators, gas heaters, or BBQs indoors for the first time; five by falling from a roof or a transmission tower while trying to remove ice; one from being crushed by falling ice; one from slipping from the sidewalk into the path of a snow-removal vehicle; one when the driver of a tractor was decapitated by a low-hanging electrical wire that he did not see; and one from a head injury.[74] "In one accident that horrified the nation, a technician who was trying to de-ice a transmission tower died after he fell about 200 feet."[75]

The accuracy of risk perceptions was important on the collective level as well. Quebec, Ontario, upper New York State, and New England were well prepared for the risk of low-intensity, short, localized freezing rain of the sort they had experienced in the past. But the long, high-intensity, widespread ice storm of January 1998 caught them all by surprise. Freezing rain of this intensity and duration was not foreseen by meteorologists, power companies, or governments; hence structures were not built to resist it, and emergency measures organizations were not set up to deal with it. The result was the most expensive disaster in the history of this vast region, a number of deaths and injuries, and a cold, frightening experience for a huge number of people. If the risk of extreme weather of this intensity, duration, and scope had been perceived and acknowledged, measures could have been taken to mitigate the problems. More underground electrical distribution lines could have been built in urban areas. More-resistant lines, poles, and transmission towers could have been mounted. Generators could have been stockpiled. Emergency measures organizations could have been better prepared. The technological and organizational expertise exists to defend the electrical system against these dynamics of nature. The risk was not, however, foreseen. The area was left vulnerable, and the result was disastrous.

In a technological society, the population expects answers from specialists holding knowledge that people have been told is sophisticated. During the disaster, victims wanted to know how long the electrical blackout would last. They had difficulty handling uncertainty. Some even believed that the lack of answers from authorities indicated a refusal to recognize the traumatic situation they were facing.[76] The population found it hard to understand that meteorologists were unable to forecast the end of the intense freezing rain and that power companies were unable to give precise, reliable predictions of when power would be restored to specific streets. Science and technology have their limits, especially when called upon to predict the future. If they are expected to be like magic, they are disappointing.

Unacknowledged Risk

A disaster researcher concluded that the "real problem is that persons refuse to recognize that they are in serious danger and need to act to protect themselves."[77] People accustomed to the normal dynamics of nature and technology that is efficient in normal weather are slow to notice risk and admit there is danger. Unperceived and unacknowledged risks become major problems when the world is not what people think it is or wish it to be. The unwillingness to recognize peril took different forms, but one was particularly worrisome in the dark, frigid weather during the ice storm. "People would remain at home even though there was no heat and no water and they were in danger of hypothermia. Police and public health personnel did identify a number of persons in danger of freezing to death."[78] About two hundred of these people were found through door-to-door calls in the rural areas around Ottawa alone; then it was realized that the most serious cases were those who did not answer the door. When people refused to go to a shelter, the family physician was called to persuade them. But even that did not always work. No one could prevail upon a disoriented woman suffering from hypothermia to go to a shelter. So arrangements were made to have a neighbour deliver three hot meals a day and a police officer and nurse to visit her regularly. The risks resulting from lack of heat in frigid weather provoked debate over whether emergency measures organizations have the legal authority to oblige people to evacuate their homes if they insist on staying: "it is very doubtful that any agency can force large numbers of persons from their homes if those persons don't want to go."[79]

The predisposition to assume normality and reluctance to recognize dangerous abnormality that threatens our comfortable way of life can characterize not only the average layperson but also experts and organizations. The Ontario Emergency Management Organization was slow to recognize the widespread danger created by this ice storm because freezing rain of this magnitude had not occurred in this region since records were kept and because the ice-loading developed incrementally. At the beginning, the power companies believed they had the situation under control and were restoring power and then another wave of freezing rain brought down more lines than they had restored. Nature itself shattered illusions of control and generated a transformation in perceptions of risk.

The Quebec commission of inquiry documented that a cultural predisposition had developed in the province which discounted nature's extreme disturbances and the need to prepare for disasters. The refusal to acknowledge risk has several components. There is the belief that being overwhelmed by nature only happens to others, especially those in "underdeveloped" countries. If a disturbance of nature occurs in a high-technology modern society, it is assumed that resources can quickly be mobilized and improvisation will be successful. Thus it is not deemed necessary to invest much in prevention, mitigation, and preparation for disasters because doing so brings the economic risk of investing in something that may never occur. Money is assumed to be better spent to attain immediate and certain goals, even if considerable amounts of financial aid have to be given during and after a disaster, because disaster is so unlikely. Many of the difficulties encountered during the January 1998 disaster resulted, the commission contended, from inadequate preparation and mitigation as a result of this cultural predisposition to discount extreme disturbances of nature.[80] I would argue that Quebec is not unique in this regard. This set of beliefs is present in other modern societies – in particular, Ontario and the United States – and these societies experienced the same difficulties as Quebec during the ice storm for similar reasons.

The commission of inquiry concluded that a more realistic perception of risk is needed than the one portrayed above. Investment in defences, mitigation, and readiness will be made only if the culture of invulnerability is replaced by a realization that disaster can occur. This does not mean that a realistic perception of risk will be easy. The commission entitled its principal volume "confronting the unforeseeable," which is a particularly difficult task. The capacity of nature to surprise even the most advanced technology, scientific knowledge, and predictions has made the calculation of risk probabilities and return times problematic. Risk perceptions will be contested: what is seen as a realistic assessment of risk by some will be criticized by others as either paranoia or wishful thinking. Nonetheless, acknowledging rather than dismissing risks is the starting point of managing them as successfully as possible. This approach is especially important for leaders because of their key positions. The study now focuses more directly on the issue of leadership in disaster.

PART THREE

Leadership in Disaster

CHAPTER 7

Worse than the Worst-Case Scenario

The difference between beliefs about risk and real, material risk can be illustrated by a pathetic case that could be a microcosm of a much larger phenomenon. "When the volcano Mount St. Helens was threatening to erupt, Harry Truman, a local resident, was interviewed by a TV newsman. 'Why don't you leave, you've been advised to evacuate in order to save yourself?' Harry responded that he had lived here for decades and never had a problem, 'how bad could it be?' Harry died in the explosion and lava flow. He is buried under the debris."[1] This anecdote shows not just the reluctance of people to evacuate their homes but, more importantly, how difficult it is for them to perceive risk that exceeds their experience. Harry Truman is not the only person who has discounted how bad it could be by extrapolating a conclusion of safety from decades of good experiences.

EXPECTATIONS OF SAFETY

Freezing rain is common in the area hit by the 1998 storm, occurring about twelve times each year. But these events are typically of short duration (a couple of hours), are not particularly intense, and are not extensive in terms of affected territory. The important question for meteorologists and for leaders is: Did they expect freezing rain of the intensity and duration as occurred in time to plan and prepare for it at least days or, better still, months in advance? Did they foresee the risk, or did they expect typical weather and therefore safety? Meteorologists claim that they did forecast freezing rain but admit they did not foresee its intensity and duration.[2] Intensity, scope, and timing are, however, the crucial elements of forecasting in mitigating disaster. So where did that leave the leaders who had the mandate to reduce the vulnerability of their communities? Expectations of decision-makers have, after all, been found to be crucial elements in avoiding, mitigating, or aggravating disaster.[3]

Florent Gagné, the Quebec deputy minister of public security, described how he and other decision-makers in Quebec expect typical episodes of freezing rain and prepare for them. The intensity, persistence, and scope of the rainfall on this occasion, however, caught them by surprise.[4] "We followed the weather forecasts, but we were sure that the next day electricity would return. There were three successive waves of freezing rain day after day, which is a very rare meteorological phenomenon, and so the government was not at all expecting it to happen. Freezing rain, yes, but not such an ice storm. No, we weren't prepared because statistically we weren't expecting freezing rain like that."

Mr Gagné clearly acknowledged that the intensity, persistence, and scope of the freezing rain exceeded anything that was foreseen even in the worst-case scenarios, and hence disaster prevention and response were inadequate. "We weren't at all prepared because an ice storm like this one had never been planned for or expected. Even in our worst scenarios it was never predicted. Therefore we didn't have the equipment to respond to such a situation." A worst-case scenario is a prospect imagined by experts in the field. It is not necessarily the worst case that could occur. Extrapolating safety from a time series based on the half-century in which meteorological data have been gathered can be quite misleading concerning infrequent but overwhelming phenomena of nature.

When asked whether he and the other decision-makers in his entourage expected this intensity and duration of freezing rain, Jean-Bernard Guindon, the director of the Centre for Civil Security for the Montreal Urban Community, answered: "No, absolutely not. In fact, the main problem concerning the storm was that we weren't expecting it. Everything that was studied about freezing rain previously had led us to believe there would be only some broken branches, some fallen wires, and some power outages but in a limited area."

Hubert Thibault, chief of staff for Quebec premier Lucien Bouchard, was asked whether leaders at his level had expected such an ice storm. "Not at all," he responded. "It was completely unexpected. At the beginning of the freezing rain it appeared to be a significant storm, but I don't think anybody realized it would degenerate into this episode."

Modern Quebec was built on hydroelectricity. In the late 1960s the electrical power companies were nationalized. Energy needs were growing.

What would be the source of the energy, and where would it come from? Nuclear reactors were being built in the neighbouring province of Ontario, in the United States, and especially in France. But the Quebec government in the early seventies, under Premier Robert Bourassa, of the federalist Liberal Party, decided not to jump on the nuclear bandwagon and instead to develop Quebec's abundant hydroelectric resources in the far north. Despite opposition from many environmentalists and Aboriginal groups, massive dams, power stations, and transmission lines were built. This development has proved enormously successful, contributing to Quebec's energy self-sufficiency, reducing its reliance on fossil fuels, providing a focus for the development of technical expertise, and greatly improving its finances. The secessionist Parti Québécois, which initially promoted nuclear energy, was eventually converted to hydroelectricity. When it later took power, its leader referred to hydroelectric energy sales to the United States as a machine to print money for the government. The public electrical utility has been a basis of pride for virtually all citizens of Quebec, regardless of their differences. It has provided a reliable source of energy. Hence expectations were high, especially for electricity for essential infrastructures.

Asked whether the lack of preparedness for an electrical blackout of the filtration plants in Montreal was the result of assurances from the electrical power utility that such an occurrence was extremely improbable, Florent Gagné answered that, "of course, we thought it was almost impossible. I'm convinced that in minds of the administrators of that equipment in Montreal an electrical blackout of the filtration plants just couldn't happen. But it did happen. So was it a lack of planning and of good decisions?" He did not answer his own question, but it is an important issue to raise.

Montreal's mayor, Pierre Bourque, described the assurances of electricity for the filtration plants and the resulting contradiction in expectations during the ice storm. "I examined the contracts that the City of Montreal signed with Hydro-Québec in 1973 when the filtration plant was constructed. In those contracts it was unthinkable that Hydro would some day lack electricity to supply the plant. It wasn't even in the contract. It was an absolute assurance that there would not be a lack of electricity." Before the Montreal filtration plants ceased functioning for lack of electricity during the 1998 ice storm, the electrical utility had always contended such a scenario was impossible.[5]

When Maine's governor, Angus King, was questioned about whether he and his advisers had anticipated freezing rain of the intensity that occurred in January 1998, he quickly answered: "No. I remember meeting on Wednesday with my director of the National Guard and emergency preparedness. We were starting to see some power failures, but I still expected to go away skiing for the weekend and never expected it to be the disaster it turned out to be." Governor King also admitted his lack of awareness of risk and of the importance of disaster preparedness before the ice storm struck. "When I first came into office and was touring the state facilities, I went into the emergency management centre and there were twenty people doing things. I asked, 'What are these people doing?' The answer I received was, 'They are getting ready for a disaster.' And I thought, 'Isn't this a waste of money.'"

Such lack of concern regarding risk prior to disaster was also true in Montreal. Jean-Bernard Guindon noted that "there were two attempts to eliminate the Centre for Civil Security of the Montreal regional government. Those attempts did not only come from politicians. Once it came from the police service and another time from the old city of Montreal. Civil security is paradoxical because it is not wanted and is marginalized, but other organizations also attempt to get their hands on the power it represents."

Unexpectedness is often used as an excuse for bad leadership, but in this case all evidence points to the fact that such intense, persistent freezing rain falling on such a large territory for such a long time was indeed unexpected by everyone. However, the fact that this freezing rain exceeded the worst-case scenario does not let leaders off the hook. Worst- case scenarios are social constructs, and nature's dynamics demonstrated that in this case the constructors of worst-case scenarios were naively optimistic. The worst–case scenario that was envisaged proved to be not bad enough. This outcome calls for a healthy dose of skepticism concerning the improbability of worst-case scenarios based on extrapolations of safety from the recent past.

FROM UNPERCEIVED RISK TO ACKNOWLEDGED DANGER: PROMPTED INTO ACTION BY NATURE'S DYNAMICS

Since the freezing rain moved slowly from west to east, people and leaders in the state of Maine heard about its intensity once it began falling in Ontario and Quebec and was headed their way. Their expectations were influenced on a daily basis by the experiences of these more westerly regions that were being communicated to them, as well as by meteorological reports that were by this time being sent out. However, it was especially the consequences of the intense freezing rain on Maine itself that led Governor King and the population to perceive risk where they had not before. "By the time we got to Wednesday night and Thursday when the power lines started to go down, it was clear we had a serious situation,"

King said. "By Friday morning it was a disaster. But we had no long, advance warning about the weather."

General Earl Adams, Maine's commissioner of defence and emergency management, described how the change from unperceived risk to acknowledged danger occurred and prompted emergency leaders to act. "The closer we came to the 7th of January, the more definitive information the National Weather Service had and finally told us we were going to get a severe ice storm that is likely to cause severe problems. About noon on the 7th of January we started to get some precipitation in the state. So we activated our emergency operation centre. The next morning the ice storm was in full swing. We started to ramp up later that day, on the 8th."

Even well into the ice storm and with knowledge of what had occurred farther west, expectations of safety prevailed and only changed when meteorologists foresaw even more dangerous weather dynamics. Governor King stated that "when the weather report came Friday around noon that it was going to fall to 9 degrees, that suddenly put an edge of urgency on it that really hadn't been there before because people can die in that kind of cold, particularly elderly people and people with disabilities and breathing problems. I recall a bolt of fear and suddenly realized (a) we had a very serious problem that could result in loss of life and (b) we really didn't have the resources to cope with it. It could be getting away from us. It could be beyond our ability to respond."

General Adams described how some of his emergency workers stumbled upon victims who failed to perceive risk. "Some of our National Guard troops went into homes and found people either unconscious or nearly unconscious from carbon monoxide and got them out of there." This near-death experience resulted in increased awareness of such risk on the part of both the victims and authorities, who then sent out warnings to people via the media. Perceptions of risk are often in a rather nebulous state until they are confirmed by physical experience.

For other leaders, it was news of the crushing of the electrical grid by the ice-loading that transformed expectations of safety into fears of disaster. Florent Gagné described what had kick-started government action in Quebec: "When pylons started to collapse on Wednesday, the government as a whole began mobilizing, in particular the political machine. Obviously nobody, including Premier Bouchard, thought it would ever be that long, but we knew it would be for several days." Even when transmission towers started to fall, expectations proved to be underestimations of the severity of the risk.

Jean-Bernard Guindon described how the news of the destructiveness of the freezing rain stirred him and his decision-makers into action, but some leaders did not react as quickly. "I was told eight transmission towers had fallen in Drummondville. Instantly I realized we were dealing with an emergency. I mobilized our people, and the next day our civil security committee had a meeting. The day after that our coordination centre was open. That was from Tuesday to Thursday. But at the Quebec civil security ministry they only started to mobilize on Saturday. By then the worse of the events was over." Nature provides prompts, but people react in different ways to such prompts, even disastrous ones.

THE INTERPRETATION OF DISASTER

The experience of this disaster led to a broad consensus concerning its character, as described by Florent Gagné. "There were three events: a natural problem that brought about a technological problem and the technological problem resulted in a problem of social disorganization. The problem of social disorganization obviously was much more difficult to manage because entire cities were deprived of electricity during long periods of time." The intense freezing rain persisted for five days, and the electrical blackout lasted a month for hundreds of thousands of people in a frigid, dark winter climate. This was a long-duration disaster rather than a brief wallop. The problem was far worse than a blackout in summer because in winter people could freeze to death in their homes. The severity of the consequences quickly went beyond the response capacities of local levels. "Municipalities normally play the role of first organizational responders," M. Gagné observed, "and they generally play that role well for ordinary disasters. But they were completely overwhelmed by this phenomenon."

Jean-Bernard Guindon interpreted the most significant impact of this intense, persistent freezing rain as "a two-pronged disaster: an ice storm resulting directly in material damage and giving a knock-out blow to an essential infrastructure [the electrical grid]. If the economic centre of Montreal and of Quebec isn't functioning, it is all of Quebec which suffers and perhaps even other provinces or Canada as a whole."

THE PHYSICAL PRESENCE OF LEADERS DURING A DISASTER

Faced with overwhelming freezing rain that was unexpected and with its destructive consequences, how did these key leaders react? What was the response to this destruction, and was the crisis well managed?

Governor King of Maine described his education in disaster management as follows: "In the United States the National Governors Association, which is the association of all governors that meets three times a year, has a school for new governors. During three days, twenty experienced governors tell you what to expect. One of the things that was driven home to us was: you cannot be out of your state during a natural disaster. You have to be there physically. That was all I had for a lesson."

Despite only this rather rudimentary training in disaster management, King sensed intuitively what to do and/or learned on the fly. "The important role for a leader in a situation like this," he said, "is to be visible and reassuring to the public: showing that somebody's in charge, that somebody cares, that somebody's working on solving the problem. So I was not in the office much at all. I went to shelters. I flew around the state three times and drove around the state. I visited line crews on the road and went to Central Maine Power at 5 in the morning when they were sending line crews out. I was on TV all the time. I was on radio three times a day and listeners would call in." Flying in a helicopter or being driven in an automobile was not pleasant in freezing rain. King concluded that leadership has an important component in addition to management. "Communication with the public was one of the most important parts of my role. It is more symbolic and psychological, which is to provide reassurance to people in a time of crisis. People want to feel that somebody's taking care of them."

In the United States this intense freezing rain fell on a large geographical area, but it was an area that was sparsely populated. Boston was fortunately missed, but not by much. As a result, the president did not get involved, but the vice-president did. In any response to a disaster, timeliness is crucial, and Vice-President Al Gore came quickly to Maine. "He came up and met with all our emergency people and listened to what was going on," Governor King commented. "His real role was to say, 'You have the federal checkbook. I'm here representing the federal government, we're on your side, the money is going to be there when you need it.' Then he did what a good politician should do in an emergency: he was visible. He and I got in a helicopter and went all around the state." State authorities did not hesitate to make necessary expenditures because they had received categorical assurance they would be reimbursed by the federal government.

Being physically present was much more demanding for the mayor of Montreal, Pierre Bourque. He was a former director of the city's Botanical Gardens and Biodome and had accepted an invitation from the deputy prime minister of China to act as a special consultant for the First International Flower Festival in Kunming, in the southwest of that country.[6] This working holiday in early 1998 was to be a temporary relief from North American municipal politics. Bourque did nevertheless telephone his assistant in Montreal daily. On 5 January she informed him that the weather in the city was awful. The next day she told him that freezing rain had fallen for two days, making things very bad, and that the members of the city's executive were quite nervous, adding, "I think you should return as quickly as possible." Mayor Bourque now sees that advice as the best she could have given him. It is understandably disappointing for a leader to have to cut short a vacation to come home to manage a crisis. But the mayor recognized that in the uncommon case of a disaster, absentee leadership does not suffice; a leader has to be physically present with his or her population and not on holiday.

So he packed his bags and made flight arrangements. Returning immediately to Montreal from the southwest of China was challenging, requiring connecting flights from Kunming to Guilin, Hong Kong, Vancouver, and Toronto. There, after twenty hours of flying, Mayor Bourque discovered that the freezing rain had resulted in the cancellation of all flights and trains to Montreal. So he called the president of the Quebec aircraft manufacturer Bombardier, who sent his personal jet. Mr Bourque described his arrival: "There were all the cameras: 'the mayor is back.' So I said to myself that I must make people feel more secure. Then I made the rounds of the shelters all night long."

The prime minister and the premiers of Quebec and Ontario had been scheduled to travel with hundreds of Canadian business leaders on a trade mission to Mexico and other Latin American countries. All three delayed their departure to oversee their respective governments' response to the storm.

Technologies of communication and transportation have shrunk space and time. Nevertheless, the physical proximity of leaders remains crucially important for significant decision-making. It is especially vital during a crisis when communication and transportation, so efficient in normal weather, often function poorly if at all. Even if those systems are intact, there is a need for war rooms, emergency operations centres, and conferences of high-ranking leaders, because face-to-face contact diminishes misunderstanding and facilitates the prompt making of decisions.

Quebec City is the capital of Quebec, but the province's premier also has an office in Montreal. It is located in the headquarters of the public electrical utility, Hydro-Québec. The company has for decades been gently trying to nudge premiers out of its building, but all have refused. They find it convenient, especially in case of demonstrations, to be tucked away in a skyscraper where demonstrators cannot find them. During the crisis, which was more than anything else one of electricity deprivation, this serendipitous proximity led to easy and close contact between the premier and his staff and the president of the electrical utility and his staff. This would have been beneficial for any premier, but it was especially advantageous for Premier Lucien Bouchard, who had difficulty getting around because he had had a leg amputated several years earlier to save him from dying from flesh-eating disease.

COMMUNICATION BY LEADERS WITH THE PUBLIC

Politicians' handlers are very sensitive about the media. Premier Bouchard's chief of staff, Hubert Thibault, noted that the media are omnipresent and like big headlines, and if that dimension is ignored by leaders, then the impact of the crisis will be aggravated. If, on the other hand, the media are used to present events in a measured way, then the population will understand what is happening and will see that decisions are being taken and acted upon, and they will be reassured. Thus he argued that it is very important to give the most complete information possible in the most rapid way. Premier Bouchard and André Caillé, the president and CEO of Hydro-Québec, gave televised press conferences daily (sometimes twice a day) to inform people about the state of affairs and what was being done, to make suggestions concerning measures they should take, and to ask for their cooperation. The population was frankly told about the magnitude of the problem and all the specifics. Each movement in the weather and its consequences were described, along with the countermoves taken in response. Mr Thibault explained this process in terms of a pre-emptive strike against potential rumours. "The media are businesses that operate twenty-four hours a day. They need news to disseminate. If you don't give them information, they will search for it themselves, and then it is dangerous. Rumours will start. It is somewhat more difficult for a journalist to allow a rumour to be broadcast if it did not have some echo in the latest press conference of the premier and the president of Hydro-Québec."

It seems paradoxical that television was used when people deprived of electricity could not watch it. Nevertheless, the rest of the population was informed by television about what was happening and what to expect, as were people in shelters, and the information was communicated by telephone (since more telephone lines than electrical lines withstood the storm), by word of mouth, and especially by battery-operated radio to the affected parties.

Since the freezing rain caused problems for half the population of Quebec – that is, for over 3 million people – communicating at the provincial level permitted a unified response and mobilized everyone. Jean-Bernard Guindon stated that Premier Bouchard was a person of strong charisma. Hence he was the man of the hour through his press conferences to reassure citizens and to advance in the suggested direction to solve problems.

Florent Gagné concluded that leaders were intentionally transparent in managing this disaster because "the worst thing for people occurs when they do not know. Communication calms people and helps them orient their actions." If the media announce a snowstorm in an area, people will avoid driving there. If they announce that the flu will spread next winter, people will get vaccinated. This response then facilitates the work of leaders. "As a general rule, I am in favour of transparency and informing people," Mr Gagné remarked. "During the ice storm disaster, we gave daily communiqués to all journalists, held press conferences, etc. We had a press centre within our organization that answered questions directly all day."

Vera Danyluk, the president of the Montreal Urban Community, added that authorities were very careful in their communication with the population to avoid mixed messages from different sources: "the attempt was made to limit the spokespersons for official messages to those two people [Premier Bouchard and the CEO of the power utility, André Caillé]. We were very prudent that the message was clear, simple, and only one message." Thus the aim was to achieve the contradictory goals of transparency and strict control over the message communicated to the population and over the choice of messenger. One uncontrolled message did get out and misled people. On Thursday, 8 January, a television anchorman understood from a colleague that there would be a complete electrical blackout of the island of Montreal at 3 p.m. This information was broadcast at 1 p.m. But the anchorman had misunderstood his colleague, who had said that radio stations were ready to take over if there were a blackout of television. A hypothetical possibility was communicated as a fact and caused confusion. The error was corrected at 1:39 p.m. Hydro-Québec was quick to assert control, using this episode to emphasize to journalists and the public that the best way to avoid error was to rely on its official information.[7]

The elected chair of the Ottawa regional government, Bob Chiarelli, also stressed the importance of open but non-contradictory communication with the public: "when you look at Katrina, you had a general speaking in one press conference, and the mayor speaking at another, and they had different facts that they were operating under and different strategies. We had just the opposite here. We had the three levels of government and the three levels of technical people at one table, and we were all on the same page." Mr Chiarelli described in detail how coordinated communication with the public was carried out: "We had press conferences where the whole team was exposed twice a day to the media for questioning, which were carried live on radio and television. That went on for eleven days. I was merely the chair; everybody was required to be ready to respond. It was coming out of those meetings that we knew where we had to tighten up, and it helped to direct us significantly. It was very calming to the community to know they were communicating exactly what was happening. It's lack of information that creates panic and fear in the community."

Press conferences at which senior political and organizational leaders were forced to answer questions had the manifest function of providing the public with information, but they also had a latent function. "Having the leaders available for questioning was an impetus for the provincial people at the table to get their act together," Mr Chiarelli said. "Otherwise they would be embarrassed. If they couldn't provide answers then it would look very bad. The press conferences were covered on province-wide television and national television. So Hydro had no choice but to get its act up to speed." This pressure forced the laggard leaders of the Ontario electrical utility to improve an initially inadequate response to the disaster.

Mr Chiarelli drew the following conclusion about disaster management from his experience with this disaster: "If I did anything right, it was two things. Number one was to recognize we had a strong team and let them do the job. I just had to make sure they were coordinated. Number two was to communicate with the public and have the public be part of the solution. Those were the key things." This assessment was confirmed by an independent study. "The decision to hold twice-daily news conferences ... was perhaps, the best decision made throughout the incident. It guaranteed that the RMOC [Regional Municipality of Ottawa-Carleton] would not only be open with its residents but would be seen to be open ... Studies show that the public can handle the truth but it finds confusion intolerable."[8]

The president of Central Maine Power, David Flanagan, also stressed the importance of avoiding contradictory messages from well-intentioned spokespersons. "You don't want to have the local manager, who ought to be focusing on getting troops deployed and work done, giving a picture when he's got incomplete information. We're above the trees, he's down in the leaves. People can inadvertently distort the picture. You want a consistent message out there to the extent you can." Mr Flanagan argued that it is better for a leader to err on the side of caution and have things fixed surprisingly fast than to unduly raise expectations and then have repairs slower than promised.

One of the conditions that enhances trust in leaders occurs when they experience problems similar to those faced by the public. The population can then sense that everyone is in the same boat, and the captain has as much need to steer safely to shore as anyone else. Maine's Governor King stated that "it helped my communication on this when we lost power here where I lived. People then liked that I could relate to their outage. People have a funny thing about leadership. They want you to be dignified and an authority, but they also like it when you wear blue jeans just like other citizens and your power goes out."

Even simple gestures can boost a leader's reputation for caring. One night Governor King brought his ten-year-old son to the shelters. "We stopped first at Dunkin Donuts. I said that I'd like all of the donuts. The guy behind me in line said, 'Can I just have one chocolate cruller?' So I took ten dozen, went to the shelters, and handed out the donuts. It wasn't a calculated 'consult with your staff: should we bring donuts?' kind of thing. I wanted the message: we're caring about you, and that was very important."

INTERACTION BETWEEN FEDERAL AND PROVINCIAL OR STATE LEVELS OF GOVERNMENT

As noted earlier, freezing rain is common in this region, but it is typically of low intensity, short duration, and focused on only a small area at a time. Hence neighbouring communities provide help if power outages are beyond the response capacity of any one utility. The intense, persistent freezing rain in January 1998 fell over a huge territory. No help was forthcoming from neighbouring communities because they, too, were severely affected. If aid was needed, and it certainly was, it would have to come from afar, which created practical problems when the needed resources consisted of line crews and their bucket trucks.

Emergency management has been planned using a bottom-up response in both Canada and the United States. The local municipality makes the initial response, but if it cannot deal with an emergency, then the province or state takes over. If the latter is overwhelmed, then the federal government is expected to be ready to supply personnel, equipment, advice, and especially money. The municipality and then the province or state take the decisions, with the federal government playing a supportive role. There is, however, often the potential for friction in intergovernmental relations, especially if different political parties are in power at the various levels. Did the provincial or state governments work smoothly with the federal governments during the ice storm, or was the relationship conflictual?

Vice-President Al Gore, a Democrat, came promptly to the devastated region and toured the state of Maine with its governor, an Independent. On his way back to the airport, Gore said, "Let me know if you need anything." When the president of the power utility heard this, he asked Governor King to request some big transport airplanes to fly crews and bucket trucks up from southern states. General Earl Adams admitted that the request was made out of desperation and was "unheard of. When we were first talking about this, people were saying forget it; it's not going to happen." King called the vice-president's office and immediately got through to him. Within twenty-four hours, huge transport planes were landing at the Brunswick Naval Station full of electrical crews from North Carolina and their bucket trucks.

Quality leadership came from the top – the vice-president and the governor – and the unusual needs of the state of Maine during this disaster were quickly met by the federal government in Washington. The American federal government and FEMA did not hesitate to use innovative methods to provide timely aid to Maine. The state and federal levels worked well together to solve problems encountered in this disaster. There were no turf wars between them. This was the golden age of FEMA: after it had learned lessons from its mistakes in responding to Hurricane Andrew in 1992 and before it had been subordinated to Homeland Security, where it forgot those lessons, as demonstrated by its feeble response to Hurricane Katrina in 2005. Unfortunately, lessons learned are sometimes unlearned.

SECESSIONIST/FEDERALIST CONFLICT IN CANADA: DID IT AFFECT THE RESPONSE TO DISASTER?

In Canada there is great potential for conflict between the federal and provincial levels of government, especially in Quebec. Chronic political tension exists over the possibility of Quebec seceding from Canada. At the time of the ice storm, a political party with secession as its first priority was in power in Quebec. The leader of the Parti Québécois was the charismatic Lucien Bouchard. He had previously established a secessionist party called the Bloc Québécois at the federal level that campaigned for seats only in Quebec and decided issues solely in terms of whether they were in the interests of that province. The Bloc Québécois held the majority of federal seats from Quebec. In 1995 secession had been narrowly defeated in a Quebec referendum, but the secessionists sought to construct the "winning conditions" for a future referendum on the question. Premier Bouchard would hold another referendum, but only when those winning conditions arrived. The head of the federal government was another Québécois, Jean Chrétien, leader of the Liberal Party of Canada. He held a seat in Quebec and was resolutely in favour of the province remaining in Canada. He did not want another referendum, but if one had to be held, he hoped that the conditions would be such that the people of Quebec would vote for a third time in favour of remaining in Canada and end the secessionist threat for a very long period.

The disaster occurred in this political context of repeated referendums on the secession issue. Did the leaders on either side make decisions concerning the disaster that would help to establish the winning conditions for their particular option? Did this division influence the response to the disaster? Was there conflict between francophones and anglophones during this crisis?

Quebec's deputy minister of public security, Florent Gagné, answered: "No. The partisan political dimension was never in any way a hindrance. I never saw any unease, neither from the Canadian army, nor from the Quebec government having to do work with the armed forces. I never felt uncomfortable working with federal authorities. Everybody mobilized to deal with the crisis we were experiencing, and political considerations were put out of our minds." I then asked whether Mr Gagné felt the crisis was so serious that no political party dared to try to profit from it for fear that would hurt their society. He answered quickly: "Exactly. I would say exactly that. I don't think anybody tried to benefit from the disaster."

The director of the Centre for Civil Security for the Montreal Urban Community, Jean-Bernard Guindon, agreed that neither side in the federalist-sovereignist debate tried to take advantage of the situation to further its ideology. André Brunelle, assistant director of the Montreal Fire Department and coordinator of emergency measures, answered a similar question categorically: "No, no. It never came up. Not at all."

Montreal's Pierre Bourque asserted there was no friction between anglophones and francophones during the ice storm and added that "being confronted with such a problem created solidarity. Each and every one of us was affected." Often, tension develops later when bills have to be paid. But even on this delicate question of payment Mayor Bourque was pleased with the response of the federal government and its cooperation with other levels of government. "The government of Canada was outstanding. The ice storm cost the city of Montreal $110 million. I spoke to Mr Massé after the storm: 'The bills have to be paid.' He was very understanding about the drama we had gone through. My hat's off to the government of Canada because they paid a very high proportion of the cost."

Premier Bouchard's chief of staff, Hubert Thibault, concluded that relations between the Quebec and Canadian governments during the disaster were "very good. Very correct in the sense that the federal government offered its full support. The federal government understood – and deserves credit for it – that the crisis should be managed by the Quebec government. In that sense there wasn't the conflict that occurs in normal times concerning files that are more political. That wasn't an issue in the disaster-response operations." I asked Mr Thibault specifically whether there was the feeling that one group or the other was trying to profit from the situation because of sovereignist/federalist divisions. "No, not at all. The federal government is not equipped for this type of crisis. Their people understood that. They offered us the greatest possible support. As a general rule, that has been true in all the crises I've dealt with."

Next I pressed him about the bill afterwards. Was Quebec compensated satisfactorily? I added that if I remember correctly, there were some tensions, especially about paying Hydro-Québec. Thibault responded that there was some negotiation about certain elements of the bill, given the magnitude of the damage. But that was minor compared to the total cost. "I have no criticisms to bring up concerning payment. The program is well made and does what it was designed to do. Astronomical costs are spread out on a pan-Canadian basis. It is a good insurance policy for each of the provinces."

Vera Danyluk, president of the Montreal Urban Community, summarized neatly what all the others also concluded: "It was the best example of how people in a crisis put aside partisanship and work together. I never, ever saw partisan political issues in this disaster: separatist, *indépendantiste*, or whatever. People worked together. It was fantastic."

Some of these leaders are secessionists; some are federalists. But they all answered that there was cooperation rather than conflict between these two camps during and concerning this disaster. Their conclusions are confirmed by other evidence; studies have not found significant conflict along these lines.[9] This major fault line in normal times was transcended during the disaster by leaders and the population.

About 15 per cent of the people in the Ottawa region have French as their mother tongue, and as the capital of an officially bilingual nation, Ottawa has a responsibility to function in both English and French. News releases during the storm were put out in both languages, and news briefings held by the Ottawa-Carleton Regional Police Service were bilingual. However, one problem did arise. After the declaration of a state of emergency on Thursday, the Regional Municipality of Ottawa-Carleton held twice daily televised news conferences at which the chair of the region, the chief administrative officer, the chief of police, the regional fire coordinator, and other spokespersons kept the population informed about recent developments. By Saturday, however, complaints began to come in that no one was speaking French at those conferences.[10] That problem was subsequently corrected. Asked whether there was anything that could have been done better in communicating with the population, Bob Chiarelli, chair of the regional government, indicated that "one lesson was that we didn't get our bilingual component up quickly enough." Other than this factor, linguistic conflict was not a problem during the disaster since people had other, more pressing concerns on their minds.

PREPARATION BY ADVANCE PROTOCOL

When technical specialists, such as electrical line repair crews, are employed during a disaster, they do not work as a charitable activity. But pay scales vary greatly between companies, states, provinces, and countries. What are they paid? Do they all get paid at the same rate for the same work? Does everyone get the wages of the highest paid on the principle that the highest paid would never work elsewhere for less than in his or her home state or province? David Flanagan, president of Central Maine Power, responded to these questions by stating that all the utilities are heavily unionized and have been aiding each other for a long time. Hence standard protocols have been negotiated. "What you do is pay the foreign crew at their rate, not your rate," he said, "and you pay them according to their rules for overtime, not your rules. Of course, there are inequities. We love to work with our friends from New Brunswick because they're always productive and know our climate, but they get paid less than their counterparts from some other states who are far less productive." Salary inequities exist, but rates are agreed upon in advance. Mr Flanagan made an assessment of Hydro-Québec workers that addresses an allegation by Quebec premier Lucien Bouchard that Quebecers do not work as hard as Americans: "When Quebec's storm was finally over, we did get some Hydro-Québec crews in, and I love those guys. They just work like dogs."

IMPROVISATION IN DISASTER MANAGEMENT

Disasters typically involve the unexpected. That is why they are so difficult to prepare for and deal with. Take a dramatic illustration. City planners in New York City sought to construct their Emergency Operations Center in a central location close to valuable structures and highly populated areas so that the response would be rapid if an emergency were to occur. Hence they rationally decided to place it on one of the lower floors of the World Trade Center. They did not foresee the 9/11 terrorist attack, which caused the collapse of the twin towers and crushed New York

City's Emergency Operations Center. The emergency communications equipment was totally destroyed. Fortunately, all emergency personnel were able to evacuate the building in the interval between the time the planes struck the skyscrapers and their collapse. Emergency specialists quickly set up a new Emergency Operations Center in a boat in the harbour and brought in new equipment. This incident demonstrates the necessity of expecting the unexpected during a disaster and of being prepared to improvise. The quality of the emergency response depends not only on the excellence of planning but also on the necessary improvisations.

The intense, persistent freezing rain that unexpectedly crushed the electrical system in January 1998 also created a need for improvisation. There was great concern that people would die in their unheated homes in this frigid, dark winter climate, especially the elderly and the handicapped. But there were not enough police officers, firefighters, and even National Guard troops to knock on every door quickly. Maine's Governor King "started calling all the radio and TV stations and told everybody to call your neighbour, particularly if they are elderly or disabled and see how they are doing. Or if the phone isn't working, go over and see how they are doing. The response was enlisting everybody in the state to be part of the team, and it worked."

One of the casualties of the ice was Maine's emergency alert system. It was based on the Public Radio Television Network, but the network's towers went down and the system went off the air. General Adams described the improvised solution: "Many radio stations simply went on almost twenty-four-hour call-in operation where people were calling in problems, and we were monitoring those stations and being prepared to respond based on that information."

Governor King confirmed that radio took over when television and computers failed, a step back in technology at a time of crisis. "Many people had battery-powered radios, which was very important in this crisis. There is one talk radio station in Bangor. They essentially went to twenty-four-hour-a-day ice storm coverage. They struggled up the hill with propane tanks to power a generator to stay on the air. They became a clearing house. People would call and say they were out of water, and somebody would bring them water."

During the extreme weather disaster, which could also be accurately referred to as an electrical-grid collapse, there was a great need for huge generators. Most municipalities could not afford to buy $100,000 generators and keep them in storage in case of a disaster. A solution was found, nevertheless, as described by General Adams: "many municipalities have large businesses in their municipalities that do have huge generators. We called on them during the ice storm." This solution was partly improvised and partly planned. It is important to institutionalize, in preparations for a future disaster, an up-to-date list of emergency resources in the vicinity that can be called on in times of crisis, even if those resources are found outside government. The experience of the disaster has shown that individuals and organizations, both public and private, are willing to cooperate in times of disaster.

Mayor Pierre Bourque of Montreal devised an impromptu solution to the unexpected problem of returning to his city during the deluge of freezing rain. "In Toronto they told me that nobody was going to Montreal: no train, no bus, no plane, everything was closed. So I called the boss at Bombardier and asked that he send me a jet to Toronto." Jean-Bernard Guindon described his improvisation after the freezing rain had ceased and pressure had built to reopen "the schools on Wednesday. I had to intervene with Mme Marois, who was minister of education, because I knew her personally. I told her we couldn't reopen the schools because the roads were iced over, the children would kill themselves, and the school buses couldn't circulate on certain roads. So on Wednesday the schools did not reopen." This anecdote demonstrates the importance of prior personal friendships among key leaders for the management of a crisis.

A WELL-MANAGED DISASTER

How did these leaders assess the overall response to these crushing dynamics of nature? The premier's chief of staff, Hubert Thibault, stated: "Premier Bouchard was making his phone calls personally. Hence everyone was collaborating well. People sensed there were things that needed their full attention, and that's hard work. There were certainly some organizational difficulties, but overall they did not stand out because we were functioning well." Vera Danyluk, president of the Montreal Urban Community, concluded that "the organization was rather good. Given the circumstances and the fact that nobody expected all that, people reacted very well to the situation. All collaborators were at the emergency table within twenty-four hours." André Brunelle, the city's coordinator of emergency measures, observed that the difficult circumstances brought out the best in people. "There was more solidarity. People made themselves more available to help than usual. Some residents in areas with electricity took others into their homes. Some of the electrically deprived went to stay with relatives, friends, or neighbours. There was a great deal of mutual assistance."

Mayor Bourque felt that the disaster response was good. "There were no deaths in Montreal. I was so pleased about that. There are so many people who are alone and fragile. We reacted well. There weren't any robberies or looting. Communication was very well done." The Canadian military helped with many essential activities, such as checking on the elderly and the handicapped, clearing debris, and providing equipment. Soldiers were also on duty to prevent looting, but they had almost nothing to do on that score because, as the Montreal police chief stated publicly, the crime rate decreased remarkably.

The crisis brought out the best in people in the United States as well. Maine's General Adams stated that "a number of our National Guard troops were used for security when businesses were vacated in downtown areas in most of our municipalities. But we experienced very little looting." Neither Adams nor Governor King could recall any turf battles. The governor concluded that "our performance in the ice storm was really good. I don't necessarily take credit for it, but everything worked pretty well. You could go back afterwards, and we did, and say, 'What could we have done better?' But by and large, given the fact there was no way to specifically prepare for this problem, it worked pretty damn well."

Every one of the key leaders I interviewed agreed that the response to the intense, persistent freezing rain was generally good. These conclusions could be discounted as self-congratulations by those involved in managing the crisis, but all evidence indicates that their assessment is correct. Independent studies also concluded it was a well-managed crisis.[11] My research in examining videos of televised press conferences, in addition to my interviews, demonstrated that these key leaders gave the public information in a calm, orderly fashion so as to elicit its cooperation. Overall, they communicated openly with the population, transparently transmitting the bad news as well as the good news in the confrontation of their respective societies with the changing dynamics of the weather. Most importantly, they conveyed a sense of caring and competence. Despite the fact that freezing rain of such intensity, duration, and scope took everyone, including meteorologists, by surprise, leaders, acting partly on the basis of pre-designed plans and partly through improvisations, made many excellent decisions and took appropriate action to minimize suffering, fatalities, and damage. They were faced with a massive, overpowering force of nature that affected the largest number of people in the history of Canada and the state of Maine, yet they handled the crisis well. All indicators point to the assessment that, overall, this was a well-managed disaster, paradoxical though that expression may be. The leaders' response was able to prevent even worse social and material repercussions.

The political leaders were physically present as much as possible and communicated all information to the public in a way that avoided mixed messages. Electricity was restored in a systematic fashion according to a set of priorities. Numerous shelters were organized. The most vulnerable members of the population (the elderly, the handicapped, the poor, and immigrants who spoke neither English nor French) were monitored and given aid. People were not left to fend for themselves; nor were they left in ignorance of what was happening and what was being done. In a disastrous situation frustrations and dissatisfactions are always present, but there were no major breakdowns in the authority structure or any turf wars that visibly impeded response to the disaster. In fact, communication and cooperation between different levels of government were generally excellent. The Quebec government, the federal government and army, and the city and metropolitan governments of Montreal worked well together, as did the governor of Maine, the National Guard, FEMA, and the federal government in Washington. So, too, did the mayor of Ottawa and the government of Ontario. In Canada secessionists and federalists resisted the temptation to use the ice storm to further political objectives. Instead, they cooperated in this struggle with a major disturbance of nature.

It is important to learn from what leaders and key decision-makers did right when this extreme weather struck two modern societies. It is also necessary to learn from how the management of even a well-managed disaster could have been improved. These questions will be addressed in the next two chapters by examining the exceptions to this successful management.

CHAPTER 8

From Openness to Secrecy as the Crisis Deepened

It is commonly believed that disaster leads to a breakdown of order: communal norms that make possible life together collapse into a war of all against all. Price gouging, looting, riots, beatings, rape, arson, and selfish behaviour are prevalent. People are too shocked to act rationally and are incapable of functioning. They desperately attempt to flee. All this becomes contagious. Hence chaos can only be avoided and order maintained by the army: martial law is necessary. If leaders inform the population about how bad the crisis is, then people will panic and act according to individual interest even if their behaviour makes the collective situation worse. Truth is alarming in a crisis, so leaders should withhold bad news from the population.

SHOULD TRANSPARENCY BE LIMITED?

The "research literature has established that these beliefs are largely myth, thus they are collectively characterized as the *disaster mythology*."[1] The media look for such behaviour to display, find some, and focus on it as if it were typical, rather than reporting the omnipresent but mundane normal behaviour. This response fosters a belief that behaviour in a disaster is deviant. The myth has perversely real consequences. In cases where people should evacuate, many insist on staying to protect their homes from looting.[2]

When New Orleans was struck by Hurricane Katrina, the situation was portrayed by authorities and the media as an example of panic and looting, but later evidence showed that this description was greatly exaggerated. "Mayor Ray Nagin said in the midst of the crisis that thousands crammed into the Superdome were 'watching hooligans killing people, raping people.' No murders or rapes were ever confirmed in the Superdome. Former New Orleans police chief Eddie Compass said on *The Oprah Winfrey Show* that babies were being raped and tourists were being beaten; he later admitted that he was repeating unconfirmed reports."[3] Such media distortions led the director of Columbia University's Project for Excellence in Journalism, Tom Rosenstiel, to affirm: "We often let our presumptions jump ahead of the facts, ... we let our preconceived notions guide our news judgement."[4]

If mythic belief was replaced by knowledge of real typical behaviour and if the actions of citizens were accurately foreseen, the response would be more efficacious.

Research into disasters has made a number of important findings.[5] During a disaster people do not want to leave their homes and have to be convinced that the threat to their person is great. Even when they are told to evacuate, a large proportion will not do so, many for fear their homes will be looted. People hurriedly and purposefully avoid danger, but panic is rare. Typically, there is much less stealing from one another and panic than the public, the media, and the authorities believe. Victims will take from stores such necessities as food, water, and medicine, which they need to survive, but that is not the same as looting valuables in an opportunistic fashion. Crime generally decreases, and "individuals tend to become altruistic during times of disaster. Survivors share their tools, their food, their equipment, and especially their time. Groups of survivors tend to emerge to begin automatically responding to the needs of one another. They search for the injured, the dead, and they begin cleanup activities."[6] The victims themselves are the first responders, and it is only later that the professionals arrive. The behaviour of citizens during and after a disaster has been found to be generally quite rational and characterized by a great deal of mutual aid.

Disaster research has found that leaders often have an exaggerated fear of panic by the population, leading them to make poor response decisions. Authorities too often make inappropriate choices out of fear of looting, and scarce resources are allocated to prevent it rather than to deal with real problems. Research has shown that information about dangers should be given to the public so that citizens themselves will have the most accurate knowledge necessary to make decisions that are suitable, but that emergency management officials tend to withhold it "because of a fear that people will panic."[7] However, such "a fear insults those who should be served, increases the likelihood of harm to potential victims, and casts such officials in the role of the all-knowing, yet ineffective Big Brother."[8] Hence the organizational response is frequently chaotic. "Accurate information is difficult to obtain; decision-makers often have a hard time communicating with one another; the written disaster plan (if one exists) is often ignored during a real emergency; the pre-impact designated disaster coordinator may not emerge as the actual post-impact leader."[9] Organizations carry out their self-defined mission without coordination, decisions are based upon disaster myths, and turf battles break out between emergency organizations.

Disaster research has also found that, typically, emergency personnel presume that because information was given out, the intended recipients have received it and responded in the way that was planned, but that this presumption is usually wrong.[10] Many people will not have heard the information, others will not understand it, and still others will not follow the suggestions. Warnings will be taken seriously and acted upon if they are clear and specific, telling the public what the hazard is and what to do; consistent with other messages; frequent; given by a credible messenger; follow accurate previous warnings; and in an area where the disaster agent strikes frequently.[11] The media look to political and emergency management leaders as their primary information source, and failure to provide a steady flow of information results in a vacuum in which the media seek their own information from any source in order to avoid being "scooped." "Failure to provide a steady diet of information will result in the feeding frenzy turning on decision-makers, consuming them in the process."[12] The public's expectations are often unrealistically high, and if there is not an honest dialogue between decision-makers and the public, blame is fixed on particular leaders.

From these findings a scenario can be outlined for how key leaders should manage a disaster, which can then be choreographed into specific planned movements. Well-designed disaster plans attempt to counter a belief in disaster mythology and to respond to the actual behaviour their community will demonstrate during a disaster. A key element involves the calm, orderly communication of information and conveying a sense that competent authorities who care about the population are in charge, are transparent in their dissemination of information, and are doing everything humanly possible to serve the interests of the population. By competent authorities are meant key political representatives and administrative and technical experts. These authorities should be present and visible. They must communicate significant information – both good news and bad – to the population so that citizens can take appropriate action. Citizens need accurate information to better their chance of responding effectively as individuals and as families; they can then improve the mitigation of the disaster in the community. For authorities to micro-manage all action is too burdensome and probably impossible, so the communication of the best possible information to all citizens is essential in order that they can take action which does not have unintended, perverse consequences. If bad news is communicated calmly by decision-makers together with plausible plans to solve the problems collectively, the population will not panic and will cooperate and act appropriately. Political decision-makers and emergency organizations should give out information through a highly trained spokesperson. Information and instructions should be provided clearly and repeatedly, specifying precisely what is being suggested.

Trust is a crucial ingredient of this relationship. The population must trust that the authorities are competent and will not mislead them, yet they must allow for human limitations and error on the part of the authorities when confronted by powerful, only partially understood forces of nature such as an extreme weather event. Rational action, not magic, is assumed. For their part, the authorities must trust the population not to react inappropriately to bad news and to the admission of mistakes. They must believe that the population consists of mature citizens who will act rationally and not panic upon hearing dreadful reports. Most demanding of all, the authorities must trust that individuals will follow the authorities' counsel on what is best for the collectivity and not descend into individual actions that provide a temporary personal advantage, to the detriment of everyone else or to the resolution of the crisis.

This trust is fragile on both sides. If the population has experiences or sees evidence that runs counter to what the authorities are saying, then rumours that exaggerate the problems will begin to spread. The authorities, for their part, will be tempted to fear that non-expert lay people will panic or seek individual advantage in a zero-sum situation, adding a second-order disaster to an already catastrophic situation which they have to manage. The devastating situation can entice authorities to let the ends (minimizing adverse consequences) justify the means (lying or/and withholding information about how critical the situation is), but disaster researchers argue that this response is typically counterproductive because citizens cannot take fully informed, appropriate action, the true situation eventually gets out, exaggerated rumours start, and trust is undermined. Leaders, by virtue of their positions in the decision-making network, have more knowledge and expertise than the public, and they should work in partnership by sharing their knowledge. Disaster researchers argue that decision-makers should put aside the fear of scaring the public, a fear that insults those who should be served and increases the chances of harm.

After the ice storm disaster, the Quebec minister of public security, Serge Menard, forcefully emphasized the need to communicate accurate information to the public. He asserted that it was almost criminal to intentionally understate the impacts of a hazard and thereby lead the population into error because poorly informed citizens would then not take the means necessary to ensure their security in time. But he claimed that it would be just as irresponsible to exaggerate a situation, such that panic would take hold of the population.[13] Capturing the reality of the situation, rather than minimizing or exaggerating it, is the challenge.

Was such a model of openness and trust followed when this extreme weather event was confronted? To what extent and under what conditions were there deviations? For half a century disaster researchers have been documenting that panic rarely occurs in disasters and have published their conclusions in scholarly books and journals. Has the message been received by emergency management and political leaders who have the practical role of leadership when a disaster strikes? Has mistrust of the population by leaders found its way into their decision-making? Can we trust our leaders to tell us the truth, the whole truth, and nothing but the truth in a timely way when a sudden disaster occurs or a slow-onset calamity probably lies ahead? It is against this backdrop of research that the management of the January 1998 disaster can best be understood.

THE TEMPTATION OF SECRECY AS THE CRISIS WORSENED

In general, the ideal script described above was almost always followed during this disaster, almost everywhere. Hence this crisis was generally well managed, as documented in the previous chapter. The only exception occurred when the situation became particularly grave: important information known to key leaders was intentionally withheld from the public and even from other authorities and elected leaders. This most serious episode of the disaster happened in Montreal.

Just before noon on Friday, 9 January 1998, the head of the fire department secretly told Montreal's mayor, Pierre Bourque, what the mayor later referred to as the "most terrifying, most appalling news: he said that in four or five hours there would be no more water in Montreal." That morning all electrical transmission lines to Montreal Island except one were crushed by the ice-loading and no longer functioned, including the lines to the two water-filtration plants. There was no energy to power those plants and their pumps, and their operation stopped. Water had been stored in high reservoirs, so the supply continued to flow by gravity, but it was rapidly being depleted. As the level in the reservoirs decreased, the risk grew that sediment from the bottom would be sucked into the lines and contaminate the water supply to homes and businesses. The worse danger was fire. Live local electrical wires supplying power from the one remaining major transmission line had fallen on top of trees, houses, and other objects. Generators were being widely used and would overheat if there was no more water to cool them. People were overusing fireplaces and chimneys for heat, as well as using candles for light. Any of these actions by any of the almost two million inhabitants of the island of Montreal could result in fire, but there would be no water to extinguish the fire, and it could spread through the city, particularly if there was wind. The prediction at the time was that water would run out when people return home after work because they then use a great deal. San Francisco had burned down in a three-day fire after the earthquake of 1906, and a similar fate threatened Montreal as a result of the heightened risk of fire and the insufficiency of its water supply following the collapse of its electrical grid.

Montreal Island included the city of Montreal, which was by far the largest municipality, twenty-seven smaller municipal governments, and a regional government appointed by the province. Water was supplied to the smaller municipalities by the city of Montreal. The commission of inquiry that subsequently studied this disaster in Quebec wrote a five-volume report but dedicated only the following short paragraph to this most dramatic moment of the crisis. "The commission ascertained that only the premier, the mayor of Montreal, and the top management of Hydro-Québec were warned about the possible lack of water. The president of the Montreal Urban Community was not informed about what was happening until the end of the afternoon on January 9th. Neither the mayors of the other cities supplied with water by Montreal, nor the affected population were informed until the public notice to boil water was issued [which occurred after the filtration plants were restarted]. Although the commission understands the reasons that led to these decisions – namely, fear of panic – the selective mode of disseminating sensitive information should at least be re-examined."[14] That is what will now be done here.

It must be emphasized that the population was eventually fully informed. Alain Michaud, director of the Montreal Fire Department, appeared on television at 9 p.m. to tell the public about the electrical blackout of the filtration plants. This was nine hours after the filtration plants had stopped functioning and almost three hours after they had returned to continuous operation. He asked the population to reduce its consumption of water to a strict minimum. At the same time he instructed the public to boil water before consuming it. The filtration-plant stoppage was announced in other venues around the same time. In a television interview the police chief requested the population to reduce its water consumption because Montreal was in a crisis situation. A Montreal

Urban Community representative asked the public to lower its water use by 50 per cent. Television journalists also instructed the population to diminish the consumption of water. The issue is not whether the population and the mayors of the other affected cities were eventually informed; they were. Rather, it is whether they were informed in a timely way so that they could take appropriate action and not make the situation worse by acting in ignorance of the problem.

Hubert Thibault, chief of staff for Quebec premier Lucien Bouchard, claimed that the worst moment of the crisis occurred when the head of the power company gave a report to the premier. "There was only one wire left bringing energy to the island. The wire was already galloping by about ten feet because of the ice-loading on it, and if it broke or fell, the whole island of Montreal would no longer have electricity for anything at a time when the bridges were closed because of ice that was falling on the vehicles. That was a situation which was difficult to manage. Everyone was trapped: the population was trapped on the island of Montreal. And it was at that moment when the risk of disaster was strongly felt." (When a line shakes, it makes a sound like horses galloping and indicates that the risk of snapping or falling is high.)

André Brunelle, assistant director of the Montreal Fire Department, had been told by experts that water contamination because of the sediment was not an immediate problem but would become dangerous if the filtration plants were shut down for many hours. Of more immediate concern was the risk of fire, which could break out at any time. There would be little water pressure and eventually no water to extinguish fires, which would result in a terrible crisis in a metropolis, especially if the population was trapped on an island. Did fires actually break out? "There were more fires than usual," Mr Brunelle said, "but we were lucky; no major ones occurred during the ice storm. However, the longer the situation lasted, the more it became critical because the water level in the reservoirs was dropping." Although the mayors of affected municipalities were not informed of the shutdown of the filtration plants, Brunelle stated that the fire chiefs were told there was a water supply problem which was being solved, that twenty-five water trucks had been parked at strategic places in the city in case of fire, and that these fire trucks would be sent if there were a water shortage.

Sylvain Tremblay, a manager in the Quebec Emergency Operations Centre, declared that when the filtration plants ceased functioning, arrangements were made to have water trucks fill up at swimming pools throughout metropolitan Montreal in case of fire. This decision was kept ultra-confidential so as not to scare the citizens. He also emphasized that finding a solution to a problem, especially if it has to be done quickly, is not always possible for technical reasons. "I had to supply three immense generators to the filtration plant to try to get it working again. The trials with the engineers, the specialists, were negative for restarting the filtration plant. That was a shock to realize that restarting a filtration plant was quite an adventure and was not something that could be done from one day to the next."

Montreal's mayor, Pierre Bourque, described what happened next as follows: "The top managers of the city of Montreal all tried to communicate with Hydro-Québec, but were not able to get through. They then told me the only solution that they foresaw – and I was completely shocked – evacuate a million people, ask people to stockpile water by filling their bathtubs or buying all the bottled water they could. I listened to their strategies and I thought about sicknesses, epidemics, water, the health of the people. I thought about the bridges and the people who would flee. I imagined the panic. I saw Montreal upside down. I was the mayor of this city, and I was the only one responsible who had the mandate to make such a decision. I told the top managers that the solutions they were proposing were not possible and did not make sense."[15] Mr Brunelle confirmed that the final decision about what to do for the city of Montreal was made by the mayor after consultation with the coordinator of emergency measures. The evidence I compiled, however, refutes the notion that any decision-maker wanted people to store extra water or to evacuate Montreal Island. All leaders were, on the contrary, afraid that people would do exactly that and worsen the situation. Mayor Bourque phoned Premier Bouchard at around 12:45 p.m. and explained the gravity of the situation. Lower channels of communication were slow, so the most effective means of getting action was at the very top: mayor to premier. Bouchard told Bourque that he would get back to him.

Jean-Bernard Guindon, the director of civil security for the Urban Community of Montreal, said that in the early afternoon of 9 January "the coordinator of the city of Montreal, Mr Michaud, called me and asked if I would be in agreement with the idea of not announcing to anyone that there would be a shortage of water because he feared the consequences [of informing people] – namely, that everyone would fill their bathtubs and jars with water, and that would lead to a premature shortage of water. We have a system that operates with seven immense reservoirs. We would have water for, according to him, four hours if we told the people and the other municipal authorities, and we would have water for eight hours if we didn't tell anyone. He felt that he had the assurance from Hydro-Québec that it would re-establish the electrical current to at least one of the two filtration plants within eight hours. That was the bet he made. I confess that I thought at the time it was a good bet and I consented."

The assumption was made that if even only the mayors and authorities of the twenty-seven other municipalities served by the filtration plants were told, information would leak to the general public, panic would ensue, and water would be depleted more quickly. The goal was to survive for eight hours to permit the power utility to reconnect electricity to the filtration plants.

Florent Gagné, the deputy minister of public security for Quebec, stated that he and his staff quickly realized that a metropolis such as Montreal could not be deprived of water, especially in winter, for more than a few hours without causing a catastrophe: hospitals, cooking, hygiene, fire suppression, and many other functions all depended on it. News of the shutdown of the filtration plants "was not communicated to the population because a panic was feared where people would immediately fill their bathtubs. And if everyone filled their bathtubs, obviously there would be no more water in Montreal." Mr Gagné concluded that to his knowledge no one decided or ordered that the public not be informed, but neither did anyone take the decision to inform the public. "When information like that comes in, we ask ourselves whether the public should be informed. And then the decision is postponed; it is postponed and postponed until the information arrives that the situation has come back to normal – whew! The decision was delayed in order to wait for more information." These leaders were paralyzed by fear of a panicked population.

There were no leaks of information from the very few leaders who had that information, and no rumours began to circulate. Why not? "In such a serious situation," Gagné observed, "I think that people quickly realize there must not be a leak. The information must be kept secret. One must not go to the media." The worse the problem that leaders are keeping secret, the more likely it is that no whistle-blower will dare come forward and divulge it.

Mayor Bourque had a televised interview scheduled for 1:30 p.m. with the best-known journalist in Quebec, Bernard Derome. When it took place, "the atmosphere was tense," Mr Bourque recalled. "I knew the truth. I knew that it was very serious, that there would be no more water in a few hours. That meant more than a million people would be in trouble. I asked the population to remain calm, to stay at home, to wait until the worst was over. I declared that we had the situation under control, even though we had nothing under control."[16]

The tape of that interview shows Mr Derome repeatedly questioning Bourque as to what the worst scenario that Montreal could face would be, and the mayor again and again said that he would not speculate about scenarios which were too hypothetical. Bourque was wise to avoid alarming the population by not revealing the worst outcomes that were going through his mind. Nevertheless, hypothetical scenarios of a panicked public response if it was told about the filtration-plant shutdown were a large part of his thinking, leading him to withhold information. It is one thing to speculate publicly about the worst possibilities one could imagine might happen. Bourque was certainly correct not to do so. But it is quite another not to reveal that the filtration plants had already been shut down.

Mr Bourque waited anxiously all afternoon for a call from Premier Bouchard. The level of water in the reservoirs was decreasing by the minute, aggravating the risks of contamination and diminishing the water pressure needed to fight fires. The higher parts of the city were about to experience a water shortage. The physics of water flow were close to sending non-verbal but visible information through the water taps that contradicted reassuring messages that the situation was under control.

When the president of the Montreal Urban Community, Vera Danyluk, finally learned about the filtration-plant problem, she and her emergency planners began negotiations to purchase bottled water. A system of rationing was worked out, and municipalities were told that if there were a shortage of tap water, then bottled water would be distributed. Because supply was limited and delivery difficult on ice-laden streets, a fixed amount was promised immediately and more a few days later. Some municipalities accepted their rations and made their own attempts to supplement them. Others, however, demanded a greater share, which would have diminished the water available to other municipalities because the total amount was limited.

Evacuation was briefly considered at the Montreal Urban Community level but quickly dismissed because bridges off the island were closed, there was nowhere for almost two million people to go, South Shore communities were in an even worse condition, and people wanted to stay near their homes. Ms Danyluk stated that, "concerning the question of evacuation, the conclusion was that, no, it is not a solution."

The premier's chief of staff, Mr Thibault, did not recall any specific decision to hide the problems at the filtration plants from the public. He suggested that, since the press conferences of the premier and the head of the power company were to be given in the early evening, the premier had pushed the power company to find a technical solution in the meantime and to make all the necessary verifications before anything was announced. Mr Thibault concluded that that there was no alternative. A very small number of key political, administrative, and technical decision-makers were informed in a blow-by-blow fashion to help them make rapid decisions, but the population and even most mayors of the affected municipalities would be informed only after all technical remedies had been attempted and the scheduled press conference began. As noted earlier, the fact that the premier had his office in the headquarters of Hydro-Québec physically facilitated communication between him and the head of the electrical company.

Fortunately, a temporary technical solution was found in about six hours. But sometimes it takes time to implement such a solution. What would have happened if the filtration plants had stayed down for ten hours, or twelve, or twenty-four? When would the population have been informed about the problem? Thibault answered that "it would have been necessary to take rapid decisions. Which ones? God knows what would have been done? That night, when there was only one remaining power line, we were on the threshold of a catastrophe of great scope. The decisions would not have been easy, so what would have been done? At our office of the premier we did not reach that point. The situation on the island stabilized sufficiently quickly [when the only remaining power line was shifted to the filtration plants], so we turned to the other places in crisis, such as on the South Shore."

In this crisis so many essential infrastructures were being crushed by the freezing rain that no one of them stood out in the media, not even the most perilous. Thibault remarked that Premier Bouchard had told the population about the loss of all electrical lines to the island of Montreal except one, but that the media and the population only perceived the dangerousness of this situation after it was over because of all the other problems reported in the media: for example, total loss of electricity to communities on the South Shore. Disaster overload can occur in media reports whereby the most salient issues get lost in the volume of bad news.

Deputy Minister Florent Gagné found the question of evacuation under these ice-storm conditions almost unthinkable. "Can the evacuation of a city like Montreal even be imagined? Even today I almost tremble thinking about it. Thank God we did not have to take those decisions." Was evacuation of the island of Montreal given serious consideration? "In my mind and that of the premier and several people," Mr Gagné responded, "perhaps in the four or five hours when it [the filtration-plant shutdown] was occurring, but [we were] hoping that it was only a nightmare and that it would be a matter of hours until another solution was found. But not otherwise, not in an organized, discussed way. It wasn't really discussed." In addition to the physical challenges of evacuating two million people from an island covered in ice, there were other good reasons why evacuation could be used only as a last resort, as Gagné observed. "They [people] prefer to remain in their home even if it is 3 or

4 degrees Celsius with a little stove in the basement or no heating at all. I have found people to be very tough, very resistant. And I realized after – it is one of the things that I learned – getting people to leave their homes is very difficult. The crisis has to go really to its limit." By virtue of their positions, leaders in charge of public security have to examine every possible solution when a catastrophe such as that is anticipated, but examination can result in rejection. "I said: 'If Montreal has to be evacuated, Oh my god, how would that be done? How would it be organized?' But then I concluded: no, it is impossible."

The director of civil security for the Montreal Urban Community, Jean-Bernard Guindon, stated that evacuation of the city of Montreal was never given serious consideration in this case and would have been stupid "magical thinking" because of all the above reasons. In fact, many places in nearby shelters were left unused. He argued that people should only be evacuated when there is an imminent danger from a hazard that would destroy shelters and when evacuation conditions are optimal (functioning roads and bridges, possible relocation, etc.). Otherwise, remaining in homes or shelters nearby is preferable. If, however, a major fire had started with no water to extinguish it, then evacuation would have had to be seriously considered, despite the atrocious driving conditions. But, fortunately, fire did not break out, so the possibility of evacuation was not given much thought. The feasibility of the response scenario depends on the character of the natural disturbance. Evacuation to nearby arenas, shopping centres, and other facilities on the island that had heat from generators was being planned, but not evacuation from the whole island of Montreal. This response was also based on the assurance given by the power company that electricity to the filtration plants would be promptly restored. Unlike news about the filtration-plant shutdown, evacuation plans were not withheld from the population because they were never proposed.

At around 4:30 p.m. Mayor Bourque received a phone call from Premier Bouchard informing him that Hydro-Québec was attempting to transfer the only electrical transmission line that remained for the island of Montreal from downtown buildings to the filtration plants. This choice of essential infrastructure – water over everything else – slowly brought the filtration plants back into service. André Brunelle, assistant director of Montreal's Fire Department, pointed out that starting up the pumps when electricity was reconnected to the filtration plants was risky because the whole system could blow again. Prudently, one pump was started, and when it had attained its usual speed, a second was started, then a third, a fourth, and so on. Thus it took time to get everything working at full speed. That was achieved by about 5:30 p.m., but the side effects included a blackout in downtown Montreal. Residents thought that the situation had become much worse because electricity for downtown buildings and the subway had been shut off, but Mayor Bourque judged that the situation had improved because water was deemed more crucial than the infrastructures which were sacrificed and water was now available. Since the water in the reservoirs had dropped to a very low level, restarting the pumps threatened to immediately stir up the sediment and contaminate the supply. So the doctor in charge of public safety requested the mayor to issue a boil-water advisory, which was done. When the premier appeared on television to give his update, he appeared pale. As Bourque put it, "people never really knew what had happened."[17] He added that Montreal had a very close call.

WAS SECRECY NECESSARY AND UNDER WHAT CONDITIONS IS IT REQUIRED?

Why did Mayor Bourque say that authorities had the situation under control when they did not? "Why? If I had announced the shutdown of the filtration plants, it would have led to panic. It would have been impossible to manage. I have seen panic worsen a situation. That would have frightened people. There would have been all sorts of accidents because eventually people would have fled. Stores with water would have been vandalized. I didn't want that."

Mr Bourque himself raised the following important question: "Should a political leader tell everything to the population or not? Should he hide things? That is an ethical question. I feel that I made the right decision. I have never regretted it. If the premier had not called after four or five hours, I would have been uneasy with my conscience. I would have been obliged to develop a plan B with him: evacuate, I don't know. I don't even dare think about how or what would have been done. When the engineers called me from the filtration plant and told me, 'Monsieur Mayor, there is now water,' it was a grand victory, perhaps the greatest victory over the freezing rain." Thus Bourque concluded that his decision to withhold crucial information was the correct one, given the speed of restoration, but that it might have been wrong had the electrical company not succeeded in restoring power to the filtration plants quickly. The legitimacy of political leaders withholding important information would therefore depend on the technical skill of the electrical engineers.

Under what conditions would a political leader reveal to the public the gravity of a problem such as this one? Mayor Bourque gave two answers: first, "when the problem is solved," which was what happened in this case; or second, when the problem could no longer be hidden for physical reasons, and this situation was rapidly approaching. "There were already streets that were beginning to lack water. It was beginning to be known that there was a scarcity of water. We would have been forced to inform the population at 6 p.m. or 6:30 p.m., that's for sure. The filtration plants came back on at 5:30 p.m. Fortunately, no rumours had started. It could not be hidden for a long time. It would be impossible." Bourque felt that "I had one scenario; there were no others. I had one way out. It was the one I chose. I made the right decision, but there were no others. It was a very bad day for me until the water returned. I foresaw a hellish scenario of panic. I didn't have much choice. But that doesn't happen often in a lifetime."

Why did he not at least inform the president of the Montreal Urban Community about the shutdown of the filtration plants? "I didn't think a second about calling Ms Danyluk because I was the authority for the city of Montreal." This response was a product of a major turf war that existed during the disaster, which will be examined in the next chapter.

Vera Danyluk did eventually learn about the shutdown of the filtration plants in the course of this critical afternoon, and she subsequently participated in discussions concerning whether to announce publicly that there was a danger of the complete loss of water. She and her group also chose to withhold the information. "Participants said, 'If it is announced, it is sure that people will panic. They will start to fill their bathtubs and pots with water.' We tried to limit the number of people who had that information because we were so afraid that someone – one of the individuals at the table – would decide to tell it to someone. Then someone would tell it to journalists. We knew what happens in situations like that."

Transparency would replace secrecy only when withholding information was no longer be possible, that is, when people were on the threshold of physically seeing diminished water pressure in their taps: "the only time when I think that it would have been necessary to have a press conference and announce the bad news to the population," Ms Danyluk commented, "is if it were certain that the production of water would stop. Then, certainly, one would have been obliged [to inform the public]. We were not yet there, but almost."

With the benefit of hindsight, was withholding information from the public about the shutdown of the filtration plants the right thing to do? "Definitely," Ms Danyluk responded. "I would say yes because I am certain that the reaction of the public would have immediately been to go fill their jars. I thought at the time that it was the right decision. I would say that I have thought a great deal during the following years about that subject, and always on that specific point – the water – I concluded that it was the right decision."

Florent Gagné also concluded that not informing the population was best. It was in the public interest because such news would have gravely upset the population and undermined social order. "The benefits of transparency would have rapidly been wiped out by a much greater harm," he remarked. "I don't dare even imagine the situation of panic that it would have provoked in the population. In addition to the multiple crises already existing, we would have had to manage a further crisis: a frightened population in panic." Telephone lines would have been jammed. People would have descended to the streets to flee as best they could.

Mr Gagné argued that in the case of 90 per cent of information, decision-makers should be transparent and the public informed, but transparency can have perverse effects if the news is too upsetting for the public. Hence in these infrequent cases the information should be controlled. Doing so means that it should be withheld and explained to the population only after the crisis has ended. "When it has a traumatizing and socially destabilizing effect," he said, "I am convinced, and I would never counsel the political authorities to divulge such information that would be prejudicial to the population."

The assumptions upon which withholding information was based were widely shared by all the leaders during the crisis. Gagné felt that there are only two conditions under which decision-makers should reveal this sort of information. The first is when leaders decide that the population must do something which requires that information. "Informing the population should be delayed as much as possible until one is convinced that there is an action that the population has to take. Then there is no more choice." The second occurs when it becomes physically impossible to withhold the information: for example, when people started noticing the reduction in water pressure in their taps. If information was not given out at that time using the media, the population would receive the missing information non-verbally from their water taps. "The technical dimension would have been the determinant," Gagné believed. "If there would have been no problem for three days because of the [size of the] reservoirs or diesel pumps [if they had worked], probably we would have gone to this limit of technical capacity [before informing the public]. I suspect that is how the situation would have been managed." Days would have gone by before the public was informed if the water reservoirs were big enough. Political transparency would be inversely proportional to reservoir capacity.

Mr Gagné gave another reason for not informing the public quickly: "You can't just say at a press conference: 'Dear friends, dear fellow citizens, there is no more water and we don't know when it will return and we don't know what advice we should give you.'" It takes time to plan what will be said and to make suggestions to the people concerning what they should do. The question of *what* the population should be told in terms of a response to an emergency is as important as the question of *when* it should be told, and the former requires planning.

Gagné was well aware of the long-term perverse effects of withholding information, yet he concluded that not informing the public was best in this case. "I know that people might imagine that the government will hide things the next time. And perhaps that will be the case." He refused to draw any general lesson from this case about whether information should be divulged or withheld in the future. "It is really ad hoc in each case. The best was done this time. Perhaps in another case the same would be very bad. Why was it good this time? Undoubtedly, we were lucky. We were lucky that in four or five hours things [the filtration plants] came back to normal." This observation implies that leaders cannot learn lessons from one crisis that would be useful in future ones, that they have to reinvent the wheel, so to speak, in each disaster, and that the outcome is determined by either good or bad luck. These are not particularly reassuring conclusions.

Sylvain Tremblay, of the Quebec Emergency Operations Centre, agreed with the decision to withhold information from the population. He was less categorical about withholding it from other municipal leaders, but even there he emphasized that having to make decisions in the heat of the action is different from making them calmly after the fact. He is certainly right to draw attention to the complexity of making sense of a never-before-experienced situation as it unfolds and the difficulty at such a time of making the best possible decision rapidly. In this present research we are, however, examining leaders' assessments of their decision to withhold information from the public and from other leaders six years after that decision was made. These leaders could have concluded after six years' reflection that it would have been better to divulge the information to the population or at least to the leaders in other municipalities affected by the stoppage of the water-filtration and pumping systems. But almost all of them did not reach that conclusion and instead continued to insist that secrecy was best to avoid panic in the population.

Some made partial changes. André Brunelle, assistant director of Montreal's Fire Department, stated that "with hindsight, we concluded that in another case we would inform the mayors of municipalities accurately about what had occurred, what the actions were that we were taking, and how we anticipated solving the problem." Thus he has now retrospectively reached the decision that the leaders of affected communities should have been informed. But should the public have been informed as well? "When Hydro-Québec decided to go in front of the television cameras to explain the electrical blackout, there was already a blackout, there was no electricity. We, on the contrary, had water and did not want to exhaust our reserve. That is why we acted in a different way." This comment sounds very much as if leaders informed the public when things became physically visible and they had no choice; otherwise they did not.

Leaders also gave out only one type of information. "I have the impression that they said in front of the camera what they felt was certain," Mr Brunelle observed. "When something was only probable or was problematic, they didn't want to talk about it publicly." But taking this approach resulted in those who were privy to the backstage discussion of probabilities hearing a discourse that contradicted what they heard from leaders on television, who "gave general information to reassure people. However, we sometimes heard the perspective of the power company experts when they were together. They raised specific problems that worried us but were not yet talked about on television. We had two versions, so we thought: 'How can we function?'" Since certainty is not abundant during a disaster, it is important for the emergency management leaders to have the best information available in order to facilitate planning, even if it consists only of knowing that some occurrences are more probable than others. But the electrical company divulged probabilities only in private to its own personnel. Listener-planners from outside the company had difficulty making their plans because they needed to know in advance the gravity of problems and what was being done, but they were not officially told.

Mr Brunelle learned from his experience about a better way to avoid causing panic. "We are more inclined now to think that people should be informed, and informed as well about the consequences if they do not follow the instructions that are given."

Jean-Bernard Guindon, of the Montreal Urban Community, agreed at the time with the decision not to tell the population that the filtration plants and pumps were not working for lack of electricity, but he has since changed his mind about the best way to manage such a crisis. He learned from the post-mortem and specialists that it would have been preferable to inform people and solicit their collaboration. "We would have had as delinquents only society's usual delinquents, who would have taken a surplus of water, but we would have made an important saving of water because the vast majority of people would have better been able to collaborate. Today, with what I have learned, with this experience that I have acquired, it is certain that I would do my best to influence the authorities not to keep it secret."

Mr Guindon now believes that assumptions of panic on the part of the population are prejudices that hinder the management of crisis situations because "all that borders on magical thinking. Just like it is claimed in magical thinking that there will be more vandalism during a disaster. Statistics have shown beyond all reasonable doubt that there is not more vandalism during a disaster than during normal times. Magical thinking like that is harmful because it leads to myths about the population that make the management of the population much more difficult." Told that the other key decision-makers still believed that panic would have resulted if such information had been divulged and that withholding it was necessary, Guindon concluded that this was because they assumed "it was certain that electricity would be re-established before having a real scarcity of water. That was the error. Certainty was not possible in a situation as catastrophic as that. Therefore we should have acted as if we might lack water for much longer. I, too, believed it at the time when told that Hydro-Québec gave a formal assurance that the filtration plants would be working within eight hours. Today I would not have believed it and would have announced that there would likely be a shortage of water and requested that everyone conserve water. In hindsight, I think that even with certainty, the public should have been informed."

Guindon added the following comment: "Reflection on what happened enables us to learn from such situations. Being in error is not ideal, but it is less serious than persisting in error." Claiming six years after the disaster that secrecy was best is seen by Mr Guindon as leaders "persisting in error" and failing to learn an important lesson about the need for transparency.

SOLIDIFYING TRUST BY REPLACING ASSUMPTIONS OF PUBLIC PANIC

Panic on the part of the population does occur during some disasters, but it typically results from an inadequate response from leaders and from the disastrous conditions in which people are forced to live under normal weather. Take two examples. The first consists of the looting and panic that occurred in New Orleans when Hurricane Katrina resulted in breaks in the levees and in flooding, although much of this looting and panic was exaggerated by the media. This behaviour during the disaster was closely tied to the extreme poverty, addiction, and vulnerable situation of some sectors of the population of New Orleans before Katrina came ashore, as well as to the inadequate response of municipal, state, and federal leaders in aiding defenceless groups during the disaster.[18]

The second example involves the explosion of the state-owned PetroChina plant in Jilin province on 13 November 2005. On that occasion, the Jilin authorities emptied 100 tons of benzene compounds from a reservoir into a river in order to dispose of it in an emergency.[19] They then waited six days before informing authorities in downstream Heilongjiang province about the potential disaster. Those authorities next shut off the water supply to the city of Harbin and told its population of 3.8 million that they were carrying out repair and inspection work. This well-intentioned lie triggered a panic because leaders had withheld information in the past and because no one believed that the first days of winter would be chosen to do a routine inspection of the whole water system. The population suspected a major problem but did not know what it was, and rumours began to circulate that an earthquake would soon strike (perhaps because an earthquake had recently hit Pakistan). The city of Harbin descended into chaos, hoarding of bottled water began, and thousands started to flee.

The governor of the province told a meeting of four hundred officials to keep on lying until the top leaders in Beijing gave permission to reveal the truth. They finally did so nine days after the explosion, when the national leaders realized that the poisoning would be impossible to hide. In addition to the risk that their own people would inadvertently become poisoned was their belated awareness that attempting to conceal this impending calamity, which could not be concealed, would eventually result in a diplomatic conflict with Russia, where the river was flowing with its now-toxic contents. Those top leaders then condemned the deceit of the lower-level leaders and proclaimed they would bring the liars to justice. This way of managing a disaster is a far cry from the model advocated by disaster researchers, and it shows the chaos that can result when a biophysical hazard strikes if there is a history of leaders withholding important information from the public. Trust in leaders is like credit in the bank. It can be accumulated by leaders as a result of transparent, competent leadership. On the other hand, a trust deficit can develop if leaders fail to share crucial information with the public.

During the 1998 ice storm disaster in northeastern North America, bad news as well as good was, for the most part, transmitted to the population by leaders, as described in the previous chapter. In their dance with nature's movements, these key decision-makers announced to the public in a calm, reassuring fashion each episode of freezing rain and its destructive, disorganizing consequences as well as their socially constructed response. Their decisions were often reactive but sometimes proactive, often improvised but at times scripted according to plan. In order to diminish the possibility of contradictions, the messages were controlled and communicated by a limited number of spokespersons, who were in a position to see the big picture. Care was taken not to raise expectations unduly. The key political leaders had a visible presence in managing the disaster, thereby reassuring the population that someone competent who cared about them was in charge and giving people the information they needed to act rationally. The population was trusted by leaders to act in a reasonable way, which eliminated the burden of micro-management on decision-makers. In short, key leaders, in their dance with these meteorological movements, largely followed the choreography of open, transparent communication with the population scripted by disaster researchers.

However, when the freezing rain had its most destructive consequences by crushing all power transmission lines except one to the 1.8 million people living on the island of Montreal and bringing to a halt the operation of filtration plants and water pumps, assumptions of irrational panic by the population came to the fore in the minds of these leaders. They were led to assume, further, that information about the filtration-plant shutdown, which was certain knowledge not requiring further verification, would have a traumatizing and destabilizing effect on the public. As a result, the leaders concluded they had no alternative but to withhold information about the perilous problem while measures were attempted to restart the plants.

Their secrecy has been applauded, on the one hand, by those who presume that the population will panic if told the truth and who presuppose that the public needs false assurances that everything is under control on occasions when the situation is out of control.[20] Even the Professional Federation of Journalists of Quebec agreed that it was acceptable to withhold information for a time to avoid worsening a situation. It warned, nevertheless, about the risk that secrecy leads to rumours, which make the situation much worse. The federation also insisted that credibility requires that the facts be given out as soon as possible after the emergency has ended.[21] On the other hand, the withholding of significant information by these leaders at the worse moment of the crisis has been criticized as a paternalistic affront to the intelligence of the public and as a violation of the rights of citizens in a democracy to learn important and pertinent information in a timely fashion; that is, it has been condemned as a violation of the right to know.[22]

There is also the more pragmatic issue examined here. The findings of disaster research have largely refuted this assumption of panic and shown that the vast majority of the population tend to act rationally in a difficult situation if the problem and solution are explained calmly and if they see tangible signs that competent leaders are in charge doing their best to meet the needs of the population. During this ice storm disaster, there was no sign of panic or looting. The Montreal police chief stated that the crime rate went down, a pattern that was typical of other communities in Canada and the United States during the disaster. People remained calm and helped one another. One could claim that in Montreal this response was because the public was not told that the filtration plants had stopped working. Had it been told, according to the panic hypothesis, all hell would have broken loose. But this is pure speculation, lacking any evidence or reasons to back it up. Montreal did not have the extreme poverty, drug problems, and violence of New Orleans. Nor did it have the history of authoritarian secrecy of China. On the contrary, it has managed disasters in a transparent way, including this one until the filtration plants ceased operation.

It is reasonable to conclude that transparent management of disaster could have continued even at the worse period and would have made for better management. There was an alternative to withholding information. If the problem of the looming water shortage had been revealed calmly by a competent leader – for example, Premier Lucien Bouchard – it would not have had a traumatizing and destabilizing effect. The public would have been asked to help conserve water in the city's reservoirs more effectively and thereby lengthen the time available to the power company to find a technical solution to restoring electricity to the filtration plants: perhaps making water available for twelve hours if the population was informed of the problem and instructed to conserve water, instead of eight hours if the information was withheld and consumption continued at habitual levels. The public would not have been required to stop drinking water, only to postpone showers (for those who still had a means of heating water), washing laundry and dishes, and perhaps toilet flushes. These are not particularly painful sacrifices. But these decision-makers did not perceive this alternative.

Instead, they assumed that the population would panic and hoard water, thereby aggravating the impending shortage. Hence the leaders withheld information about the seriousness of the problem from the public and from the leaders of other affected municipalities until the problem was solved. If the electrical utility had not succeeded in restarting the filtration plants so quickly, then household water pressure would have decreased first in the neighbourhoods most remote from reservoirs or on the highest ground. People would have observed in their taps that something was wrong. When the authorities later told them to conserve water because the filtration plants were not functioning, citizens would then have asked: How long has this serious problem existed, why were we not told, and what else are our leaders hiding? Withholding crucial information undermines trust in leaders. These leaders took a risk of the information being leaked, which would have been a scenario much more likely to incite panic than announcing the problem. Thus there was risk of panic on both sides of the issue – divulging the information or withholding it – and these decision-makers took the risk of withholding information.

It is hard to avoid the conclusion that the gravity of the material problem had a traumatizing and destabilizing effect on key leaders, leading them to switch from transparent communication of problems and solutions to concealment. The openness mode of sharing information with the population was transformed into a secrecy mode when the problems became particularly serious and known to leaders but not yet visible to the population. The issue is not whether the population was informed but when it was informed. As long as the problem was invisible to the public, even though it was becoming more perilous by the hour, information about it was withheld even from other elected mayors because leaders based their decisions on assumptions that the population would panic and that individuals would act to advance their individual interests to the detriment of the collectivity. The possibility of eliciting the cooperation of the population was sacrificed for fear of triggering baser instincts of terrified self-preservation. Moreover, one leader concluded from this experience that the more serious the problem that is concealed, the less likely there will be a leak of information to the public. If he is correct, then whistle-blowing decreases as the gravity of the concealed problem increases.

All this does not augur well for dealing with serious but creeping environmental problems whose effects are invisible to the public at present but will potentially be grave at some point in the future. Neither the filtration-plant shutdown nor global warming brings harm immediately, but both threaten devastation after a time lag. These leaders did not want to inform the public about something that was only probable rather than certain, but the only available knowledge about environmental problems and slow-onset disasters is based on probabilities, not certainty. If certainty of disaster is a requirement for transparent communication, then no such communication from key leaders will occur. Will they fear alarming the population and play down information about the dangers of environmental problems such as greenhouse-gas emissions? If so, then this opaque form of leadership, based on a secrecy mode of withholding crucial information from the public, will result in the accumulation of a trust debt that will eventually undercut trust in leaders when it is needed in a crisis.

There is a better way than concealing how serious problems are until they become certain and/or visible, which in some cases may be too late. The population could be reassured and calmed by competent leaders while being given the information needed to act rationally for the long-term collective good. In this way false, short-term reassurances would not undermine trust in leaders over the long haul. Transparent leadership based on an openness mode of sharing information with the public would put trust on a more solid basis and result in the accumulation of trust credits by leaders which would serve leaders well in dealing with risk and crises in the future.

In defence of these decision-makers, it must be said that the period of reflection during the crisis was relatively short, a solution was being attempted, and water flow had not yet stopped or pressure tangibly dropped. They had to make sense of the novel situation quickly during a particularly grave crisis, as indicated by the fact that these leaders were put, by the freezing rain, in the position of having to choose between essential infrastructures: either the water supply or the subway and heat, light, and energy for skyscrapers in the downtown area but not both. They had no prior experience with such a dilemma and were ill-prepared.

The more worrisome part is the belief by most, but not all, of these leaders even six years later that the population would have panicked and individually hoarded water instead of following instructions to act according to the collective good. They also persist in believing that decision-makers should only reveal the gravity of the problem when it becomes visible, when it is certain, or when it is solved. These leaders have long since been removed from the heat of the action and have had time to reflect. Their maintenance of their risk assumptions must, nevertheless, be interpreted with awareness that leaders are human and have predispositions to believe they made the best decision and to admit error reluctantly. A few of these leaders have since changed their conclusions about withholding information concerning particularly serious problems and now advocate transparent communication and open solicitation of the population's cooperation, but most have not. It is only by admitting error that individuals can learn to do better next time. Since these Quebec decision-makers acted as openly and transparently as those in Ontario and Maine when the disaster was less severe, and only started withholding information when prompted by a particularly severe crisis of catastrophic proportions, there is no evidence for inferring that this thinking is peculiar to Quebec. The extreme weather simply did not put leaders in Ontario and Maine to the test to the same extent. More worrisome is the possibility that leaders everywhere will pass over a threshold from transparency to secrecy when the information becomes especially grave. Hence it is particularly important to learn from this episode about the value of open communication with the population on the part of leaders.

CHAPTER 9

Leaders in Conflict during a Disaster

Even though the ice storm crisis was well-managed overall, conflict between leaders erupted along several pre-existing lines of tension. Some of the key leaders interviewed here were central protagonists in these conflicts and had to deal with the tension. Did conflict and turf wars adversely affect the response to the crisis, or was conflict managed successfully? How was conflict resolved?

REGIONAL GOVERNMENT, AMALGAMATION, AND DISASTER

When a disturbance of nature occurs, it exerts its effects in a particular socio-political and physical context. Historically, villages have grown into towns and then into cities, and neighbouring cities have evolved into large metropolitan areas. An awareness has developed that there has to be some form of cooperation and coordination between cities in a metropolitan area for transportation, emergency services, and other utilities. Therefore a regional structure of governance has been developed.

In Canada's capital of Ottawa, this took the form of the Regional Municipality of Ottawa-Carleton under the leadership of an elected chair, Bob Chiarelli, who had been voted to that position a month before the ice storm. By the time of my interview with him almost eight years later, that structure of governance had been replaced by amalgamation into one big city of Ottawa, and Bob Chiarelli had been elected its mayor. Was there conflict between the chair of the regional government and the mayors of the various cities, especially the large city of Ottawa, during the ice storm disaster? Mr Chiarelli answered that "there was no tension at all. I didn't think that the two levels of government impeded the response at that time. I thought the response was at a very high and satisfactory level."

The absence of tension was the result of an accommodating spirit on both sides to deal with the crisis. The regional government saw the cities "as part of the solution," Chiarelli said. "It was very cooperative. We tried to be totally inclusive and at most of the press conferences we had the mayors or the municipal representatives available in the room to raise any issues on behalf of their municipality." For their part, the cities, including the city of Ottawa, ceded authority over the disaster response to the regional government. "The lower-tier municipalities accepted the reality that it was the regional government that had the disaster response responsibility. And there was operationally good communication." Nevertheless, Chiarelli concluded that "decision-making is easier in every manner with one level of government. There's no question about that."

The Montreal area went through a similar evolution in its structure of governance: from autonomous cities to a regional umbrella government superimposed on the cities and then to an amalgamated big city several years later. The regional government existed at the time of the disaster, and the amalgamated big city at the time of the interviews. But the interaction between the regional and big-city levels of governments during the disaster was much more conflictual than in Ottawa.

Montreal is on an island that contained twenty-eight municipalities and about 1.8 million people in early January 1998. One of these municipalities, the city of Montreal, had a much larger population (about 1 million) than any of the others. Each municipality had an elected mayor and councillors. A regional government, called the Montreal Urban Community, with a president appointed by the provincial government, had been created to coordinate the actions of the various municipalities. In this structure of governance, the fire department was controlled by each city, whereas the police department was the responsibility of the regional government. The city of Montreal had authority over the filtration plants (clean water in, which was subsequently distributed to the other municipalities on the island), while the regional government was responsible for the purification of sewage before it was dumped back into the river (clean water out).

Both these levels of government had a legal status. There was a huge rivalry between the regional government, led by its appointed president, Vera Danyluk, and the elected government of the largest city on the island, Montreal, led by Pierre Bourque. Mayor Bourque was not satisfied with this political structure and launched a campaign to amalgamate all these municipalities into one city of Montreal using the slogan "One island, one city." Vera Danyluk was a staunch opponent of such unification and argued in favour of local identity but regional coordination. Opposition to amalgamation was centred largely but not entirely in anglophone municipalities fearful that their local identity would be lost if they were swallowed up in one big francophone city. Wealthy municipalities were also not pleased with the tax-sharing implications of amalgamation. The freezing rain fell on an already chilly relationship between the city of Montreal and the Montreal Urban Community, in particular, between their respective leaders, Mayor Bourque and President Danyluk. How did they manage these tensions during the disaster, and did the tensions affect the emergency response? It must be remembered that, although Mayor Bourque had proposed the amalgamation of municipalities before the ice storm, it was implemented after that event but before my interviews. A different party took power at the provincial level, and it permitted local referendums to approve amalgamation or end it, which some municipalities did. Nevertheless, that provincial government imposed a level of integration exceeding that which had existed prior to amalgamation.

Bourque pointed out the duplication, even triplication, of emergency management which resulted from the parallel structures of municipal authority that existed on the island of Montreal at the time when the freezing rain fell. "There was, unfortunately, an emergency operations centre at the regional government level as well as at the police station and here in the city." He described the absence of interaction between the president of regional government, Vera Danyluk, and himself when he was told about the shutdown of the filtration plants. "I didn't think for even a second about calling Ms Danyluk because I felt I had authority over the city of Montreal. For us, the mayors, particularly for me, Ms Danyluk was a civil servant. We were six from the city of Montreal and six mayors from the suburbs, and she was working for us in the final analysis. At the time there was a great rivalry. There were two authorities: regional government and the city. The regional government had a life of its own. It's true that we had our emergency and they had their emergency. That isn't normal. There should have been one government. If there is one thing that favours having one city government for the island, it is the unified command structure."

There were no elected officials at the emergency management table of the regional government. Instead, key administrators from each municipality were present – for example, the fire chief for the city of Montreal. Each municipality had a person responsible for implementing emergency measures, and every day the president of the Montreal Urban Community communicated with that person or the mayor of the municipality. There was, nevertheless, one municipality different from the others. Ms Danyluk stated that "the city of Montreal, by far the biggest city in the urban community, functioned somewhat on its own because its size."

The regional government began negotiations for items that were needed in the emergency, such as beds, in order to have a coordinated approach. Some municipalities did not wait and began negotiations on their own, which Ms Danyluk felt was a waste of money and energy. Which municipalities did that? "Especially the city of Montreal. Mr Bourque decided he was going to order one thousand beds. We told him, 'Don't do that because we are negotiating with the army to quickly obtain portable beds.' But Mr Bourque said, 'No, no, no.' He wanted to place the order. After the crisis was over there was the issue of who was going to pay all those bills. At the urban community, I said, 'This isn't our bill, we didn't order this.' I don't know who paid the bill." Since the federal government paid most of the expenses of the city of Montreal, as Mayor Bourque stated, the bill likely ended up being paid in large part by the federal government, and this factor kept a lid on the municipal conflict.

President Danyluk and Mayor Bourque agreed about one thing: the need for a clear command structure in times of emergency. As Ms Danyluk put it, "during the ice storm the emergency measures office of the regional government was responsible for coordinating everyone on the island of Montreal. But it didn't have the power to tell Mayor Bourque or Premier Bouchard that the way to function was specifically described in the emergency plans and then work as a team. That is where the Quebec government caused a problem of fragmentation because the lines of authority were not clear concerning a situation like that." It is one thing to agree on the need for a clear command structure, but quite another to achieve consensus concerning where the enforced commands should originate. Danyluk and Bourque did not agree about who should command and who should obey. According to Bourque, commands should come from a centralized city-wide government in which all municipalities would be amalgamated. For Danyluk, the regional government should be given the power to impose its coordinating functions on all the municipalities, including the recalcitrant city of Montreal.

It is clear that Quebec premier Lucien Bouchard had the ultimate power and authority in this crisis. But beneath him the command structure was ambiguous, especially on the island of Montreal. The emergency measures office of the regional government employed experts on risk and emergencies. It was supposed to have the authority to coordinate emergency measures, yet the city of Montreal, the public power utility, and the Quebec government were not obliged to share their information with that office. The overall emergency response worked well in spite of this structure, rather than because of it.

President Danyluk was not ready to admit that one big city was the best way to have a clear command structure. She contended that size is more a problem than a solution. "Small municipalities were able to open shelters rapidly and offer services to their population. The bigger a city, the more difficult it is to respond quickly." But would smaller, poorer municipalities not have had fewer resources to confront an emergency? Ms Danyluk argued that regional government existed precisely for the purpose of offering services to help them, and they were well-organized thanks to those services. Smaller municipalities were also more familiar with mutual aid and had reciprocal agreements with neighbouring municipalities for help in times of emergency. She insisted that a regional government coordinating local municipalities was better than one big amalgamated city to confront an emergency, assuming that the command structure was unambiguous, which was not the case here. "There weren't clearly defined lines of authority with the obligation to communicate to ensure that everybody who was part of the decision-making process had all the information. That was the greatest weakness."

Jean-Bernard Guindon saw the issues from both sides because at the time of the ice storm he was the director of the Centre for Civil Security for the Montreal regional government and at the time of my interview, after the amalgamation of municipalities, he had become director of the Centre for Civil Security for the enlarged city of Montreal. Will amalgamation help to confront such a crisis in the future, or was it better with the previous structure? "Viewed from a public security perspective," Mr Guindon replied, "I can conclude that amalgamation into an enlarged city of Montreal was a solution blessed by the gods. It was the best thing that ever happened." Previously each municipality had the legal power to apply emergency measures, and the regional government had to coordinate all of these not by law but by virtue of its other powers. "That was not sufficient to act effectively. We were just a team of advisers and experts on civil security for the cities. Only rarely did we carry out operations. Now we still play the role of experts and advisers, but we also play a very operational role at every level. Amalgamation has been wonderful from the point of view of public security, but not necessarily when viewed from other angles."

Previously, if a major fire broke out in one municipality, then it would call in help from other fire departments on an ad hoc basis. Now after amalgamation the operational efficiency has improved greatly: "a unified procedure and a unified approach have been implemented. Even if all else were dismantled in a de-amalgamation, it would be necessary to keep everything that relates to public security unified: police, the fire department, and civil security," Mr Guindon argued. "In matters of security the organization must go beyond political boundaries to attain the efficiency necessary for the protection of citizens."

Florent Gagné, Quebec's deputy minister of public security, stated that everyone knew there was tension between the city of Montreal and the regional government, and he had witnessed it himself. He concluded that the fact that civil security was placed in regional government in the previous structure was an indication that even then it was accepted that "civil security is best organized on a geographical basis. That responds to practical criteria better than political borders. So in that sense an amalgamated city is better able to meet civil security needs than a compartmentalized city." Mr Gagné described the interaction between his ministry and the two warring municipal organizations as follows: "Municipal civil security at that time was handled by the Montreal Urban Community. Our relations were much more direct with the Montreal Urban Community than with the city because officially in the division of labour the former was supposed to address questions to us."

Interaction at the highest levels of leadership was, however, opposite to these officially prescribed communication channels, as described by Hubert Thibault, Premier Bouchard's chief of staff. "I remember relations with Mayor Bourque well, but I don't recall discussions or telephone calls between the Office of the Premier and the Community during the management of the crisis." Informal personal contacts between leaders at the highest levels circumvented pre-established channels in establishing communication networks during the crisis.

André Brunelle, assistant director of the Montreal Fire Department, worked for the city of Montreal during the ice storm and now works for the amalgamated city. He contended that in the previous structure coordination did not work well. Two dozen municipalities with all their services are difficult to coordinate. There was duplication during the disaster. For example, the police decided to check on people in an area to make sure they were not in difficulty but did not communicate this decision to the fire department. The latter only found out when they checked the same homes. This duplication of effort because of poor coordination wasted scarce resources. Mr Brunelle concluded that these problems had been largely solved by amalgamation. "It is a major advantage because there are decisions that must be taken, and they are coordinated more easily if they are centralized." He judged that hierarchy and centralization are necessary for an efficient response to a disaster.

The structural division between the powers of a large city government and those of regional government, as well as the tension between the leaders of the region and its principal city, were far from ideal for an emergency response. Nevertheless, these leaders managed to overcome these obstacles so that the response was still very good.

Another source of municipal tension occurred in less populated areas. Many towns wanted to operate their own shelters even if they had few people who would use them and even if they did not understand the complexity of running a shelter: for example, protecting against contagious diseases. The emergency management leaders had to choose which towns would be the location of a shelter for their area, to the chagrin of the other towns. Thus Earl Adams, Maine's commissioner of defence and emergency management, stated that "we had difficulty in almost forcing them and saying: 'We will set up a shelter at this location, for this area, and that will be the shelter.' Because of the scarcity of generators, we just could not afford to put generators in every place that wanted them." General Adams specified what was learned from these experiences: "Once the storm was over we did reviews and concluded the necessity to pre-designate shelters, make sure they are equipped and have trained people to manage them."

TENSION BETWEEN AN EMERGENCY MANAGEMENT DIRECTOR AND A POLICE CHIEF

Emergency management organizations often have a very close relationship with the police, to the point of cohabiting the same building, as is currently the case in Montreal, in Ontario, and in New York State. There are good reasons for this association since the police have many resources for the maintenance of public security. It does, nevertheless, constitute a chronic takeover threat of emergency management by the police. The expertise necessary for the analysis of risk from natural and technological hazards, as well as preparing for such disasters, could be subordinated to organizations charged with the mandate of protecting against criminal activities and terrorist threats. The poor response of the Federal Emergency Management Agency (FEMA) to Hurricane Katrina in 2005, compared to its excellent response in the case of the 1998 ice storm, is indicative of the problems that resulted from its subordination to Homeland Security. Protection against both natural-technological hazards and terrorism is important. Subordination of one objective to the other can, however, diminish the effectiveness of response to the subordinated threat. Did a takeover of emergency management by the police occur during the ice storm? If so, how did it occur, and did it affect the emergency response?

Jean-Bernard Guindon stated: "I experienced a major problem concerning leadership because of the presence of a police services director who decided to exercise hegemony that surpassed and didn't conform to the planned emergency measures. He gave solo press conferences and took initiatives that even the president of the regional government had difficulty controlling, even though she was the director of police services' boss."

Mr Guindon admitted that this problem had to do with the particular personality of the police chief and his pursuit of visibility to launch his campaign for mayor a few months later. It was not typical of previous or subsequent police chiefs in Montreal. But when such an issue occurs during a disaster, it is highly problematic because the Police Department commands powerful resources that enable its chief to impose his wishes on others. What does an emergency management director do when the police chief pushes him aside, violates the protocol that had been established to deal with a disaster, and takes control? Guindon argued that this should not be allowed to happen again. "It caused many problems because his own police officers thought their boss was blazing the trail for them to follow. So the officers had an attitude comparable to that of their boss. I had received the mandate from the Quebec Ministry of Public Security to control the deployment of the armed forces in Montreal. But Mr Director of Police Services seized control of the armed forces because of his friendship with the general."

Guidon gave the example of his night replacement, who had scheduled a meeting with members of the armed forces. However, a police officer placed himself in front of the door where the meeting was to be held and insisted the man was not authorized to enter there. Guindon observed that communications became poor because the police intercepted them and prevented the municipal emergency measures organization from communicating with the army. The police directed the army according to needs perceived by the police, which were not the same as those perceived by municipalities. The latter complained the army was doing work that was not necessary and was not doing needed work. This was because, Guindon argued, the army was being deployed by the police instead of by the municipalities or by regional government, as specified in the emergency plan.

How did the police manage to take over control of the emergency response when that was not what had been planned? A report prepared by the Montreal regional government implies that the seeds of the conflict between the municipal emergency measures organization and the police were sown higher up the hierarchy.[1] It is, however, expressed in such bureaucratic jargon that everyone can evade responsibility. In my interview, Mr Guindon explained the problem and its source in plain language: "Near the beginning of the crisis Premier Bouchard called into his office a certain number of leaders he knew personally: ex-directors of the Quebec police force, the director of the Montreal police services, a deputy minister here and a deputy minister there. He gave them each a mandate: you take care of this, and you deal with that. He told the director of the Montreal police services: you will control Montreal. The police services director came back that evening from the meeting all puffed up with the power given to him by the premier and said: 'I'm the person who is in charge in Montreal.' That greatly undermined the procedures which had been planned."

Why did that happen? "Mr Bouchard knew absolutely nothing about the civil security plan for the province," Mr Guindon responded. "He had not been concerned about it before the disaster. Suddenly he was forced to face the music. Instead of using organizations inside his government such as departments, he preferred to choose people on the fly and give them ad hoc mandates. He completely destabilized the predetermined procedures at that moment, undoubtedly with good intentions and goodwill but in complete ignorance of the emergency plans." To avoid this type of problem, it is important that top political leaders, who will have to make decisions rapidly in a disaster, take the time to fully inform themselves about the best-practices planning that has been created to cope with a disaster. Otherwise those top leaders will inadvertently upset emergency plans and invent new plans in circumstances not propitious for well-thought-out solutions.

In the case of the ice storm, the ad hoc decision of Premier Bouchard led to an inversion of the power hierarchy that exists in normal weather. The police chief, who worked for the regional government and was under the authority of its president in normal times, was suddenly catapulted to a position of power over the regional government and its president, in clear violation of the plans that had been elaborated for an emergency. The police chief had, by virtue of being named by Premier Bouchard, "a political legitimacy, but he didn't have a place in the operational legitimacy."

The police chief then proceeded to act arbitrarily. Without receiving a request for aid from the municipality or the regional government, he sent police assistance based on his presumption that perhaps the municipality needed it. He referred to this process as being "proactive." However, in some cases such aid was not needed, and in other cases it would have been better used where it was requested.

Mr Guindon confirmed he had to go several times to the deputy minister of public security to check his mandate because it was in contradiction with the mandate given by Premier Bouchard to the police chief. Guindon felt that the deputy minister, too, was bothered by the evident contradiction between pre-established mandates and the ad hoc mandate given by the premier to the police chief. Thus a second-order crisis of authority had been superimposed on the ice storm crisis. "It was hardest to manage what I would call the epiphenomenon of human interaction as expressed through politics, the municipalities, etc.," Guindon remarked. "When you have disorganized power, no matter what the source of power, it affects the response to a disaster. It becomes the source of a crisis caused by the fact that the human staff in charge of responding to a disaster are not well controlled or there is confusion or ambiguity in their roles. Hence actions that need to be taken are not taken with the same effectiveness."

Guindon gave a nuanced evaluation of the role played by Premier Bouchard during the disaster: excellent political leader but poor chief executive officer. By that assessment he meant that "no one could criticize this premier as a spokesperson. But when he intervened to manage, not only over the head of his public security minister, but also over the head of the whole civil service apparatus of coordination that had previously been put in place, that government machinery could not help but function poorly. Despite his goodwill, the premier was largely responsible for the incoherencies that occurred in the government's response. He was visibly not someone who was ready to manage a crisis. But because he is a man with strong charisma, he was also paradoxically the man best suited to reassure citizens in order to create a unanimous movement, especially in press conferences, to go in the desired direction, that is, toward solving the problem." That bipolar result was indeed an immense paradox.

Mr Guindon noted that now relations between emergency management and the police are relatively good, but that the police remain the police. An effort is needed to prevent police hegemony from returning because, even if the police director is reasonable at the present time, not everyone in the police hierarchy has the same level of wisdom.

Guindon's boss, the president of the Montreal Urban Community, Vera Danyluk, agreed that according to the emergency plans the Office of Emergency Management was to be the centre of coordination for the response, but that the police service became the de facto centre because of its 5,000 employees and the personal contacts between its chief and high-ranking army officers. But her evaluation of the change differed greatly from Guindon's. "Neither the president of the Urban Community [herself] nor the director of emergency measures, Jean-Bernard Guindon, were personalities. It was more important for the population to have a Jacques Duchesneau in a police uniform as spokesperson in a disastrous situation. When the population feels threatened or vulnerable, right away people think that the only ones who can help are police officers and firefighters. So Jacques Duchesneau was perceived like the rescuer, the saviour."

A fully equipped room had been prepared in the Office of Emergency Management to respond to a disaster, but the decision was taken to move to a war room in the police headquarters because it was closer to the police, to downtown, to City Hall, and to the action. This relocation contributed further to the impression that the police had taken over control of emergency management in contradiction to what had been planned. Other actions reinforced that impression. Roads were barely passable, so President Danyluk was driven to work from home and back by police officers. Instead of participating in the press conferences of the Office of Emergency Management, the police held their own press conferences. Ms Danyluk commented: "Police Chief Duchesneau was there. Mr Guindon and I were at one end of the table. Nobody noticed us. It did not bother me at all because I told myself: if the response to the needs of the population is good, then that's the way it should happen. My biggest problem was my director of emergency measures [Jean-Bernard Guindon] because he did not agree."

Ms Danyluk sought the best possible outcome, and if the police chief had taken over, she was willing to sacrifice the emergency plan rather than fight the chief during the disaster. She saw her role as that of a conciliator between her two warring subordinates: Emergency Management Director Guindon and Police Chief Duchesneau. "It was necessary to find a way to reconcile it all and to ensure that we accomplish what was most important, the most urgent, under the circumstances. I think we succeeded pretty well in the end."

Things did work out reasonably well, but the deviation from what had been planned raises the question: Why bother with the expense and time involved in preparing an emergency plan if it is thrown out the window when a disaster occurs, or with emergency management experts if they are pushed aside by the police, or with an Office of Emergency Management if it is replaced by the Police Department? There was the potential for a debilitating turf battle here, but it did not take place because the president of the regional government played the role of conciliator and because the emergency management director, despite his irritation at the violation of the prearranged plan by the police chief and the premier, yielded in the interests of an efficient response. There is no reason to believe that the response would have been less effective had the emergency plan been followed, with the police playing an important role but not an overbearing one. On the contrary, the response would likely have been even better with less duplication of effort and reduced tension.

CIVILIAN EMERGENCY MANAGEMENT AND THE MILITARY

The civilian emergency management leaders praised the excellent contributions of the military during the ice storm in both Canada and the United States. There were, nevertheless, some tensions at the interface between civilian and military practices in both countries.

Jean-Bernard Guindon signalled a lack of consultation prior to the disaster. "The armed forces had always refused to make plans with us prior to disasters. Everything was improvised. We didn't know what the armed forces could do, and the armed forces didn't know what we could do. So when we asked them to do this or that, they did not necessarily have the capacity to do it. It seemed to them that we expected them to be scouts who could be asked to do anything. Members of the armed forces didn't appreciate being treated that way."

The military's refusal to join in advance planning for disasters with emergency management organizations proved to be bad for the military as well as for an emergency response. Why did the military refuse to participate in joint training sessions with civilian emergency management? Mr Guindon responded: "The armed forces told us that their role is not to do civilian support but rather to protect and defend Canada and do overseas missions on the orders of the government of Canada." The military focuses on priorities such as fighting foreign wars, which it views as more important and more in line with its mandate than disaster response at home. It has, nevertheless, always helped in times of domestic emergencies. Rather than refusing to plan together and then having to improvise when an emergency occurs, Guindon suggested that "it would be much better to plan in advance, and in the planning the possibility of a refusal would be foreseen. At least then the mutual capacities would be known, the lines of communication would be established, and the procedures would be put in place."

Mr Guindon, who is a civilian, found the military commander and some other high-ranking military officials arrogant during the disaster response. "He behaved like an officer who believes the world is divided into two categories: the military, who have the right to truth and to life, and the civilians. He spoke to us as 'you other people, the civilians,' with a contemptuous and condescending tone." Guindon noted that relations and communication between the civilian emergency management and the military are much improved since the experience of the ice storm.

Quebec's deputy minister of public security, Florent Gagné, worked closely with the armed forces during the ice storm and when he had previously been director of the Quebec police. He noted that the military is well-organized and has a rigorous approach to planning. That is its strength but also its weakness. When asked to help, the military "replied that they would have to make a plan first, that generally it takes thirty days to mobilize a battalion. In thirty days we wouldn't need anyone here. The military is excellent at planning. But don't let war break out this afternoon. There needs to be a thirty-day delay before declaring war." Nature does not accommodate such waiting periods before its hazards cause natural disasters. Mr Gagné admitted that the civilian security system is not as well-planned as the army's, but he claimed its response is more rapid: "If a disaster occurs today, by the end of the day civil security will already be in place with the means to help." He suggested that the strengths of each are needed and the response should be two-staged. "Civil security has a more immediate response but is less organized than the army. So the well-organized army arrives for the second phase." Gagné's comments were somewhat exaggerated to make his point: the army was not as slow as he implied, and civil security was not so rapid. But his more general point is well taken. It is important to recognize that different traditions and forms of organization, both military and civilian, are in play in managing a disaster. These have to be negotiated, preferably in advance, in order to have a coherent response.

In the United States each state has a National Guard, which is the oldest military force in the country. It is a dual-mission organization. One mission is to respond to the state and its governor if needed. The other is to respond to the president if the nation needs it. The president can override the governor. Normally, there is a cooperative agreement between the federal and state governments. It is primarily the federal government that pays for the National Guard in each state, except when the governor calls on it; then the state must pay for the troops, the equipment, and everything used for the emergency. During this disaster, the governor of Maine called on the state's National Guard to help with the emergency, and it provided crucially important assistance.

General Adams, who was head of both emergency management and the National Guard for Maine, gave an important example of the relationship between emergency management and the other uses to which the military is put. He argued that if a disaster such as the ice storm occurred at the time of the interview, the capacity to respond would be greatly diminished because half of Maine's National Guard and its equipment was not available, and the same is true of neighbouring states that would normally be in a position to help Maine. "The primary unit that responded to the ice storm in 1998 is now over in Mosul. Many of our other troops are in Afghanistan and Iraq and some other places related to the war on terror. All that manpower is gone; their equipment is gone; the engineering equipment that was so vital in the ice storm is mostly over in Iraq. There are a number of troops left, but they're going to face a much harder battle than we faced back in '98, when I had all of our troops, all of our equipment, right here in the state of Maine, and was able to get helicopters and generators from other states."

Adams made this insightful comment in an interview on 12 December 2004, about nine months before New Orleans was devastated by Hurricane Katrina, in part because the National Guards of Louisiana and the other Gulf states, as well as their equipment, were largely tied up in Iraq. Following the Hurricane Katrina calamity, the general's insight reads like a premonition. It is a pity that his foresight was not more widespread before Hurricane Katrina struck. His experience with the ice storm led him to a conclusion that was not disseminated and acted upon and has only become widely shared after the tragic experience of New Orleans. Because of the ineffective emergency response to that hurricane, state governors became more aware that the use of their National Guards and equipment to fight foreign wars deprives the states of an important resource for internal emergencies. This has now become a major complaint voiced by state governors in the United States.[2] They now realize there is a fundamental contradiction inherent in the dual purpose of the National Guard. If the federal government embarks on a foreign mission, then the challenge is to diminish that contradiction as much as possible by planned compensatory measures to make personnel and equipment available to each state in case of need.

In both Canada and the United States there was a serious salary inequity between military forces and electrical power workers, but it did not result in conflict in either country or diminish the effectiveness of the disaster response. General Adams explained the problem well: "I had my Guards out there making eight dollars an hour working alongside line workers making, God knows, fifty dollars an hour. These folks were working twenty-hour days and getting extremely low wages compared to people who were responding to the same event all around them."

TENSION BETWEEN POLITICAL LEADERS AND THE ELECTRICAL COMPANIES

There was also a problem of communication between political and emergency management leaders, on the one hand, and the electrical companies, on the other. Those companies were centrally involved in this disaster because the freezing rain crushed the electrical transmission and distribution system, but it was not easy to extract information from them.

Jean-Bernard Guindon agreed that it was difficult to get information out of the public electrical utility, but he contended that its chronic secrecy diminished rather than increased during this crisis. "Hydro-Québec has always been an organization that cultivates secrets and withholds information. Contrary to such habits before and since the ice storm, it was more open during that disaster." Communication between Montreal's Office of Civil Security and the electrical utility was better during the crisis because of daily telephone conferences and because that office succeeded in placing one of its members full-time at the electrical utility's coordination centre. The company's culture of secrecy has had the perverse, unintended consequence of mistrust; thus it does not receive credit even when it opens up. "It is easy to accuse Hydro-Québec of everything under the sun because of its well-founded reputation of not communicating. They haven't created a relationship of trust, so even when they behave well, their good behaviour is not recognized. People say instead: 'They must be hiding something else.'"

In the neighbouring province of Ontario, there was also tension between, on the one hand, political leaders and emergency management directors and, on the other, officials of the electrical utility. Bob Chiarelli, chair of the Regional Municipality of Ottawa-Carleton, described the actions, during the disaster, of Ontario Hydro, a company with a misleading name because most of its electricity is now produced by nuclear reactors and coal-fired plants. "It took me calling the premier of Ontario and saying: 'We can't get in touch with them. We can't coordinate anything with them.' And it took the premier calling Ontario Hydro to demand that they get somebody senior from Hydro down onto our disaster team. The first effort at responding to the premier's wish was to have somebody come down here who had no authority and who was almost as unplugged-in as we were. So I had to go back to the premier a second time, and then he got a very senior person to come down here and sit at our table." The senior director of the electrical utility was propelled into action by having to answer complaints at press conferences and being embarrassed when his utility's response had been chaotic.

What was the reason for the poor disaster response of the electrical utility? Was it because the public utility was being deregulated, partially privatized, and split into subunits? Mr Chiarelli answered in the negative : "It's just that Hydro was big and cumbersome and wasn't ready for that much freezing rain. They had written off this type of eventuality."

REPLACING A MINISTER OF PUBLIC SECURITY JUDGED INCOMPETENT DURING A DISASTER

One of the important qualities of a good leader is to choose competent people to carry out tasks. Another is the capacity to correct one's errors. A premier or governor has to select individuals to head the various ministries of the cabinet according to many criteria, such as equity on the basis of region, gender, race, and ethnicity. But competence to do the job must remain an essential qualification. Experience is an important quality; nevertheless, everyone has to acquire experience somewhere. In a cabinet there must be place for new blood. Which ministry will be given to beginners to acquire experience, in many ways by trial and error? Of all the politicians voted to office by the electorate, did the premier choose the right newcomers for the cabinet? What happens if the premier has chosen poorly? Will the error be corrected, or will loyalty and image trump competence? All of these questions loom particularly large in the demanding challenges for leadership during a disaster.

Prior to the ice storm Premier Bouchard had chosen a young, inexperienced politician as his minister of public security. This decision already says a great deal about the lack of importance attached at that time to public security in the government's hierarchy of priorities. If the chosen person made mistakes because of youth and inexperience but nevertheless demonstrated potential, ability, and good judgment, then the learning curve would be steep, but at least learning would be occurring. The problem in this case was more serious. President Vera Danyluk, of the Montreal Urban Community, stated that the individual "was a young minister who didn't have any experience and who was very mistake prone. I know because I had to work with him on all the Montreal Urban Community police files." Jean-Bernard Guindon, of the community's civil security centre, confirmed that description. "He was a minister who didn't inspire any respect, not from the director of police services, nor from most other people for that matter. The director of police services was almost mocking him openly."

The freezing rain had produced an extreme crisis. Decisions had to be made rapidly and correctly. The population had to be reassured that everything possible was being done and done well by the most competent people possible. If a mistake had been made in selecting a person into the cabinet and placing him in charge of the Ministry of Public Security, which the freezing rain had transformed from a secondary ministry into a key one, this was not the time to persist in the error. Faced with such a dilemma, Premier Bouchard was decisive. Mr Guindon described the premier's actions as follows: "This minister was rather quickly pushed aside, explicitly and publicly pushed aside. It was pitiful. The premier squeezed him out and took over the role that the minister of public security should have played."

Ms Danyluk noted that the dismissal of the minister of public security occurred on television and was symbolic of the immediate transfer of direct control of the crisis to Premier Bouchard. "A news report on television one evening had the minister of public security of Quebec, Premier Bouchard, and Mr Caillé, the president of Hydro-Québec. Viewers saw these three leaders, and then all of a sudden the cameras cut the minister of public security from the picture. Only Mr Bouchard and Mr Caillé were left. The message was clear. The leader and spokesperson for this file was not the minister, even if it was his responsibility. It was Mr Bouchard and Mr Caillé, and that remained the case for the remainder of the ice storm crisis during the next several weeks. The minister of public security was no longer seen." The premier could have taken the lead role yet have his minister of public security carry out an important supporting role, but that did not occur, probably because Premier Bouchard and others did not have enough confidence in the minister's abilities. Thereafter the young minister of public security was not on the radar screens of the population at all, even though the management of a crisis such as this one was formally his job. The political career of that young cabinet minister was finished as well.

Although we can sympathize with the cabinet minister, quick action by the premier was necessary in the crisis to correct his own error in choosing that person for the cabinet. This rapid, decisive change can be compared with what occurred when Hurricane Katrina struck New Orleans in 2005. As the disaster was unfolding there, it became obvious that FEMA was responding poorly. The director of the agency, Michael Brown, chosen by President Bush, was being criticized for its substandard organization. Instead of correcting his error and replacing Brown immediately, President Bush made his infamous comment 'You're doing a great job, Brownie' at the very time everyone else knew Brownie wasn't. President Bush had to backtrack several days later and replace Brownie when even he saw the ineffective work of the department Brown headed.

BUSINESS PRESSURE ON EMERGENCY MANAGEMENT DIRECTORS TO REOPEN A CITY

There was another decision that remains shrouded in secrecy, namely, the speedy reopening of downtown Montreal. It is the second largest city in Canada, is the economic engine of Quebec and of a large part of eastern Canada, is integrated into just-in-time production with facilities elsewhere, and houses head offices for many corporations. The ice storm shut down all this: freezing rain made roads impassable, grounded flights, stopped trains, covered roofs of tall buildings with ice that then fell in chunks threatening pedestrians, and crushed electrical power lines. If electricity was reconnected too quickly, power surges could cause fires or make the system collapse once more. Was downtown Montreal reopened when it was safe to do so?

By Monday of the second week of the disaster, electricity had been reconnected to Montreal, but the power company had not yet rendered the electricity of the downtown core secure. Jean-Bernard Guindon assessed the decision to reopen the downtown as the worse one made during the ice storm, and it led to the most dangerous phase. "There was a desire on the part of all the economic players to reopen the downtown. Hydro-Québec was pressured and decided to take the risk of opening the downtown despite opposition from the people responsible for safety. Those economic players succeeded in forcing us to accept their decision. It was lucky no one was killed or hurt. An unparalleled risk was taken in the name of the resumption of economic activities in the downtown. We came close to real chaos. If the same situation arose in the future, I would be completely firm because of the evidence of danger for the citizens. Avoiding a day or so of economic losses is not worth risking so much." Vera Danyluk confirmed she was subjected to similar pressure to reopen the downtown. "All the retailers and banks told us: 'It's time to reopen because the economy has ceased functioning in the downtown.' But we were afraid to say yes because of the fragility [of the infrastructure]."

Fortunately, the resumption of activities occurred without injury, although not without some disorganization. Mr Guindon concluded, nonetheless, that it was a reckless decision. "It is not because someone does something completely stupid and gets away with it that it is less stupid. Because we were precipitated into that decision, we did not have the capacity to plan correctly. We tolerated it and we let it pass. I admit that I would not have recommended to the president [of the Montreal Urban Community] to accept it if I had not myself been subjected to extraordinary pressure. I remember several telephone calls I received from major employers and from several top managers of Hydro-Québec that really put strong moral pressure on me that was difficult to withstand. Behind all that there were certainly stakes that were never known or revealed which enabled those pressures to occur." Where did the pressure come from? "The major banks, large commercial establishments, owners of big buildings, holders of large real-estate investments in the downtown, and large commercial investors who were becoming impatient because economic activity had ceased for over a week in the downtown. I believe that the safety of people was dallied with in an unjustifiable way."

André Brunelle, assistant director of the Montreal Fire Department and coordinator of emergency measures, also witnessed these pressures from Hydro-Québec and the business sector. His team resisted but eventually succumbed to the pressure. "We delayed the opening by about two days because it was too dangerous for the population. We learned a lesson because we never thought we would be subjected to those pressures. From now on we need to have more documentation to be able to explain why it shouldn't be done." In addition to creating a dangerous situation, the secretive commercial pressures that led to the premature reopening of the downtown violated principles of transparency.

EVERY CRISIS HAS A SCAPEGOAT: BLAMING AN INEFFECTIVE RESPONSE RATHER THAN INEFFECTIVE PREVENTION

The Quebec Ministry of Civil Security was blamed for a slow, ineffective response to the crisis. The criticisms came from many sources: the Canadian military, municipalities that themselves were not prepared, Quebec politicians, civil security organizations in Montreal, and so on. Florent Gagné, Quebec's deputy minister of public security, whose department was the target of the criticism, responded as follows : "In any crisis, a scapegoat is needed, and the Quebec Ministry of Civil Security was effectively something of a scapegoat in this crisis."

The electrical power company was extremely clever in its public relations, presenting itself in the best possible light and explaining to the media all the good things it was doing. It succeeded in deflecting attention away from the fact that the vulnerability of its electrical system was the cause of the disaster. Mr Gagné commented: "It is true that civil security did not do everything. In a context like that, obviously people look for a black sheep or someone to blame, and we had very bad press. There were essentially two dimensions. Hydro-Québec was technologically the source of the problem. But those who had to deal with the problems were instead the ones who were blamed. There was irony in that. I experienced that feeling, and so did a large part of my team." If the food in the shelters was not good enough, or the blankets not warm enough, civil security was blamed for not taking care of its citizens. Gagné concluded that in such a crisis people should act in solidarity with their government and help it to solve the crisis instead of continually blaming it.

CONFLICT MANAGEMENT

Tensions between regional and city governments, between emergency management and the police services, between emergency management and the army, and between emergency management and the electricity utility in a disaster of electrical deprivation had the potential to develop into multi-faceted turf wars. People were blamed, an incompetent minister was replaced, and business pressures were exerted to open the downtown prematurely. All these very human conflicts between leaders and between competing institutions with different interests were worked out more or less satisfactorily. Those involved, irritated though they may have been, did not become stubborn or allow the tension to undermine a collective response in the crisis. This event proved to be an example of effective conflict management. The resulting overall response was not perfect, but it was nevertheless very good.

CHAPTER 10

Making Sense of Disaster and Its Management

Many specific technical and organizational lessons were learned from this disaster that have been enumerated in other publications.[1] There are, in addition, general lessons that can be learned through the experience of leaders who were directly involved in the management of the crisis. How do they make sense of this ill-defined crisis and define the problems several years later? Can they provide insights as how better to deal with the risks from such hazards in the future? This chapter investigates the meanings these key decision-makers and leaders gave to their experience with the disaster and the general lessons they learned from it.

COORDINATION: A CRITICAL PROBLEM IN AN EMERGENCY

Hubert Thibault, chief of staff for Quebec premier Lucien Bouchard, observed that there is one aspect of crisis management above all others that is particularly challenging. "What is difficult in this kind of crisis is coordination. There are many leaders involved. At any given moment things happen so rapidly. Therefore it wasn't simple to find a way to make sure that everybody could function together without being overwhelmed by consultations or discussions or the mechanics of decision-making."

He confirmed that even in normal times crucial decisions of government are made by fifteen or fewer people: the premier, three or four important cabinet ministers, two or three high-ranking civil servants, the secretary-general, and two or three advisers. Meetings of twenty-five people who debate details are not productive, especially in an emergency. But participants must be at the top of the decision-making structure, and information and decisions must be communicated clearly to all others who are needed to implement these decisions. "We can't always count on people at the next level [who were not at the meeting] to understand the decisions," Mr Thibault observed. "Participants in these meetings have to take responsibility themselves, each in their sphere of influence, to make the decisions and information go down the chain of command correctly."

As the premier's chief of staff, Thibault contacted cabinet members who were not present at the meetings. "It is necessary to make sure that everybody understands the decision that has been taken, why it was taken, that follow-ups are done, and that the marching orders are applied." Effective coordination and communication are serious challenges in a crisis. The top leadership cannot be tardy or disorganized in responding to these challenges, as seems to have occurred when Hurricane Katrina struck New Orleans in 2005.

DECISIVENESS OR PERFECTION IN A CRISIS?

It is important that leaders make the right decisions during a crisis, but the situation evolves rapidly. Windows of opportunity can close if decisions are not made in a timely fashion. How do leaders weigh the importance of correct decisions with the need for swiftness? Mr Thibault noted that a disastrous situation "isn't a time when we can allow ourselves to indulge in never-ending discussions. At a certain point in a crisis it is more important to make a decision, even at the risk of it not being perfect, than to make sure that all the i's are dotted and the t's crossed, thereby delaying the decision." The goal of perfection promotes hesitancy, which is inappropriate in a situation requiring swift choices. Decisiveness is essential in managing a disastrous situation that is changing rapidly. Effective leadership recognizes that choices are fallible but can be corrected or improved later if necessary. The population, too, has to understand the time constraints leaders are working under in a crisis.

Mr Thibault compared decision-making during a crisis with decision-making in a commission of inquiry after a crisis. "It is easier after a year of investigating, after having heard all the witnesses speak, after all the time possible at your disposal to do studies, consult with experts, double-check, revise and correct the reports. In real life, events happen a little more quickly than that, and it is more complicated." Leaders in a real-life crisis must satisfy exceptional requirements for speed of reflection that are not imposed on commissions of inquiry and researchers, including the present author.

Florent Gagné, the Quebec deputy minister of public security, pointed out a serious problem. "When decisions are made in a context like that, they are taken necessarily with the information that is available, which is not always all the needed information." David Flanagan, president of Central Maine Power, stated, "I've always operated on the theory that the perfect is the enemy of the good. I'd rather have some reports from the field than none." The requirement of speed in decision-making and the scarcity of necessary information are serious constraints on the quality of decisions in a disaster. The management of this ice storm calamity was particularly impressive when judged in this context.

THE ADVANTAGES FOR DISASTER RESPONSE OF CLOSE PERSONAL RELATIONSHIPS AT THE TOP

Disaster preparation involves a great deal of advance planning, exercises, and other activities. There is, however, another dimension that is crucial for the effectiveness of the response but is more difficult to ensure. Maine's governor, Angus King, argued that "one way you prepare for an emergency is to have a smoothly functioning team in advance of the emergency, who are comfortable with one another, friendly, and work well together." This is an admirable goal, but not all teams work well together and have members who are comfortable with one another in routine times. What determines whether a smoothly functioning team will be in place? "I was very close to my cabinet," Governor King responded. "We met all the time. We went on retreats together. So when something like this happens, we had pre-established, close personal relationships. The head of the DOT [Department of Transport] and the head of the National Guard knew each other closely, first-name basis, easy communication, no bureaucratic lines of 'You can't make that call and you have to go through staff.'"

During a crisis, routine bureaucratic channels are too slow and cumbersome. Decisions and communication are often most effectively made leader to leader. Hence it is important to have occasions in normal periods to bring top leaders together so that they can become comfortable with one another, because they will have to trust one another and cooperate during a crisis. This requirement does not presuppose that leaders are angels without a trace of jealousy, credit hogging, or authority-seeking behaviour. Rather, it suggests that the groundwork for trust and cooperation can be laid before a disaster occurs. Tightly knit personal relationships connected the governor to private utilities, and that linkage facilitated the management of the disaster. "I was already previously on a first-name basis with a bunch of people at the power companies that I had known for years. Coincidently, I had been in the energy business, so I even knew them before I was in government." Personal contacts developed in normal times can be very helpful in extraordinary disastrous times.

Governor King contended that smallness breeds efficacy in disaster management. A much broader study would have to be undertaken to determine whether his hypothesis receives support. Smallness of scale and close personal relationships are not only characteristic of Maine's government and companies but also of the state itself, with its population of only 1.3 million people. "Maine is a big small town. It has a community feeling to it. That makes it easier to manage a disaster like this because there aren't bureaucratic barriers. It's pretty easy to go around the chain of command to get things done. We could work on things without fancy meetings with my staff and your staff. David Flannigan [the president of Central Maine Power] would call me at home at ten o'clock at night."

King admitted that trusting and feeling comfortable depend on the personalities of the people, "but it's also a function of the place. It is more likely to have those relationships than it would be if it were a place with – what's the population of Quebec: seven million? That's a whole different scale. But it [this disaster] could have happened in Boston."

If close personal relationships among leaders depend on the smallness of the population, how can larger population centres cope? The freezing rain just missed Boston, with its large population. Would it have been unable to deal with the extreme weather as well as the smaller state of Maine? The experience of Montreal gives an indication of how a larger metropolis handles a crisis. Montreal's mayor, Pierre Bourque, confirmed that disaster management was consolidated at the highest levels when communication through the lower levels proved to be inadequate. "It was done chief to chief: the mayor of Montreal with the premier. And he gave instructions to the CEO of Hydro-Québec. The command structure is very important. In this case the political authorities really did their duty. If Mr Bouchard had not responded to my call, it would have been a catastrophe. The middle-level people at City Hall had difficulty getting through to the bosses at Hydro."

Communication and decision-making among a metropolitan government, the provincial government, and public utilities are carried out in normal times by a middle-level staff. During this crisis in Montreal, the staff forming the middle layer were skipped over, and communication went directly from the head of the city government to the head of the provincial government to the head of the electrical utility. It was as if the disastrous extreme weather had shrunk the province of Quebec and the metropolis of Montreal into a big small town like the state of Maine as far as governance was concerned.

UNIMAGINABLE TROUBLES IN A CRISIS

The worse moment of the crisis for the premier's chief of staff, Hubert Thibault, occurred when CEO André Caillé of the electrical utility announced that all the transmission lines to the island of Montreal except one had been severed by the freezing rain, and that last line was covered with ice and galloping about ten feet. If it fell, the island would have no electricity whatsoever at a time when the bridges were closed because they were too dangerous. "That was a situation we had obvious difficulty managing. We were all trapped; the population was trapped on the island of Montreal." In a crisis like that, the balance of probabilities for occurrences cannot be calculated, especially not with sufficient accuracy and speed. When there is such great uncertainty, no one knows what will happen. How do leaders make sense of such an ill-defined situation?

We now know in hindsight that the freezing rain stopped abruptly at the worse moment of the disaster and that this change, more than any decision or human action, was the beginning of the end of the calamity. If the freezing rain had continued, then the last remaining ice-laden, galloping line would most likely also have collapsed, so attaching it to the filtration plants would not have accomplished anything. I pushed these leaders to imagine what they would have done if a technical solution had not been found quickly. What decisions would have been taken if electricity for the filtration plants had taken twelve hours to re-establish, or twenty-four hours, or more? Montreal's Mayor Bourque answered: "Whew! ... I don't dare imagine that. It would have been hard to live with. It was hard on me that day until the water came back on." He felt that the constraints on his choices were already very tight. Decisions would have been extremely difficult had engineers taken longer to find a way to supply electricity to restart the filtration plants or if the one remaining line had also been crushed by the ice-loading. André Brunelle, the assistant director of the Montreal Fire Department, answered that evacuation of the island of Montreal might have been considered in a second phase if living conditions had greatly degraded, but evacuation of the island on impassable roads in icy weather was clearly a terrible alternative.

Hubert Thibault specified that decision-making was already very difficult and would have become more so. "It would have been necessary to make quick decisions. Which ones? God knows what we would have done. And that is why that night in particular – when the last ice-loaded transmission line was galloping – we were on the brink of a large-scale catastrophe. Then decisions would not have been simple, so what would we have done?" Mr Thibault did not answer his own question. It is admittedly hypothetical to imagine a scenario of the freezing rain or the filtration-plant shutdown lasting longer or the one remaining line falling, but the hypothetical situation came within an hour or two of occurring. The island city of Montreal was hanging by only one galloping, ice-laden transmission line when the freezing rain serendipitously stopped. That might not occur next time. Thus it is important for future leaders to imagine dramatic scenarios such as this one, to make sense of them, to define the problems, and to envisage solutions in order to prepare in advance for an event that almost happened this time.

Consequences resulting from unforeseen risk lead to short-term improvisation. This in turn is followed by the attempt to transform improvised sense-making into preplanned decision-making in order to prepare better for the next big risk. "Best practices" consist of reflective social constructions, but they also are typically prompted by the experience of nature's dynamics – in this case and many others, by a disastrous experience. They involve learning from erroneous assumptions of safety in the past. Response planning will, nevertheless, always be partial when done in advance because the precise character and timing of nature's hazard is uncertain. Improvisation cannot be planned out of existence.

DOES MODERN SOCIETY DEPEND ON LUCK IN ITS INTERACTION WITH NATURE?

When the water-filtration plants on the island of Montreal lost their electricity because the transmission lines had been crushed by the ice-loading from freezing rain, the risk of fire was great, yet the means of extinguishing it was rapidly being depleted. Strange as it may sound, rain threatened to cause fire. By interrupting the rationally planned supply of water, freezing rain threatened to produce the same result in Montreal in 1998 as the earthquake of 1906 in San Francisco had: the burning of a city. Worse still, in Montreal the population was trapped because the bridges off the island were impassable. Deputy Minister Florent Gagné admitted that "without a doubt we were lucky. But will luck be there the next time? God knows!" He is fundamentally right to acknowledge that the successful resolution of the crisis did not depend only on technical and organizational work and skill but also and mainly on good fortune. At that worse moment of the crisis on Friday, 9 January 1998, the freezing rain unexpectedly and suddenly stopped. A leader of emergency management in an advanced modern society must feel disappointed to have to admit that luck played an important role in mitigating disaster, but his admission is, nevertheless, an honest and candid assessment. "Luck" is an oblique way of expressing modern society's dependence on autonomous constructions of nature that are benign rather than destructive.

THE PERVERSE EFFECT OF FEAR OF LAWSUITS

Prior to the ice storm, the state of Maine had a cooperative agreement with the neighbouring Canadian province of New Brunswick that specifies how things will be done during an emergency, how mutual aid will be paid for, and so on. Paradoxically, Maine did not have such an agreement with neighbouring American states. Earl Adams, Maine's commissioner of defence and emergency management, described the difficulty that resulted as follows: "When we first called them [Massachusetts] and requested their assistance, their immediate response was 'We'd be glad to help you.' But then the lawyers got into it and said, 'You can't let the workers go until we've got these agreements. What's going to happen if somebody gets hurt or killed? Who is going to pay?' It bogged down the operation for a while. I called my counterpart down there and said, 'Let the lawyers work this out, but we got to get some help up here in a hurry.'"

Fear of legal conflict can retard or paralyze an emergency response, so it is important to have prior agreements in place specifying how aid across political boundaries will be delivered and financed in a disaster in order to eliminate conflict and remove legal obstacles to a timely response. The experience of this extreme weather disaster prompted American states to create agreements to respond more effectively in a disaster.

IS MANAGING NORMALITY MORE CHALLENGING THAN MANAGING DISASTER?

Hubert Thibault made an insightful comment that at first glance seems surprising: "The most difficult decision was not in the immediate management of the crisis. In times when the population needs to be mobilized, when it must become responsible, it does that quite well. There are always exceptions, but that is an important tendency. I would say that the most difficult decisions are when the immediate crisis diminishes. Then decisions have to be taken to ensure it doesn't happen again."

Mr Thibault did not remember any decision during the crisis that was difficult to make, and he felt supported by the population and by its willingness to cooperate. Decisions were urgent but not difficult. A cooperative spirit of mutual aid and understanding reigned. If leaders did a reasonable job of managing the disaster, the population was satisfied. "By the end of the ice storm crisis, the level of satisfaction in the government of Mr Bouchard rose to 90 per cent." he observed. A self-confessed political animal, Thibault advised Premier Bouchard to call an election immediately after the disaster. They both decided, however, that it would appear too opportunistic. So they waited. The opposition party chose a new leader, a strike by nurses broke out, and their governing party lost 25 points in the polls. Normality was not as kind to the political party in power as disaster had been.

The decision during the disaster to switch the only remaining electrical transmission line from the downtown to the filtration plants was not difficult because it was clear that the most urgent and important need was for water. Thibault concluded that "once again these are so obvious. It is more complicated to approve the plan for the extension of the electrical grid after the disaster than to decide to unplug the downtown during the disaster in order to plug in the filtration plant." When the crisis was over, the cabinet decided to construct a new transmission line to reinforce the electrical network. At that point, however, the NIMBY (not in my back yard) syndrome and usual opposition against new electrical transmission lines emerged, and the project, which he felt easy to justify rationally, became impossible to sell to the population.

Mr Thibault also concluded, contrary to popular belief, that the ice storm disaster did not lead to a crisis of confidence in the electrical utility. On the contrary, its CEO became a public personality because of his television appearances, and the utility found a credible way of communicating with the population during the disaster. It was after the ice storm ended that decisions by Hydro-Québec concerning location of transmission lines and windmills, conservation of energy, and other policies were criticized, and it seemed to lose its capacity to communicate.

Thibault then provided an analogy to illustrate that leadership is easier in times of crisis than in ordinary times. "Managing budgetary difficulties is much easier than managing surpluses. When you don't have any money, you say, 'No, I can't.' When you have money, you are obligated to say, 'Not your project, his.' Then it is much more complicated. In a certain sense a crisis is easier. The crisis itself proves the need for a decision. Solutions are, as a general rule, relatively obvious. Only exceptionally are they harder to make. Hence surprising though it may seem, the everyday management of a complex, modern society is much harder" than the management of that same society during a disaster. A crisis, much like a deficit, creates conditions propitious for sympathy, shared priorities, and a direction for the government if it can seize the opportunity. Mr Thibault added that management was also easier because there were none of the struggles between the federal and Quebec governments that are common in normal times. On the contrary, solidarity reigned during the crisis between the two governments.

The same was true in the United States. David Flanagan, president of Central Maine Power, described how the spirit of solidarity made his work easier in the crisis. "Our guys got into the spirit of working these long hours, and there were no complaints whatsoever. The foreign crews picked up the fact that our guys were working hard and that they were getting tremendous public support and the press was in there being supportive, the legislature, the governor. So they too put in all the extra time. It was really tremendous for moral." The support that workers received from the population led them to work far harder in atrocious conditions than they would in much more favourable routine working conditions.

Florent Gagné drew attention to another aspect that made managing this disaster easier than matters in normal times. "We didn't have any financial problems. We spent all the money we wanted. Human resources were sometimes lacking, but we had a great deal of financial resources." Unlike in normal times, when they faced financial constraints, governments and taxpayers provided all the necessary money to deal with this disaster.

This conclusion that managing normality is in many ways more difficult than managing a disaster was confirmed as well by Maine's governor, Angus King. He agreed that decisions made in this disaster were not as difficult as some he had to make in normal weather. "There were lots of decisions, but they weren't hard. They were obvious. If you decide to call out the National Guard when half of your people have lost their electricity, that's not a hard decision. I don't remember agonizing about decisions. It was more a case of executing."

The state's commissioner of emergency management, General Adams, concluded, "I had more political turmoil and attempted influence in minor emergencies. Sometimes in those events you do get politicians that would try to say, 'I'm a senator and I want this in my area and I want it now.' During the ice storm we really didn't [experience that]." The gravity of the disaster prompted senators and representatives to avoid pressuring this emergency management director for special favours for their areas. For the wider good of the state in crisis, they accepted a negative response to their requests.

Why was there more political turmoil in minor emergencies than in major ones? "Prior events were more localized," Adams responded. "You had competition between the municipalities in a small geographic area of the state. In this event everybody was involved. Everybody knew that we had a terrible situation on our hands." The worse the experience of disaster, the less often the "me first" syndrome appeared. Far from interfering, legislators helped emergency management leaders in handling this extreme crisis.

The disastrous consequences also stimulated greater awareness of another important element, as Ms Danyluk, president of the Montreal regional government, explained: "As a result of this crisis, especially when we were having meetings to analyze everything that happened, people realized the importance of collaboration on the part of all the partners and all the leaders. But did this sense of the importance of cooperation remain with people? That I don't know."

Maine's Governor King was not criticized for on-the-spot decisions during the ice storm, but he was denounced for a difficult choice he made a month later. "The federal government asked: Will the state reimburse people for generators they bought during the ice storm? It was our call. I said no because it struck me as beyond the responsibility of the state. Also, there were people who had bought generators themselves years before to be prepared. I didn't see why we should be rewarding somebody who wasn't prepared and penalizing somebody who was prepared. I took a lot of heat for that decision, but I felt it was the right decision. That wasn't a crisis decision. It was a policy decision made afterwards."

In times of crisis, leaders receive more support than in normal times and are given the benefit of the doubt because no one wants to weaken the disaster response. Thus solidarity in fighting extreme weather is similar to the collective solidarity that emerges during a war. The material context of grave danger augments support for leaders during a disaster or a war. In both cases, however, there is much criticism if things go wrong and much second-guessing when the threat ends and the situation returns to normal.

SELF-RELIANCE OR DEPENDENCE ON THE NANNY STATE?

Deputy Minister Florent Gagné claimed that the comforts of life in a wealthy, well-organized society have led the public to lose the initiative for self-defence. "The population has developed the reflex, when a disaster occurs, of relying on the government, which must provide everything. That was what happened [during this disaster], and it was the main problem." He gave several examples of the lack of individual autonomy and the dependence on the state. "I visited shelters where people asked me:

'Mr Gagné, there is nothing to do here. We have been here for eight days. I have three children. Could we have some entertainment?'" So the government had to buy large-screen colour televisions for the shelters and organize leisure activities. But that was not all. "Condoms were provided in the shelters because there were complaints that people are in shockingly close quarters, and sexual relations have been observed. My God, is it the responsibility of the government to take care of such problems of intimate human behaviour that people decide to have? Well, believe it or not, the answer from many people is yes. It is up to the state to take care of all problems." Gagné saw this attitude as one element of a much larger problem of lack of autonomous initiative that results in enormous demands on the state during a disaster.

André Brunelle, the assistant director of the Montreal Fire Department, also concluded that a spirit of individual self-reliance would make government more efficient during a disaster than if the state has to provide everything. "People have to be made more aware of the need to be as self-reliant as possible: to stockpile a basic minimum of essentials, to be able to manage on their own, and not depend only on public authorities."

Maine's General Adams told municipalities that "we can help you some, but we're a huge state [geographically] with a lot of towns, and there's no way we could ever hope to satisfy all their needs." State organizations and professionals provide backup help during a disaster, but victims themselves have to be prepared to be the first responders.

EMERGENCY MANAGEMENT AND THE MILITARY

In the state of Maine one person heads both emergency management and the National Guard. Is this structure typical of all states in the United States, and is it effective? At the time of the January 1998 ice storm that person was General Earl Adams, who answered that "a little more than half the states are set up the same way we are. I would recommend it highly because I know some of the other states where they're not that way. So when they get an emergency like the ice storm, there's a butting of heads between the emergency management director and the National Guard commander. We don't have the trouble of butting of heads because we have it in one hat that controls them both."

General Adams is undoubtedly correct to argue that in small states, where emergency management has very few resources, a structure has to be set up that does not have to beg for resources when disaster strikes. Thus for lightly populated states such as Maine, the joint management of emergency measures and the National Guard may well be a reasonable structure. But the experience of the more populous Canadian provinces of Quebec and Ontario demonstrates that an emergency management structure independent from the military does not result in a butting of heads in a way that diminishes the efficacy of an emergency response. Whatever conflict occurred was minor and did not hinder the response because the command structure was clearly defined: the non-military emergency management was decisional, and the military was supportive. This evidence disproves the hypothesis that a militarization of emergency management is necessary. Even in small states such as Maine, the merger of emergency management and the military provided a successful response because it was under local control of the state. It involved Maine's National Guard, rather than the US Army, and thus benefited from intimate knowledge of local conditions.

A CULTURE OF SECURITY?

General Adams observed that during the disaster "so many people got out their battery-powered radio only to find dead batteries in it. Then they couldn't get to the store to get new batteries. So people learned." The experience of disaster prompted changes because it made inadequacies visible.

However, disaster does not determine that long-term social change will necessarily take place because it can be dismissed as such a rare occurrence that nothing needs to be done. Adams observed, "Since the ice storm was listed as a one-in-a-hundred-year event, some people are saying, 'I'm not going to be around to experience the next one,' even though that thinking is not the way it should be." The shock from a disaster acts as a prompt, but by itself the disaster does not produce social change and can be shrugged off as not likely to occur for another century.

Asked whether disasters such as the ice storm and 9/11 shocked the population and authorities into a greater sensitivity to the issue of disaster protection, Montreal's Mayor Bourque responded: "Exactly, exactly. Without that, people essentially live without stress. Why give yourself unnecessary stress?" But he added a warning: "all that [security] is expensive. It costs a fortune. Often it's a lot of money wasted." Leaders and the broader society have the difficult challenge of making sense of the new situation subsequent to a calamity, of distinguishing between real risk and false claims of risk, and of deciding how to respond. A calamity often prompts *a* response, but leadership decides *the* response. As time passed after the disaster and as the negative side effects of new transmission lines to make the electrical network more secure became evident, a NIMBY reaction emerged and leaders became hesitant. Hubert Thibault noted: "Security isn't just about buying fire trucks or just about having 30,000 beds in storage in case something happens. It is about many decisions that taken individually can be very annoying, but that have to be taken." Ideals of safety run up against bothersome side effects caused by the means required to attain those ideals in the cases of both protection against disasters and solving environmental problems. Particularly courageous and effective leadership is required to reconcile the two and find the proper balance.

During a disaster, all-too-visible material consequences stimulate a spontaneous culture of security. But after the disaster ends, other priorities once again come to the fore. Mitigating potential disasters in the future then seems less urgent than other problems. The issue becomes whether the spontaneous culture of security during a disaster will be ephemeral. In routine times, disaster preparedness and solving environmental problems can seemingly be put off to the distant future.

Vera Danyluk, of the Montreal Urban Community, concluded that the ice storm and other recent disasters led to a cultural change in that they undermined expectations that nothing bad would happen. She pointed, however, to a fundamental discrepancy between cultural ideals and real, material practices when there is no crystal ball to tell us what kind of crisis will occur and when it will happen. "When there is a change in culture, it's very impressive because people realize they need to accomplish things. As soon as they have to pay for those cultural changes, however, that is something else entirely. There is now a cultural change, but if someone tells us tomorrow to pay more taxes because we need to be prepared [for disasters], people would scream bloody murder because they don't want to pay." Discourse should not be equated with behaviour. A culture of security in words cannot be assumed to have produced a culture of security in practice. Nature's dynamics interact with material practices and not with words in the production of either preparedness or vulnerability, safety or disaster.

POPULAR, POLITICAL, OR INSTITUTIONALIZED CULTURE OF SECURITY

A disaster tends to arouse a temporarily heightened awareness of the need for a culture of security, but the issue of whether it withers away as nature's prompt becomes more distant in time has a nuanced answer. Florent Gagné argued there are different social sites of a culture of security and that decline or maintenance of such a culture varies according to the site. "Among the professionals of public security in the government, the culture [of security] does not fade away. On the contrary, it is being organized more and more. But on the political level and on the level of the population, I don't see any indication of a more developed culture of civil security now than before the ice storm. It is strictly the business of professionals in that area [of emergency management] where definite and considerable progress has been made." The political leaders at the time of the disaster are no longer in power. The population turns to other priorities, and so do the new political leaders. Gagné's conclusion was that the only learning from disaster is on the part of disaster professionals and is institutionalized in organizations, particularly emergency management ones.

ACTING UPON NATURE'S PROMPTS

When extreme weather triggers a disaster because of social and material vulnerability, the population and, in particular, key leaders attempt to make sense of the new situation. In doing so, they are influenced by both signs from culture and prompts from the material consequences of the autonomous dynamics of nature. The socially constructed outcome is determined by the social *and* material contingencies within which it is embedded. There is a decay curve of sensitivity to risks and hazards by which the increasing time after normality has returned tends to lead to the fading away of the acknowledgment of risk. A major challenge of leadership consists of preventing this decline from occurring.

Social change takes place when a prompt is acted upon to construct new structures and practices. The disastrous ice storm was the impetus that led the government of Quebec to construct new laws and procedures for civil security, a point that Montreal civil security director Jean-Bernard Guindon underscored. "After the ice storm, the minister of public security seized the momentum and proposed to the National Assembly a draft bill concerning civil security." The new law approaches civil security systematically, transcending the previous ad hoc fashion. It "enables risk analysis in the territory to be done in an integrated way, to make an action plan for all vulnerabilities related to these risks, and to consult the population. We will have to implement the actions and be held accountable to the minister and municipal authorities." Laws and procedures concerning vulnerability reduction and disaster preparedness will be similar to those regarding environmental protection and fire prevention and will be included as specific dimensions in the overall urban development plan for the city of Montreal. This will require huge investments.

Many of these social changes would not have occurred without the shock of this extreme weather disaster or a similar one. But they would also not have occurred without the active intervention of leaders and/or the population. The enhancement of disaster mitigation and of preparedness was developed from the combination of the prompt from nature's construction of intense, persistent freezing rain and the way that leaders and the population chose to respond. The interaction effect between biophysical dynamics and socio-cultural processes is important in explaining changes in beliefs and practices concerning risk. This conclusion raises a significant issue. Can social change to mitigate creeping environmental problems occur when there is no prompt from an acute disaster, and if it can, how can such change occur? As Montreal's Mayor Bourque stated, without a disaster, people live without stress. So why should they stress themselves by thinking about risk and its mitigation when they are not forced to do so by a disaster? The next part of this book will explore this question in terms of the slow-onset problem of global climate change that has not yet produced disasters but could cause them in the future.

PART FOUR

Learning for a Future with Global Climate Change

CHAPTER 11

Preparing to Avoid Disaster or Preparing for Disaster

The extreme weather event of January 1998, producing intense, persistent freezing rain over a wide area, resulted from unusual warming caused by El Niño. Such ice storms affecting any one particular metropolitan area in North America are improbable, even under global climate change, but they are possible and not necessarily infrequent over a large number of cities dependent on long electrical transmission lines, as the insurance industry concluded.[1] Since many European cities are farther north than Montreal and are dependent on the Gulf Stream for their warmth, they too could be vulnerable if the Gulf Stream is rendered less effective as a conveyor of heat. The electrical grid could also be crushed in many metropolitan areas by hurricanes or cyclones, which recent research suggests are becoming more intense because of global warming.[2] If such warming results in rising sea levels, heat waves, drought, and more intense extreme weather, defences will be put to the test. This freezing rain disaster exposed the vulnerability of even modern developed societies. Although it was not caused by global warming, it could well be a harbinger of hazards that will occur in that emerging climatic state.

THE RISK OF GLOBAL CLIMATE CHANGE

There are four fundamental questions concerning global climate change. First, is it occurring – that is, is it a real, material risk or merely a standalone discourse without a physical referent? In the early 1990s skeptics denied it and claimed that it "may exist only in computer simulations."[3]

Second, is climate change the result of only nature's cycles, such as changing energy coming from the sun during solar cycles, variations in cosmic rays, and modifications in the Earth's orbit? These have affected climate throughout our planet's existence. Or are human activities, particularly greenhouse-gas emissions, adding a new dimension to global climate change?

This second question then leads to a third: Will climate change have a harmful or a beneficial effect? There may be winners from global warming: some Canadians and Russians are already looking forward to warmer winters. There will, however, likely be a great many losers. Most of humanity does not yet experience the consequences of climate change, which appears to them a remote possibility. But the Arctic Inuit already see melting of the permafrost and reductions in the ice floes upon which their way of life has been based.

Take one example of the mixed effects global warming can have. The Arctic ice cap is melting. This change is opening up the Northwest Passage in the Arctic to shipping, reducing by 7,400 kilometres the distance for cargo between western Europe and the Far East compared to the present route through the Panama Canal. The melting of the Arctic ice cap by global warming resulting from fossil-fuel emissions also increases the possibility of extracting more fossil fuels and thereby accelerating global warming and doing even more damage. This process intensifies the risk of environmental degradation in the Arctic, which is already occurring for other reasons: toxic materials are already carried by air and water currents from distant industries to this sensitive region. Oil spills become a threat as exploration, extraction, and shipping come into being. These activities would especially occur if the passage were to be declared international waters in a way that prevented any country from enforcing environmental protection laws. Such risks create the need for legislation and policing, but which countries have jurisdiction in this region? These possibilities and dangers have already led to tensions between Russia and Norway on the European side of the Arctic Ocean, between Canada and the United States on the North American side, and between Denmark and Canada.

There is a fourth fundamental question, and it is the one that concerns us here. Should societies practise business as usual, adaptation to global climate change, mitigation of it, or both of the latter? Adapting to the new situation and preparing to cope with these adverse changes would include preparing for the more frequent and/or more intense hazards, and hence disasters, that may result from global climate change. The response could instead be an attempt to diminish activities that result in global warming. Hazards could be mitigated by reducing greenhouse-gas emissions. Or an attempt to both adjust to climate change and diminish it could be made. Or neither could be done, and instead we could go full speed ahead, with increased demand for fossil fuels resulting in more drilling for oil and more use of coal.

These political options are related to the general theoretical perspectives presented in chapter 1. If leaders believe that nature can be socially reconstructed by the market and technology in a timely fashion, then business as usual will be their decision. Or they may conclude they have no choice because modern society is on a treadmill of production and consumption whereby economic growth is the only possibility. The scientific evidence of potential harm resulting, paradoxically, from the successes of modernization could, on the contrary, prompt leaders to take into account environmental risks and fundamentally modify market practices, regulations, and taxation in order to modernize ecologically rather than suffer market failure in ecological terms. Or perhaps leaders could suggest other ways of reflectively dealing with the risk of global warming. How do they frame their discourse concerning this emerging danger produced by modern risk societies?

Arriving at a scientific consensus about whether global warming is occurring, whether human activities cause it, which specific activities are the causes, and whether it will have adverse or advantageous effects on humans and other species is a difficult task for science to achieve because the questions are so global in scale and require projections into the future. Since science is fallible but self-correcting, it is possible that the answer it gives at one point in time will prove wrong later. And we cannot forget that scientists are human, influenced like other people by cultural predispositions and economic interests. The latter do not predetermine findings, but they do incline scientists toward some problems and answers (hypotheses) rather than others. Moreover, scientific consensus does not mean scientific unanimity. This characteristic is beneficial for the development of science because dissenting scientists raise new questions and can propose new explanations, which can then be either disproved or substantiated. Dissenters to a scientific consensus create the possibility of planting the seeds of a scientific revolution, which is rare but part of the process of scientific development.[4] Thus achieving a scientific consensus is a complex matter.

In addition, there is a second-order level of complexity. The acceptance of a scientific consensus by political leaders and the broader population as a basis for decisions and for modifying existing social practices is not straightforward or unproblematic. A scientific consensus can be rejected, and the theories of scientific outliers (often referred to as "contrarians") can be accepted if they fit better with existing cultural and lifestyle predispositions.[5] Scientific consensus that implies changing our consumer habits, and perhaps even sacrificing some of them, is not as comforting as the theories of scientific contrarians which reassure leaders and the population that they can continue business and consuming as usual. Dissenters who lie outside a scientific consensus can attract unwarranted attention in the media and thereby mislead the population and political leaders. This distraction can politically paralyze the attempt to use science as a basis for improving the interaction between humans and their natural environment. It is extremely rare that dissenters to a scientific consensus create a revolution in science. Most often they are simply wrong. The scientists who claimed they had found a revolutionary new process of "cold fusion" illustrate this point. But the existence of scientific contrarians enables political leaders to cherry-pick whatever science suits their predispositions: consensus science or contrarian science.

Despite the difficulty for science to come to a consensus about the enormous question of global climate change that involves a projection into the future, decisions must be made by non-scientific leaders. Even doing nothing amounts to a decision when alternative courses of action are available. So how do political and emergency management decision-makers analyze, interpret, and make sense of such a problematic situation? Leaders have to make sense of scientific findings within a broad context of political, economic, and lifestyle priorities in order to arrive at decisions. They are not scientists, but their understanding of these scientific issues is particularly important because of the positions they hold or have held. Their interpretations speak to the issue of whether the best available science will be used to benefit society in its decisions or whether science will be locked away in the ivory tower when it brings troubling news.

The United States has become the largest greenhouse-gas emitting country in the world because of its population size and high rate of emissions per capita. Despite similarly high per capita emissions, Canada escapes such notoriety only because it has a tenth of the population. Both these countries have had high per capita rates of emissions for a long time and are wealthy, so development to escape poverty cannot be used as a justification. The political and emergency management leaders interviewed for this study live in the belly of the fossil-fuel–emitting beast and in a region that has experienced an extreme weather disaster as a result of warming but not warming caused by human activities. How do they frame their interpretation of the issue of global climate change and what to do about it? By virtue of their experience in their positions and in managing this disaster, can they provide helpful insights into the related but more general issue of global climate change? It is important to learn how these key decision-makers view issues such as these about a phenomenon that has been called the most fundamental change of our time.

There is a widespread assumption, even among researchers, that transforming the taken-for-granted modern beliefs of invulnerability to nature's forces will require the experience of disaster.[6] Have these key decision-makers been prompted by their experience of an extreme weather disaster to adopt an unequivocal concern about the risk of global climate change? Or do they dismiss that risk and the possibility of another extreme weather event as highly improbable and unworthy of serious preparations for the rest of the century? There is a current of thought which argues that disasters alarm the population into being overly concerned about highly improbable possibilities and incite politicians into wasting money and time preparing for unlikely events.[7] From this point of view, the proper response is to suppress the social scare and moral panic because they consist more of paranoia than reality and to return quickly to business as usual. Do the political and emergency management leaders interviewed for this book interpret warnings about climate change as false alarms, as climatophobia, or perhaps as distractions from more urgent concerns?

If they judge there is a real physical problem, do these leaders nevertheless perceive proposed solutions such as the Kyoto Protocol as doing more economic harm than climatic good? They are the ones who must decide what actions to take in a context of incomplete evidence, uncertainty, vested interests, lifestyles that have become habitual, and culture.[8] Some authors view climate change as a technical problem with a technical solution: for example, sequestering greenhouse gases deep underground so that they do not enter the atmosphere. If this could be done, not only oil sands and shale but also the massive reserves of coal could be used in a more sustainable way.[9] Do these decision-makers frame the question in technical terms? Or do they view the issue instead as a political one requiring imaginative leadership to overcome resistance to mitigation and/or adaptation? Or do they see it as an ethical issue whereby non-emitting societies suffer as a result of the greenhouse-gas emissions of uncaring emitting societies?

DEFERENCE TO SCIENCE

The leaders interviewed here had to arrive at their own interpretation of risk and of science and make sense of the various affirmations about climate change in order to decide what to do. Maine's governor, Angus King, gave a response typical of the other leaders. He perceived a scientific consensus but not unanimity, which nevertheless provides a basis for action. "You try to go on the science, and the science, the majority view of the science, is clearly in the direction that there is a human contribution to this problem that's significant." None of these political and emergency management leaders denied that climate change was occurring, that human activities were at least partly responsible for it, or that it would have harmful effects. They all deferred to more knowledgeable scientific experts concerning the issue but relied on their own understanding of science for their conclusions. They were willing to change their mind if scientific conclusions altered.

PREVENTING GLOBAL WARMING OR ADAPTING TO IT

Despite risks resulting from global warming, consumers and taxpayers are reluctant to spend more money for prevention and protection. For example, people in wealthy countries are very attached to their automobiles, which are one of the causes of global warming, and people in developing countries make automobiles one of their first purchases if they become more prosperous. Factories pump out a large proportion of the greenhouse gases that go into the atmosphere, and their owners threaten to close the factories, with resulting unemployment, when they are required to reduce emissions. Implementing the Kyoto Protocol is running into stiff resistance, but climatologists argue that it is only the first small step in stabilizing greenhouse gases in the atmosphere. Should we continue to practise business as usual in the hope that climate change and disasters will not occur and then try to deal with those calamities when they happen? Or should we try to prevent the conditions that lead to climate change and disasters? What emphasis should be placed on mitigating global warming, and what emphasis on adapting to climate change? Both threaten to be very costly endeavours. Even if global climate change were a purely natural phenomenon, with human activities having no effect, societies would still have the difficult and costly task of adapting to it. How do political and emergency management leaders make sense of a situation where scientific evidence is suggestive but not definitive – that is, a situation in which there are large areas of uncertainty? It is often difficult enough to make decisions concerning what is foreseeable, but how are decisions made for something that is difficult to foresee and is contested? On what basis do leaders make practical decisions in such a context.

Maine's Governor King referred in the following terms to the issue of global warming and climate change: "I think it's real. Something's happening, and we do need to be prepared. Next time it might not be an ice storm. It may be a hurricane or extreme cold. You can't afford to be totally prepared for an ice storm, or totally prepared for a hurricane. You have to have your preparations proportionate to the risk. But there is a risk and a somewhat higher risk than it might have been twenty years ago."

Quebec's deputy minister of public security, Florent Gagné, was pessimistic about the possibility of mitigating climate change, and so he placed his emphasis on adapting to it. "For prevention, nothing much can be done, but much can certainly be done to prepare [for climate change]. We will certainly experience this kind of event more frequently. Hence our civil security organizations must be sharpened. Unfortunately, that is not happening, or just a little."

Governor King was not as pessimistic about the possibility of prevention and placed more stress on it. He stated that fossil-fuel use should be reduced to prevent global climate change from happening. I reminded him that some people claim such a reduction would be disastrous for the economy and asked him whether we should let global warming happen and just increase disaster preparedness. His answer: "No, no, no. We should try to do something about it. I suspect, and I am not a scientist, that carbon dioxide accumulations and the greenhouse effect has and will affect climate."

The president of the Montreal Urban Community, Vera Danyluk, was well aware of the obstacles to the prevention of greenhouse-gas emissions in a society of mass consumption. "Leaders, elected officials, and people who have authority should prepare for the possibility of disaster. Besides that, we need leaders who are taking preventative measures. We have to work on these two fronts: have in place plans for disaster situations, but also work hard with our young people on ecology and environmental protection." Disaster preparation is part of adaptation to global climate change. Ms Danyluk did not assume that such adaptation can be so successful that mitigation of climate change would be unnecessary. Mitigation and adaptation should not be treated as mutually exclusive. She judged that a two-pronged strategy which includes both would be the most promising.

André Brunelle, assistant director of the Montreal Fire Department, also argued in favour of prevention and not just adaptation. "In a city like Montreal public transportation has to be valued. It is also necessary to build cars which generate less pollution. We need to work at various elements to succeed in reducing global warming." This comment is not surprising coming from a fire chief, since fire departments not only deal with extinguishing fires but have also shown leadership in fire prevention. Mitigating global warming can be seen as an extension of the same preventive logic on a planetary scale.

Governor King concluded that we will be forced to live with the degradation of the atmosphere we have caused by our activities because nature takes time to restore itself. "If we did reduce greenhouse-gas emissions today, it would take fifty years to dissipate what's already up there. You can take steps like carbon dioxide reduction, but in the meantime you've got to assume we're into a cycle of extreme weather." Reducing fossil-fuel emissions now would not alleviate global warming immediately because nature's processes need many decades to eliminate the carbon molecules already released into the atmosphere. Hence attempts at preventing greenhouse-gas emissions have to be complemented by efforts to prepare for the harmful effects of global climate change.

Jean-Bernard Guindon, the director of the Center for Civil Security for the Montreal Urban Community, went further by pointing out that adaptations to climate change and even preparations for disasters are necessary precisely because efforts at preventing climate change have been so feeble. "The Kyoto Accord is a minimum. It should have been even more rigorous than it was. Organizations dedicated to that purpose remain marginalized and have only a limited influence. We are stuck having to live with the consequences of these climatic extremes, and nature's mechanisms will be extremely difficult to reverse. So it is necessary to prepare in the meantime and prevent as much as possible."

These leaders concluded that the Kyoto Protocol is at best a very small step toward reducing greenhouse-gas emissions enough to decelerate global climate change. Since the internationally agreed upon reference year of 1990, such emissions have increased in most countries instead of decreased, especially in the United States and Canada. This outcome has led some commentators to claim that many governments have capitulated over greenhouse-gas emissions, particularly the American and Canadian federal governments. Their lack of action in turn makes adaptation, including risk assessment and disaster preparation, even more important. Precisely because organizations and mechanisms devoted to preventing greenhouse-gas emissions are underdeveloped, investments in preparing for global climate change have to be greater. Cutting costs in prevention increases them for preparation. Modern societies are faced with the chronic problem of trying to prepare for the harmful consequences of nature's disturbances they have exacerbated.

CLIMATE CHANGE IN GEOLOGICAL PERSPECTIVE

Mr Guindon nevertheless situated this development as a new level of danger superimposed on existing cycles of nature's dynamics. "When we look at the Earth's history, we see that there has been eternal cycles of climatic extremes and calm periods. There have been worse periods than this even without the effects of all the pollutants human activities have now made." That is true, but during those worse periods in planet Earth's history, there were not six billion humans, with most living near oceans or deserts exposed to powerful forces of nature. The development of human societies, especially since the industrial revolution, has put more humans in harm's way and has increased the potential harm from nature's hazards, a situation that constitutes a double jeopardy of vulnerability.

Governor King believed that the additional warming effect resulting from human activities must be restrained, but he too provided a reminder of unpreventable cycles of nature that interact with human activities. "We shouldn't assume that, if we did attend to global warming, that would relieve us of the danger of global climate change because it just doesn't work that way." King gave the following example of climate variability not caused by human activities: "Ten thousand years ago, where we are sitting was covered with two miles of ice. That's not all that long ago. There have been enormous changes in global climate throughout the history of the world having nothing to do with people." There are cycles of nature that modern societies will have to adapt to because they are beyond the control of even modern technology.

Nevertheless, neither King nor any of these other leaders used nature's cycles as an excuse to avoid reforming human activities that unleash additional perturbations of nature. These decision-makers had a nuanced understanding of the interaction between social constructions and constructions of nature within which contemporary climate change occurs. They rejected oversimplified assertions that attribute climate change only to human activities or solely to nature's constructions.

THE DILEMMA OF FOSSIL-FUEL DEPENDENCE AND GREENHOUSE-GAS EMISSIONS

Dependence on fossil fuels is an important cause of the environmental problem of global climate change, much as dependence on a centralized electrical grid was an important cause of the ice storm disaster. Both are infrastructures upon which modern societies rely. The United States and Canada both have large land masses, which were sparsely populated until a century ago, and they developed after the invention of the automobile. The private automobile enabled the mass choice of single-family homes and resulted in urban sprawl, which in turn renders public transportation uneconomical because of low population density and makes the automobile almost necessary. Choices at one point in time constrain subsequent choices.[10] A vicious circle of fossil-fuel dependency and greenhouse-gas emissions has developed that is difficult to escape even with the best intentions, and intentions are not always the best.

There is likely more consensus among scientists about human activities causing global warming than there is about any other issue that involves a prediction about what will occur in the future. The overwhelming weight of scientific conclusions, as indicated not only by the International Panel on Climate Change made up of scientists whose disciplines are most directly connected with the issue[11] but also by the national scientific academies (in some countries referred to as royal societies) of the United States, the United Kingdom, Germany, France, Italy, Sweden, Japan, Brazil, India, and China, is that global climate change is occurring; that it is caused by human activities, especially greenhouse-gas emissions; and that it will have many adverse effects. The Academy of Sciences of the Royal Society of Canada concluded that "the scientific evidence demands effective steps now to avert damaging changes to the Earth's climate."[12] The Millennium Ecosystem Assessment compiled by scientists arrived at the same conclusion. The Kyoto Protocol is the initial cooperative agreement among nations to reduce greenhouse-gas emissions. It was constructed on the template of the successful Montreal Protocol to control CFC emissions, whereby wealthy modern countries agreed to take the lead and provide an example for poor developing nations to follow, and the latter would join in at a second stage.

The leaders interviewed here live in wealthy nations that produce the highest per capita greenhouse-gas emissions on the planet and in a region that has experienced an extreme weather disaster as a result of warming. How do they frame their interpretations of the issue of global climate change and what to do about it? Governor King responded to these issues in the following way. (It is important to note that he was one of only two Independent governors in the United States at the time of the ice storm disaster, neither Democrat nor Republican.) "My understanding of the Kyoto Protocol is that there were problems with it, and it wasn't only the Bush administration, but the majority of the US Senate wasn't too interested in it. My concern is that the administration isn't seeking an alternative and is not taking the issue seriously. Global warming is seen as an ideological issue by American conservatives. They don't believe it: if you believe in global warming, you're a wimpy liberal. That is nonsense to me. So I don't necessarily criticize the administration for not signing Kyoto. I do criticize them for not seeking another alternative and for emphasizing drilling as the solution to our energy problems, which I think is very short-sighted."

So what are the alternatives to drilling for more oil? Governor King stressed conservation and renewable energy. "There is so much energy wasted in this country. Without a hell of a lot of discomfort to anybody, we could save a huge amount and develop alternatives. The only energy policy seems to be greater subsidies to the fossil-fuel industry to produce more, which is only a short-term solution. We ought to be doing a crash thing on hydrogen, on wind power, on conservation, on a kind of standardized nuclear option. There is lots that can be done, but not much is being done."

King emphasized that the high level of greenhouse-gas emissions in the United States has a great deal to do with the denial of climate science by American conservatives, and he came back to this issue again later in the interview without being questioned. "Global climate change has become a litmus issue for the American conservative movement, which is what is now in control of the United States government," he observed. "They don't believe it. It's become a matter of ideology as much as science. Anybody who is objective would say the vast majority of scientific opinion concludes that global warming is real and that humans are causing it. But for some reason the Rush Limbaughs of the world just don't want to believe that. It's become an issue of political correctness on the right. If you believe in global warming, you're a wimp. You're a tree hugger."

Talk-show hosts such as Rush Limbaugh have trumped scientists in influencing many Americans and the American government.[13] Conservatives dismiss conclusions about greenhouse-gas emissions causing global climate change as social scares and moral panics without a physical referent; they believe that only in computer simulations do human activities cause climate change, not in the real world. Lay non-scientific knowledge has many forms, and this talk-show form influential in the United States has led to denial of the risk of greenhouse-gas emissions. It is based on an unwillingness to change and on extrapolations of safety from the recent past, rather than on scientific knowledge. In his parody of combative conservative talk shows, the American satirist Stephen Colbert coined the term "truthiness" to describe their content. He defined truthiness as "truth that comes from the gut, not books."[14] The phenomenon has become sufficiently common that the American Merriam-Webster dictionary has accepted truthiness as a neologism. Greenhouse-gas emissions have been legitimated by truthy emissions from the gut on talk shows that have dismissed the best available knowledge.

Talk-shows hosts are but one set of spokespersons for the American conservative movement, usually the working-class variant. Studies have documented how conservative think tanks joined with climate change contrarian scientists and the fossil-fuel industry to defeat the Kyoto Protocol in the American political sphere, despite growing scientific evidence and widespread concern among the American population.[15] These groups have successfully developed consciousness-lowering strategies concerning this environmental problem.[16] Science has been brushed aside by the lay truthiness of conservative media and political elites. It is noteworthy that Governor King and the other leaders interviewed here were more influenced by consensus science than by contrarian science, which was not the case with the Bush administration in the United State.

I asked Governor King what the solution to global warming was. "The markets to some extent are taking care of it," he responded. "Gas has gone from ninety cents a gallon to a dollar ninety-five in a couple of years, and there is a nine-month wait to get a Toyota Prius. People are making those decisions on their own based upon the current price." It is true that until the recession the price of gasoline had risen everywhere in the world because of increasing demand, depletion of easily accessible supplies, and political conflicts in oil-producing countries. But gasoline in North America remains by far the lowest taxed and the cheapest in the developed world and per capita greenhouse-gas emissions remain the highest. If players in the market are not steered away from emitting greenhouse gases – by government regulation and/or the taxation system – they will continue to seek the easiest and lowest cost for transportation, electricity, and other uses, even if that involves externalizing costs that will have to be borne by future generations or by innocent victims in distant localities who make negligible use of fossil fuels. Market competition bestows an advantage on the lowest price and the least costly procedure for mass production and transportation, and hence on the lowest common denominator of environmental practices, unless governments set environmentally friendly rules and taxation practices for all competitors. Why should a manufacturer go to the expense of developing fuel-efficient vehicles if consumers buy gas guzzlers because the price of gas is low? Vehicles such as the Prius have not originated from American manufacturers in part because the tax on gasoline in North America is lower than in Japan. The market has not been successful at replacing high-emission technologies by low ones under the present rules of the North American market contest, which give free rein to emitting greenhouse gases into the atmosphere.

Montreal's mayor, Pierre Bourque, had a less sanguine view of market competition and of supply and demand as a solution for environmental problems. "Others say that man will progressively consume less per individual, gas is to be better refined, and then bi-energy. We are now supposed to be reducing the size of larger cars. But all this isn't certain, and I, too, am worried. Everything is tied to energy." Market competition by itself cannot be relied on to solve environmental problems such as global climate change. In 1977 it was claimed that market innovation would result in the widespread use of hydrogen fuel-cell vehicles emitting only water vapour in just thirty years. Now thirty years later, that transition is still "just thirty year away" despite three decades of increasing indications of the dangers of global warming, pollution, and smog.

Governor King went on to describe a cultural trait of the United States: "America has a deep strain of 'let me do what I want.' The history of environmental laws in the US is: identify major polluters and go get them, but don't bother me with my SUV. It's very easy to clean up a river if you can go to six paper companies and lay a law and tell them to clean it up. It's much harder if you have to go to everybody and say, 'You've got to do something about your septic system or something about your car.'" King is here describing a very fundamental aspect of both American and Canadian culture. Even leaders who acknowledge that fossil-fuel emissions are causing global warming and who want to solve the problem feel they must do so only by regulating big companies. Whether it is the Republican Arnold Swartzenagger in the United States on the right or the New Democratic Party in Canada on the left, the proposed solution is to identify fossil-fuel companies and regulate them, not to increase taxes on fossil fuels. However, King's hypothesis of individual resistance to environmental regulation but corporate compliance is too general. When American citizens were provided with regulations and an organized structure to practise recycling, they usually responded enthusiastically. But when regulations were proposed to eliminate CFCs, which deplete the ozone layer, and sulphur dioxide, which produces acid rain, corporations fought them tooth and nail. The oil and automotive industries are now at the centre of resistance to measures that would reduce greenhouse-gas emissions.

Economists who are concerned about climate change and other environmental problems argue the market is so powerful that environmental problems will never be solved unless the incentive structure of the market is rationalized ecologically.[17] Hence they contend that the total cost of commodities must be reflected in their price so that consumers, when purchasing according to price, are also purchasing the least environmentally costly commodity. Externalities distant in time or space can be incorporated into the price by shifting the tax structure, which could be focused on taxing "bads" (pollution) and thereby reducing taxes on "goods" (income) so that the change would be revenue-neutral and taxes would not increase overall. If the use of fossil fuels has harmful consequences now or in the future, then taxes on them could be increased in order to diminish their use and to provide capital for alternatives.

What are the possibilities of shifting the tax burden from goods to bads in the United States; that is, reconfiguring the taxation system to a polluter-pays approach? I reminded Governor King that one of the reasons for the high use of fossil fuels in North America is that the taxes on these fuels are so low. Americans and Canadians pay ten and five times less gas tax respectively than inhabitants of the United Kingdom and have few tolls on roads.[18] With gas being relatively inexpensive, there is no incentive to use alternatives that conserve fossil fuels or to develop technologies that accomplish this. Then I asked King if Americans would accept to have gas taxed as it is in Europe. His answer was categorical: "No, no. Everybody that's ever mentioned it has disappeared."

North Americans have become so accustomed to relatively cheap gas that political leaders do not dare suggest a carbon tax no matter how convincingly economists argue that it is the least costly and most flexible solution to greenhouse-gas emissions. Almost all leaders in North America perceive as political suicide a policy of taxing gas as European countries do. The Green Party of Canada proposed a 12 cent per litre tax on gas, which would amount to about a 10 per cent increase in the price, to be offset by a corresponding reduction in personal and payroll taxes so that it would not be a tax grab by government.[19] That party acts as the conscience of Canada concerning issues such as global warming, but it has no seats in Parliament or any illusions that this proposition will attract a tidal wave of popular support. Leadership depends not only on the qualities of the leader but also on the culture of the population that is led. In this sense, the population gets the leadership it votes for and deserves, for better or for worse.

King added that in the past the United States was "very lucky that we had some really good leadership in this country, and we did a lot of work on cars. But it's no accident that we haven't upped the requirement for the fleet average gas mileage in ten or fifteen years." He is undoubtedly correct to argue that leadership on environmental issues has had its ups and downs. The United States has had periods of excellent environmental leadership in the past, but the George W. Bush administration has constituted a particularly poor period. The *New York Times* reported in 2004 that auto-emission standards in China will exceed the American norm by 2007 unless the United States improves its standards.[20]

FOSSIL-FUEL DEPENDENCE, CLIMATE CHANGE, AND WAR

Despite the fact that Governor King claimed the market was taking care of fossil-fuel reduction to some extent, he nevertheless made an innovative suggestion concerning how fossil-fuel emissions in the United States could have been reduced by using the taxation system to redirect the market at the time of a historic opportunity. "One of the things that bothers me about the war in Iraq is that the president isn't asking us to pay for it. There is no sacrifice whatsoever being made, except by the families of those people that are there. If the president really thought it was critically important to foreign policy, he could have announced a victory tax on a gallon of gas, which would pay for the war, help solve our budget problems, and lead to lower dependency upon oil because it will induce greater fuel economy. He didn't do that. I think the American people will respond if asked. But nobody has asked."

President Bush did not propose such a victory tax for fear it would undermine American support for his invasion of Iraq, which was attractively packaged to the population as a victory that would be swift, decisive, and cheap. Even President John F. Kennedy in his celebrated phrase "Ask not what your country can do for you; ask what you can do for your country" was not requesting something as unusual for Americans to accept as a higher tax on gas. But Governor King is probably right. Such a victory tax to pay for the struggle against terrorism, reduce the United States' dependence on foreign oil, and diminish greenhouse-gas emissions would have likely had a good chance of being accepted by the American public in the unique circumstances after the terrorist attack on the World Trade Center, when Americans felt threatened. But nobody invited them to take up that challenge. Instead, the war in Iraq has been financed by driving the United States into debt, especially to China.

King then explained how a victory tax would have operated. "Three years ago, we could have imposed a fifty-cent-a-gallon gas tax. Chances are it would have held the price of fuel down, and people would be paying the same they are today. What we have essentially done is allowed OPEC to collect the tax. If you're paying two dollars a gallon, why not have fifty cents of it go to Washington instead of to Kuwait? But we've chosen to let it go to the oil companies and the oil-supplying nations, which is not good policy. Plus it distorts our foreign policy. We can't deal straightforwardly with Saudi Arabia because of our dependency upon their oil. So we have to be extra nice to the royal family and all those things which may in the long run not be in our best interest. If we were less dependant upon oil, we'd be more free to act in ways that might be better for our country."

Saudi Arabia was the country of origin of eighteen of the nineteen terrorists who carried out the 9/11 attacks. It was the nation where the leader of al Qaeda was raised and where much of the financing for that terrorist organization came from. It is the source of some of the most virulent preaching of anti-Western hatred. Hence the "niceness" of the United States government to Saudi Arabia after the 9/11 attacks is indeed perplexing, and it likely does have much to do with American dependence on Saudi oil. A superpower has rendered itself super-dependent.

The oil tax described by Governor King could still be implemented because the American debt continues to grow and the war goes on. However, the tax would now have to be labelled with a name somewhat less triumphant than "Victory Tax" because the invasion and occupation of Iraq have led to a multitude of perverse consequences that were foreseen but dismissed by American and British leaders: the death of many tens of thousands of Iraqi civilians, the incitement to terrorism both in Iraq and on a global level, the emergence of Iraq as a training ground for terrorists, a decrease in the standard of living and security of ordinary Iraqis, the pushing of Iraq to the edge of civil war, the Iranization of much of Iraq, and the death of thousands of American soldiers. Such a tax was not asked of the American people by the government in Washington after the 9/11 attacks. So a high level of fossil-fuel emissions, American dependence on Middle Eastern oil, deference to the undemocratic regime in Saudi Arabia, and growing American debt to finance the war in Iraq continue unabated. A multi-faceted opportunity for creative leadership in unique historical circumstances was missed.

GREENHOUSE-GAS EMISSIONS AS AN ETHICAL ISSUE

Fossil-fuel production and use, with the resulting greenhouse-gas emissions, threaten those who produce and use it. For example, Louisiana and Texas produce and use great quantities of fossil fuels. Their principal cities of New Orleans and Houston are threatened if greenhouse-gas emissions result in ocean level rise and increased intensity and/or frequency of extreme weather events because those cities are at sea level and are located in a hurricane-prone region. However, the consequences are not only local. On the contrary, greenhouse-gas emissions have a planetary and lasting effect. Carbon dioxide molecules released into the atmosphere remain there almost a century and affect areas around the globe distant from their place of origin. This phenomenon creates many ethical dilemmas because future generations and present populations with miniscule levels of emissions could suffer the harmful effects of a degraded environment as a result of the activities of current emitters. The atmosphere has become a medium through which some groups by their actions hurt other groups of innocent bystanders distant in space or time and do not cease their hurtful activities even though this effect has been known for some time.

This ethical issue was expressed most starkly by the Montreal Urban Community's Jean-Bernard Guindon. "We now have awareness on a planetary scale, whereas in the past that planetary consciousness did not exist. We have the means to act according to that planetary awareness. It is socially irresponsible not to do what is necessary – for example, to implement the Kyoto Protocol, which is in my opinion a super-minimum.

And when I see countries like the United States that refuse, I think that is a crime against humanity. We accuse other countries of crimes against humanity. But that is a crime against humanity because we refuse to take the means that could protect future generations from catastrophes. The Kyoto Protocol is the long-term way of preventing natural disasters."

Mr Guindon pointed to the United States. He probably did not mention Canada because at the time of the interview the Canadian government had agreed to the Kyoto Protocol and was trying to develop means to implement it that were being opposed by the fossil-fuel industry. Since then that government was defeated because of a scandal unrelated to the issue of climate change and was replaced by a political party that claims the Kyoto Protocol is impossible to implement. The new government halted the admittedly timid emission-reduction measures that had been introduced and/or planned and then procrastinated, implemented weak alternatives, and shifted the reference year from 1990 to 2006, which had 33 per cent higher emissions. Not only the United States but also Canada and Australia are culpable for not supporting the international treaty to diminish greenhouse-gas emissions, and other wealthy countries are at fault to varying degrees.

Guindon drew attention to the generational effect: greenhouse-gas emissions produced by humans who use large amounts of fossil fuels threaten to cause harmful changes to the environment that will be experienced by future generations, including their own descendants. He is correct to emphasize this consequence because for most of humanity the adverse effects of greenhouse-gas emissions constitute a problem for the future, even though practices in the present are causing it. But harmful consequences are propagated not only in time but also in space. In the Arctic a perverse effect is already occurring within this generation. The way of life of the Inuit, which is based on hunting and fishing, is being destroyed by the melting of the ice floes needed for such activities. Thawing of the permafrost is also resulting in the collapse of their homes and buildings. The Inuit produce negligible greenhouse-gas emissions. Rather, they are innocent bystanders being victimized by activities in distant regions of the world. The production and consumption practices of citizens in the prosperous south, who demand fossil-fuel–powered transportation, manufacturing, and air conditioning, have degraded the environment needed by innocent bystanders in the high north. Technology has shrunk the world into a global village, not only in the positive sense of innovative transportation and communication technologies diminishing the time it takes to cross space but also in the negative sense of the injurious effects of pollution being transmitted globally throughout the atmosphere.

Risk has been seen as positive in business, mountain climbing, and other activities, whereby the adventuresome choose to take the risk of loss of their own property or even bodily injury in order to attain a goal of profit, prestige, or satisfaction. These groups who decide to take a risk for themselves can rightly be called risk-takers. This voluntary behaviour is very different from the actions of risk-makers who foist risk on others who did not choose to place themselves in a dangerous situation. Harmful effects have been imposed upon the Inuit by risk-makers in the south.

The Inuit are but one example of innocent bystanders who will be affected by the fossil fuels used in other countries. Island nations and poor, low-lying countries such as Bangladesh risk inundation by rises in the ocean level resulting from greenhouse-gas emissions, despite the fact that inhabitants of these countries produce relatively few emissions themselves. The ethical implications of harm done to some nations or peoples by the activities of other nations, or to future generations by the activities of the present generation, have yet to be acknowledged and dealt with. The burden of ethical decision-making is inescapable, both by key decision-makers and by the population. No technical imperative can be used as a shield to ward off such issues.

Ethical problems then become legal problems. Companies selling or/and emitting fossil fuels are giving themselves a legal problem similar to that of cigarette companies. Lawyers for the Inuit are already drawing this parallel. They have launched a lawsuit because of the destruction of the Inuit way of life as a result of greenhouse-gas emissions.[21] A petition to this effect has also been placed before the Human Rights Commission of the Organization of American States in the name of the Inuit of Canada and Alaska.[22]

Bob Chiarelli, the elected chair of Canada's national capital region and a lawyer, stated that greenhouse-gas emissions are "already moving towards an economic and social disaster in the north, where the traditional hunting and fishing areas are being lost because of the loss of ice. It is an ethical issue and certainly a political issue. It's a public policy issue of the highest priority." He then explained that the courts can be used to develop public policy and public mores even when the chance of success in litigation is small: "the group that is going to be subjected to even a five per cent possibility they would lose the case sees that the cost – politically and financially – would be enormous, as it was in the tobacco situation. It's that type of cutting-edge, in-your-face legal issue that can bring action on the part of governments."

There is, in addition, the ethical problem of rich countries that have industrialized using fossil fuels and have been emitting greenhouse gases into the atmosphere for the last 250 years telling poor countries they must not follow the same path. Since carbon dioxide molecules released into the atmosphere remain there for over half a century, the important question becomes that of which countries have been responsible for greenhouse-gas emissions over the past half-century? The answer is clearly industrialized countries, with the United States being the one that has produced most greenhouse-gas emissions. Countries that have emitted in the past cannot now convincingly claim amnesia concerning the source of the carbon dioxide molecules they have already put into the atmosphere. The responsibility to develop a more environmentally sustainable technological path for other nations to follow is those countries that have become wealthy as a result of their past emissions. Hence it did not make sense for the United States to use Chinese emissions as an excuse for not ratifying the Kyoto Protocol. The issue is whether wealthy countries, in particular Canada and the United States, will set a good example for the developing world or a bad one.

Moreover, will the population in wealthy countries and oil companies accept a carbon tax to finance adaptation in poorer ones to the consequences of greenhouse-gas emissions coming from rich countries? At the United Nations meetings on climate change in Montreal in 2005, poor countries requested financial aid from richer ones to enable them to adapt, but their request was denied.

Emissions per capita of emerging emitters, such as China, India, and Brazil, are still far below those of the United States and Canada. Their emissions have increased in order to enable some of their people to escape poverty, whereas North American emissions have grown because of sales of heavy, gas-guzzling sports utility vehicles (SUVs), recreational vehicles, recreational trucks, and muscle cars (such as Mustangs and Camaros). In the rhetorical struggle, developing countries use per capita emissions as the standard for blame (Canada and Luxembourg high, China and India low), whereas developed countries use per country emissions (China and India high, Canada and Luxembourg low).

The genius of the Kyoto Protocol was that it fixed a gold standard reference, namely, a year. Future emissions were to be reduced with respect to emissions in 1990. Comparing a country with itself allowed for differences among countries: big ones versus small ones, hot ones versus cold ones, populous ones versus sparsely populated ones. But however logical a protocol, its success depends on whether leaders and citizens of a country are willing to implement it. European countries such as the United Kingdom, Sweden, Denmark, Norway, and Germany have been successful in reducing their greenhouse-gas emissions using the 1990 benchmark as a reference point. The United States and Canada benefit from the sacrifices and costs to accomplish this reduction made by these European countries, but they refuse to pay the price of reducing their own emissions. Hence it is difficult to avoid the conclusion that Canada and the United States, as well as some other wealthy nations, have been free riders prospering on the backs of others that either have paid to control their emissions or have suffered the consequences of North American emissions. Since high-emission structures (such as urban sprawl and dependence on sour-oil extraction) have become embedded in Canada and the United States, it will be a major challenge for leadership in these two countries to implement the ethical policies required to solve the problem of global climate change and for the population to choose and support enlightened leaders on this issue.

It is nevertheless true that the effect on the environment does not depend on who emits greenhouse gases. The atmosphere distributes emissions throughout the world. Populous developing countries are increasing their fossil-fuel emissions at the fastest rate, with China having recently overtaken the United States as the world's biggest emitter, and they will likely suffer the most from their own pollution. Hence developing countries, especially big ones, have to be brought into the system for restraining greenhouse-gas emissions. Emissions could be capped in a way that allows more space for developing countries as low per capita new emitters than for developed countries that have been long-term emitters.

SOME FACTORS PARALYZING THE MITIGATION OF GLOBAL CLIMATE CHANGE

One important set of factors obstructing the mitigation of humanly caused global climate change consists of the type of physical problem itself.

Physical Contingent Factors

Perhaps the most significant physical contingent factor is that global climate change involves primarily delayed-action effects. There is a decades-long time lag between human causes and tangibly harmful consequences for most of humanity. Climate change creeps up with adverse effects that are very slow-onset, though it may become sudden if a threshold is reached beyond which the world's climate tips into a different steady state. The populations in the countries responsible for most emissions continue to experience generalized well-being. The incremental character of the increase in risk, as well as its globally diffuse rather than concentrated quality, diminishes the motivation of the population and of leaders to treat it as a matter requiring immediate action, much like disaster preparation. The tendency of both economists and the population to discount the future results in a downplaying of the danger of global climate change. Well-being in the present tends to beget lack of concern about risk for the future because most lay knowledge consists of a simple extrapolation of present trends into the future. The absence of immediacy of physical consequences leads to political and socio-cultural neglect of the problem.

In the 1980s the United States was at the forefront of many environmental protection measures, such as the introduction of catalytic converters on automobiles and the effort to stop depletion of the ozone layer. European nations were resistant to phasing out CFCs and were unwilling to replace them with a less destructive technology. The United States was pushing them to sign the Montreal Protocol.[23] But with global climate change the roles have become reversed. Are greenhouse-gas emissions a peculiar issue that has to be differentiated from the others, or have there been recent cultural changes in both the United States and Europe concerning environmental issues?

Part of the answer to this question consists, as explained earlier, in the rise of a conservative ideology in the United States dismissive of environmental problems in general and of humanly caused climate change in particular as wimpy, liberal social scares and moral panics. But there are other factors, including physical ones.

Concerning Europe, Governor King raised the issue of geography and the dependence of that continent on the Gulf Stream for heat. "If the Gulf Stream shifted, England and Scandinavia would be uninhabitable. If you look on the map, Scandinavia is at the same latitude as Hudson Bay. It is the Gulf Stream that makes those places livable at all. People in North America forget how far north England is. London is on the same latitude as north of Halifax." Here Governor King is understating how much further north those European regions are than North American regions. Stockholm, Oslo, and Helsinki are all at 60° latitude, as far north as the northern part of Hudson Bay in North America and much further north than any North American city except small communities in Alaska and the Northwest Territories. London is at 52° latitude, much further north than Halifax at 44°. Many of Europe's great cities, such as London, Amsterdam, Brussels, Paris, Berlin, and Zurich, lie farther north than the Montreal-Ottawa-Maine area that was struck by this event of intense, persistent freezing rain. Montreal is at the same latitude as Venice in Italy. The Gulf Stream is one of the massive infrastructures provided by nature that brings heat to northwestern Europe. If it becomes less effective as a conveyor of heat because of global climate change, as some scientific research indicates,[24] then Europe could experience severe cooling and ice storms like the one that struck northeastern North America in 1998.

It is precisely the counterintuitive and complex phenomenon of global warming causing cooling of some regions that has prompted scientists to use the term "global climate change" rather than "warming." So I asked Governor King if he concluded this is one of the reasons why Europeans are much more concerned than North Americans about the risk of global climate change. "That could be part of it. If I were in England or Norway or Sweden, I would want to know the condition of the Gulf Stream every day."

He added another physical contingent factor that prompts Europeans to be more sensitive to the harm caused by fossil fuels than North Americans. "I think in part it is because they're more densely populated. They have more people in a closer area, which means they're more sensitive to these kinds of things." Urbanization after the invention of the automobile has, on the other hand, led in North America to urban sprawl, low population density, dependence on the automobile, and hence an opposition to any measure that would make fossil fuels more expensive.

These physical contingent factors have, however, complex and variable effects. The low population density, more open spaces, and lack of dependence on the Gulf Stream for heat are important contingent factors in the formation of beliefs of invulnerability to global warming in North America. But this continent is rapidly developing a high population density. The sprawl of its metropolitan areas results in more open spaces, but its urban highways are now subject to gridlock. Physical contingencies must be taken into account in any explanation of cultural differences, but they are not determinant. North America, too, has its physical reasons for fearing global climate change. The southern coast of the United States is much more subject to hurricanes than is Europe, and recent research demonstrates that global warming is rendering hurricanes more intense and hence more destructive.[25] Both continents have great coastal cities that would be vulnerable to rising ocean levels and storm surges: New Orleans and Houston as much as Amsterdam and Venice. Global warming could increase desertification as much in the Canadian prairies and the American Midwest as in Spain and intensify wildfires in California and the boreal forests of Canada as well as in Portugal.

The NIMTOO Syndrome among Leaders

The time lag between human activities that cause climate change and its harmful consequences has particular importance for leaders. Will they mitigate climate change or follow a "business as usual" scenario? Hubert Thibault, chief of staff for Quebec premier Lucien Bouchard, gave an insightful answer: "Not exactly business as usual. But one feels, when we read scientific texts about climate change, that its harmful consequences won't be before the end of the government's mandate. It will be in fifty years. And even that is in some ways contested. When we see how President Bush of the United States dares to question even scientific studies, there is no fear climate change and its adverse effects will occur in two years."

A leader assumes, probably correctly, that the destructive consequences of climate change will not occur during his or her mandate. It is a short distance from that perception to the conclusion "So I need not be bothered with it. If I as leader make the decision to introduce measures of restraint, such as a revenue-neutral carbon tax or preventing drilling for oil in nature reserves, I will become unpopular. Yet if there are benefits from my decision, then some later political leader will reap those benefits instead of me because my mandate will have ended." This "not in my term of office" (NIMTOO) syndrome acts as a powerful brake on decisions that involve a time lag between decisional pain and visible gain. It is the equivalent for leaders of the "not in my back yard" (NIMBY) syndrome among the population. Thus US president George W. Bush has rejected polluter-pays carbon taxes, cap-and-trade systems, atmospheric temperature increase targets, greenhouse-gas limits, and timelines. Instead he has just wanted to talk about the issue until his term of office ended. Canadian prime minister Stephen Harper has similarly refused to implement those effective measures during his mandate and instead has cultivated the faint hope that technological innovation without polluter-pays tax inducements in the market will reduce emissions over the long haul. They will both be long out of office by the time the oceans rise, so why should they worry?

In 1988 Conservative prime minister Brian Mulroney correctly hailed the International Conference on Atmospheric Security's call for action on climate change as "an important first step." In 2007 Conservative prime minister Stephen Harper greeted the G8 consensus to convoke another meeting to seriously consider the possible reduction of greenhouse gases by 2050 as "an important first step."[26] At that meeting in 2007 the rigorous caps and methods proposed by the Europeans were rejected by Canada and the United States in favour of more talk. If after twenty years of mounting evidence of risk, Canada and the United States are still taking their first steps, an observer would be forgiven for concluding that these two North American slow learners are so hesitant to take a second step that they might never learn to "walk the walk" on restraining greenhouse-gas emissions.

Climate change creeps up slowly and incrementally, and it threatens to cause irreversible problems before there is a shock. So it requires preemptive action whose benefits will occur well beyond the mandate of any office-holder. The very character of global climate change – namely, the long time delay between causal human practices and harmful climatic consequences – hinders an effective response. It does not determine that there will be no response; rather, leaders and a population disinclined for socio-cultural reasons to take the measures necessary to mitigate greenhouse-gas emissions sense no urgency to tackle the problem. Proactive future-oriented decisions in such a context require extraordinarily courageous and visionary leadership. That such leadership is not abundant is an important reason why very little pre-emptive action to avert slow-onset climate change and the risk of disasters is being undertaken by political decision-makers in North America.

The Interface between Multiple Levels of Government

The chair of the Canadian capital region, Bob Chiarelli, pointed out a serious structural problem faced by well-intentioned municipal political leaders. He argued that cities "can't accelerate our stance in public transit because we don't have the income base in the cities to access the money to build those transit systems that would take people out of cars. So we need a federal government and a provincial government that are prepared to work with the cities to build transit systems. We have, as a country, missed the boat. There's a lack of leadership provincially; there's a lack of leadership federally." A fiscal imbalance exists between municipalities, on the one hand, and provincial and federal governments, on the other. Important and expensive consequences from environmental and disaster problems are experienced by municipalities, but the financial resources to deal with them remain with the federal and provincial governments.

The president of the Montreal Urban Community, Vera Danyluk, also related the problem of paying for disaster preparedness and climate change to the broader issue of financing the infrastructure for metropolitan areas. "The biggest cities in our country should never depend solely on revenue from property taxes to ensure that those cities are able to deal with major risks such as these." Only about 12 per cent of the taxes collected in metropolitan areas go to municipal governments (in the form of property taxes).[27] The remaining 88 per cent go to provincial and federal governments (in the form of income, corporate, and sales taxes). Hence municipal governments rely on special projects from higher levels of government to access the taxes paid by their own residents in order to undertake major infrastructure, disaster preparedness, and environmental work. The very limited basis of municipal financing places severe constraints on what even the best-intentioned municipal leaders can do. It is a structural weakness of governance in Canada that hinders the introduction of remedies for both disaster protection and environmental problems.

Ms Danyluk argued that the federal government should have the responsibility for ensuring that the largest metropolitan areas have an adequate infrastructure. From her position in Quebec, she confirmed that this proposal runs up against the opposition of many provinces, especially Quebec, which do not want what they define as federal interference. Although Canada has its own peculiar constitutional tensions, it is likely that this structural weakness in the tax-collecting powers of different levels of government also hampers attempts to reduce vulnerability and deal with environmental problems in the United States and many other countries.

GLOBAL CLIMATE CHANGE: ADAPTATION, MITIGATION, BOTH, OR NEITHER?

Whether the climate is changing was a debated question in the 1980s and even the 1990s. A decade later, scientific analyses, as well as visible phenomena such as the melting of glaciers, Arctic ice floes, and permafrost, leave little doubt that there is a physical referent underlying talk about climate change.[28] Moreover, there is now strong scientific consensus,[29] albeit not unanimity,[30] that human activities are affecting the climate of the Earth. Those activities consist especially of greenhouse-gas emissions from fossil fuels used in transportation and factory production. Nature's cycles still occur, and in addition, human activities have become inadvertent catalysts and accelerators of them such that a tipping point could be reached by which the world would be thrown into a new type of climate.

Whatever is revealed in the future about the overall net effects, there is little doubt that climate change will have many harmful consequences. The melting of permafrost by global warming is already releasing mercury into the water.[31] If the massive land-based glaciers of Greenland and Antarctica melt because of the greenhouse effect, all the oceans are predicted to rise 7 metres if only Greenland melts, 5 metres if only Antarctica melts, and 12 metres if they both melt.[32] The result would be disastrous flooding of low-lying coastal areas, including many cities, islands, and impoverished countries such as Bangladesh. In other regions, climate change is expected to bring drought, intense and frequent heat waves, forest fires, new forms of pestilence, greater desertification, and diminished snowpacks and glaciers crucial for summer water supplies.

Particularly worrisome is the time frame. If climate change occurred over thousands of years, adaptation would be facilitated. But studies are now confirming that when a threshold was reached in the past, the climate was tipped into a different type of state in a matter of decades.[33] This upheaval would likely occur with human-unleashed climate change as well because a positive feedback loop would begin whereby melted permafrost and warmer water over the ocean floor would release even more greenhouse gases into the atmosphere, water would absorb more heat from the sun than the ice it replaced because ice and snow reflect solar radiation back into space, and so on. The physical consequences of climate change would be slow-onset and then sudden. If major climate change occurs in a short period of decades, then adaptation would be very difficult and in many cases impossible. Imagine the problems of implementing a timely response to ocean-level rise for London, Amsterdam, Tokyo, New Orleans, Houston, and Miami, not to mention the predicament of highly populated coastal areas in poor countries with little means to construct defences or to adapt.

An important issue concerning the relationship between human activities and global climate change is that of using the best available knowledge to assess risk. Knowledge about nature's dynamics is provisional, but some knowledge claims tend to be more accurate than others. The talk-show form of lay knowledge based on choosing contrarian science over consensus science concerning climate change has been influential in federal government circles in North America. It legitimizes the business-as-usual approach of increasing greenhouse-gas emissions. Those who deny the adverse effects of fossil-fuel emissions are correct to claim that science is not a popularity contest, but they are wrong to cast aside the preponderant scientific conclusions and cherry-pick the work of the rare few who agree with their preconceived notions. Even if climate change had *net* beneficial effects for humanity (a dubious assumption) and even if it were caused only by nature's cyclical dynamics (an erroneous assumption according to the present consensus in science), timely adaptation and increased disaster preparations would be necessary to deal with the adverse consequences. If climate change is caused at least partly by human activities, then decisive action could mitigate or slow down that part of it, thereby providing more time to adapt.

Scientifically, it may be premature to conclude that global warming will produce more intense or frequent extreme weather. But waiting for definitive proof of danger leads to inadequate preparations. Fact-based science is not limited to disaster-based science. Science can produce facts that provide indications of danger before disasters occur. Dismissing such facts renders science impotent to anticipate and avoid disasters. Furthermore, discontinuities and surprises from nature will likely become more frequent as a period of global climate change begins. Hence there are many reasons to err on the side of caution.

The first three questions at the beginning of this chapter now have unequivocal answers. But what about the fourth? In a political context where science is but one of many influences, will leaders choose to respond by attempting to mitigate greenhouse-gas emissions, by adapting to the new normal, by both of these methods, or by neither – that is, by business as usual?

In order to guide their own decisions and practices, the leaders in the fast-track societies on the greenhouse-gas emitting treadmill interviewed here had to make sense of the scientific evidence and distinguish bogus risk allegations from real threats in a socio-political context of limited resources and high expectations of consumption. They concluded that climate change is occurring, that it is at least in part due to human activities, and that it will have serious harmful consequences. None had confidence in business as usual or, by implication, in the "magic bullet" theory that human intelligence will necessarily invent technological solutions as soon as they are needed. Hence they concluded that both adaptation and mitigation must be undertaken, and that it is only prudent to prepare for climate change and try to reduce activities that could contribute to it. These leaders sought to diminish activities that could unleash more of nature's disturbances. Their views, nevertheless, were nuanced, and they differed in emphasis. Some expressed the greenhouse-gas emission issue in ethical terms, as a crime against humanity. All were worried that current governments in North America were constructing a mitigation and adaptation deficit concerning greenhouse-gas emissions. One high-ranking political leader admitted the importance of the NIMTOO syndrome, which functions as a brake on enacting policy that may be unpopular and not produce visible gain in leaders' term of office.

The leaders in the management of the extreme weather disaster of January 1998 concluded that the jolting experience of that event had influenced their views about risk. Their conceptions of risk and climate change were, nevertheless, also affected by their prior cultural and political dispositions and their economic interests. I do not claim that this disaster determined their views, especially about an issue as broad as global climate change, only that it is one of the many important contingencies, both physical and socio-cultural, that influenced their interpretations.

Since this study interviewed a limited number of leaders, it could not determine how typical their views are of leaders more generally. They are clearly not characteristic of the Bush administration in the United States and the Harper government in Canada. Disasters affect beliefs but are not necessarily transformative of belief systems. Deeply held attachment to beliefs[34] and to lifestyles can result in wishful thinking and dismissal of nature's indications of danger, even when the latter are visible to some lay people (through receding glaciers, melting permafrost, and other phenomena) and in scientific studies. Such dismissal was the reason why some societies collapsed in the past when they failed to recognize that they were degrading the natural environment needed for their survival.[35]

CHAPTER 12

The Acute and the Chronic

The extreme weather event of intense, persistent freezing rain associated with El Niño in January 1998 crushed the electrical grid and resulted in the most expensive disaster in the history of Canada and the state of Maine, affecting the most people in both those places. It just missed Boston. It was not solely a natural disaster or just a technological disaster. Rather, it was a hybrid techno-natural disaster resulting from both a hazard constructed by nature and vulnerability that was constructed inadvertently by humans. The building of a centralized electrical grid in northeastern North America and reliance on it for essential infrastructures, including heating in a dark, frigid winter climate, increased vulnerability to freezing rain and resulted in the manufacture of a particularly costly disaster. In the past when extreme freezing rain fell on regions where homes still used wood stoves or coal furnaces for heat and kerosene lamps for light and relied on human and animal power on farms and in cities, the outcome was not so disastrous.

Intensified human activities have led to a "new species of trouble" and have blurred "the line we have been in the habit of drawing between the acute and the chronic,"[1] between sudden disasters and slow-onset ones, and between disasters and environmental problems. "Things bite back,"[2] and one of the main reasons is because social constructions are in continuous interaction with the dynamic, emergent, infinite, and only partially understood constructions of nature both outside and inside the human body. Human activities generate greenhouse-gas emissions and unleash the ongoing problem of global climate change, which gives rise to acute problems among which more intense and frequent extreme weather events have been predicted. Climate change is but one among many environmental problems resulting from human activities, albeit a particularly important one. In this changing context of relations between humans and the natural world, how do the leaders interviewed for this study, who have experienced the acute crisis of a disaster, envisage the chronic problem of managing relations between human constructions and those of nature, of dealing with risk, and of preparing for the future?

PREPARING FOR YESTERDAY'S DISASTER

When do people tend to rush out and buy flood insurance? After a flood, of course. The most common way of preparing for the future is by extrapolating from recent occurrences. If the next dangerous event is similar, then the community will be better prepared than it was for this ice storm. Florent Gagné, the Quebec deputy minister of public security, described the importance that experiencing a disaster has for future preparations, but he also described how this response can be misleading: "After experiencing a disaster there is a tendency to prepare for the disaster that has just been experienced and not for the next one. The disaster we just lived through may not happen again for 150 years." He was not arguing that Quebec will be disaster-free for 150 years but, rather, that the source may be different: a hurricane, earthquake, epidemic, nuclear reactor explosion, or terrorist attack rather than freezing rain. Extrapolations that assume hazards will be similar to those in the recent past have been wrong. The next disaster could even be much worse. It is important to learn lessons from a disaster and to be better prepared for something similar, but a deeper lesson is that preparing for yesterday's disaster is often not sufficient.

By the end of 1997 lessons had been learned from previous blackouts, so the electrical grid was constructed to be the most robust in Quebec's history. Nevertheless, it was crushed by unexpectedly intense, persistent freezing rain in early 1998. The ice storm demonstrated that what could happen went far beyond what was expected based on occurrences since meteorological records have been kept. Worst-case scenarios are just that: scenarios. They are not necessarily the worst cases. Extrapolations from recent experience have some use, but they must be employed with caution. This is especially true as we enter a period of global climate change. Essential infrastructures such as the electrical grid have been designed according to robustness criteria based on extrapolation of meteorological data for the past sixty years or so. They are therefore designed to deal with nature's extreme disturbances over that period but not with global climate change. Thus there is an adaptation deficit if climate change is not now taken into account in decisions about robustness and resilience.

THE CLIMATIC ACUTE AND THE BURDEN OF CHRONIC PREPAREDNESS

The experience of the disaster of 1998 prompted analyses of the response to this extreme weather event with the goal of learning lessons from it and improving preparedness for disaster response and mitigation. The most elaborate analysis occurred in the place that suffered the most disastrous consequences, namely, Quebec, as explained in chapter 6. The acute experience of a disaster brings new perceptions of risk and of the chronic problems of emergency preparedness, mitigation, and adaptation. New laws and regulations to reinforce disaster preparedness were also introduced in Ontario as a result of this ice storm. General Earl Adams, Maine's commissioner of defence and emergency management, described how the freezing rain forced his state to enter into a cooperative agreement with surrounding states and Canadian provinces to facilitate mutual help in time of disaster. "It didn't take us long afterwards to get it. The northeast governors, Quebec, and the Atlantic provinces had a meeting that had been scheduled not too long after this. We prepared a cooperative agreement for Governor King to take there. The other states that were involved and the Atlantic provinces all realized the need to have that agreement, and it was enacted quite easily."

Asked whether he believed such a storm would not occur for another two hundred years and therefore no immediate preparations were necessary, General Adams responded: "Oh, no, no. We did some rather thorough after-action reviews of the ice storm, with the idea that it could happen again in the form of an ice storm or some other event that could be equally catastrophic. We refined many of our plans. When we put that new Emergency Operations Center in, we did it using experience we had during the ice storm."

Mayor Pierre Bourque of Montreal described the learning curve generated by the freezing rain as decision-makers went from unperceived risk to acknowledged danger: "We gave ourselves the technological means to make sure that if ever the water-filtration plant stops functioning, there would be a backup. That didn't exist before. We used to think that it was not necessary. Now we've learned better. We should have learned it before, but we were never able to imagine it. But that doesn't mean that nature won't hit us in a different way." This comment demonstrates the great difficulty of foreseeing risk before disastrous effects are experienced and raises a perplexing question. If it takes the experience of disaster for leaders to take seriously the risk of future similar calamities, how will they deal with the creeping, slow-onset phenomena that modern societies are unleashing by their activities, which could have disastrous effects that were never previously experienced? Greenhouse-gas emissions provoking global climate change are an obvious example.

Maine's governor, Angus King, admitted that when he was first elected, he saw disaster preparations as an area where cuts could be made if the budget was tight. Did the ice storm provoke a change in his perceptions? "Absolutely. The way to think of it is as an insurance policy," he responded. "You may not think you need insurance, but when you do need it, you really need it. They had prepared a structure all the way down to the smallest town. That's quite a logistical feat. You couldn't do that without prior planning."

Asked what he had learned from the ice storm, Hubert Thibault, the Quebec premier's chief of staff, specified many lessons on a general level. "First, one can never rule out a disaster. Second, it isn't achievable to prepare ahead of time one hundred per cent for all possible disasters. It would be too expensive, people wouldn't believe in the risk, etc. It is unrealistic to think that we can have plans that are perfect in every way when dealing with the worse. Third, the components of civil security have to be regularly reviewed, checked, and refined. Fourth, it is important to know how to count on our organizations and inspire confidence in them. We have to avoid the reflex of looking to blame someone on the second day of the crisis."

Is it necessary to experience a disaster in order to be motivated to prepare for a disaster? "There is a dimension of this that is true," Mr Thibault acknowledged. "There was nevertheless an organization before the ice storm crisis. But it's true that 'once bitten, twice shy,' as the proverb states. A disaster is excellent at inciting awareness concerning the need to be better prepared." Although freezing rain was not new to this region, dependence on a vulnerable electrical grid was indeed new and constituted a species of trouble that made the chronic burden of defending against acute disasters necessary.

EMERGENCY AND DECLARING EMERGENCY

There was a striking difference between Quebec, on the one hand, and the United States and Ontario, on the other, concerning the declaration of a state of emergency in response to a disastrous situation. Quebec did not declare a state of emergency for the ice storm; in fact, it has never declared a state of emergency in its history. States of emergency were declared in the United States in places where damage and disruption were much less. The Regional Municipality of Ottawa-Carleton in Ontario also declared a state of emergency during the ice storm, as did sixty-three other, smaller municipalities in Ontario. Is this cultural difference in dealing with emergencies significant, and has it changed as a result of the ice storm?

Jean-Bernard Guindon, director of the Centre for Civil Security of the Montreal Urban Community, answered that recent disasters such as the ice storm have resulted in a new law, adopted in 2001, by which a mayor can declare a state of emergency under specified conditions. "The justification that allows him or her to declare a state of emergency is precisely the need to exercise one of the eight powers that the state of emergency permits, which are extraordinary powers, one of which is evacuation; another is the power to requisition services forcibly if necessary. There are eight like that. If there was a huge disaster, but if the extraordinary eight powers were not needed, it would not be necessary to declare a state of emergency. A small-scale calamity could lead to a declaration of a state of emergency if the extraordinary powers were required."

Mr Guindon contended that the change was not merely to declare the obvious or to emotionally reassure the population that authorities have indeed recognized the obvious, but to grant special powers. The experience of this disaster and other lesser calamities prompted lawmakers in Quebec to make changes, but they responded in their own fashion in accordance with their history and culture. Hence there still remain cultural differences between Quebec and the United States in the declaration of emergencies.

CHRONIC SOCIAL OBSTACLES TO DISASTER MITIGATION AND PREPARATION

This acute extreme weather event made visible vulnerabilities that were not apparent under normal weather and thereby stimulated a motivation to diminish them. The desire to increase safety, however, encounters social obstacles.

The Cost of Mitigation and Preparedness

A particularly important impediment is cost. André Brunelle, assistant director of the Montreal Fire Department, emphasized that materially diminishing vulnerabilities is much more difficult than merely being aware of them and talking about them. "We learned a certain number of lessons, but solving the problems is something else. What was said should be done – namely, making the filtration plants completely independent of the electrical grid – wasn't done. This was because of the necessity of examining the cost-benefit relationship." One proposal to make the water-filtration plants self-sufficient in electricity consisted of paying an aircraft manufacturer to install giant electricity-producing turbines next to the plants, but they would be very expensive to build and would be a redundancy that would not be cost-effective in the weather normally present. The cost of disaster mitigation and preparation creates a divergence between what is desirable and what is done, with the less-than-desirable creeping back as disaster recedes to the back burners of memory.

The cost of disaster preparation in normal weather deterred some disaster preparations in the United States as well. The wishful thinking of "we can take care of ourselves in an emergency" on the part of several municipalities and counties was refuted by nature's dynamics when the emergency occurred. They then became free riders begging others for help. But such thinking returned when the cost of disaster preparation became known.

Risk Perceptions Diminished by the Experience of Normality

The president of the Montreal Urban Community, Vera Danyluk, pointed out how difficult it is for decision-makers used to normal weather to prepare for emergencies: "I will always remember how many times elected officials said, 'The Office of Emergency Measures – they don't do anything.' When budgets were being prepared, those officials used to say, 'Let's cut that. We don't need that.' But there was only one mayor who always said, 'It is our insurance policy.'"

Ms Danyluk stated that municipalities were left to decide what to do with money from federal programs, but often their leaders made decisions that were popular in the short run when all was well but unwise in the long term. "There were municipalities that took money for infrastructure and built swimming pools. Decisions were possible in the past that could have helped enormously in cases like the ice storm disaster, but unfortunately, they didn't help because bad choices were made." With no visible sign of disaster on the horizon and with other budgetary priorities, it takes foresight for a leader to go beyond a simple extrapolation from present conditions of well-being to a realization that a disaster could lie in the future and prepare for it.

Uncertainty

A particularly important obstacle to disaster mitigation and dealing in a cost-effective way with environmental problems such as global climate change involves the limits of human knowledge of nature's dynamics. Scientific forecasts of safety and rational risk assessments are valuable but fallible. The most elaborate report on this extreme weather disaster, written by a scientific commission, was entitled "Confronting the Unforeseeable."[3] How can the unforeseeable be confronted in a safe way? These leaders who have now experienced a disaster they did not foresee have developed an acute sense of the limits of extrapolations and of attempts to forecast hazards. Maine's Governor King argued that "the problem with preparing for disasters is you never know what the disaster is going to be."

Deputy Minister Florent Gagné emphasized the limitations of scientific forecasting of hazards. "Weather events can only be forecast for very short time periods: meteorologists excel in their predictions twenty-four hours before an event; forty-eight hours in advance their predictions are more or less accurate; but seventy-two hours before, it's nothing at all." He is certainly correct to draw attention to the limitations of even scientific assessments of safety and risk. It would, however, be a mistake to dismiss them as useless. There is also the paradox of science being able to predict long-term general phenomena but not short-run specific ones. For example, science has accurately predicted the presence or absence of earthquakes for particular regions within one-hundred-year periods, but it is unable to specify whether and where they will occur next week. Science can forecast that weather events will become more extreme if the oceans become warmer over the next fifty years, but it cannot specify when, where, or how the next extreme weather event will strike. Some proponents of business as usual raise the objection: How can science predict global warming for the whole planet in fifty years if it is not capable of forecasting the weather in Montreal three days from now? This excuse for dismissing scientific predictions of global climate change has no more validity than dismissing the connection between cigarette smoking and lung cancer because doctors are unable to predict who will develop lung cancer next month. It is important to comprehend the limits of science without ignoring its accomplishments.

Unfounded Assumptions of Far-off Recurrence

The need to invest in emergency preparedness runs up against the predisposition to believe that nature's extreme events are exceptionally rare, even when that belief has no scientific support. Hubert Thibault claimed that for similar freezing rain events, "experts indicated there is a recurrence once every ten thousand years perhaps or every thousand years."

In my research I could not find any scientific study that indicated such a lengthy period of recurrence for freezing rain of this intensity, duration, and scope. A few employees of the power utility speculated that it was a two-hundred-year storm,[4] but most meteorologists suggested a hundred-year storm would be a better estimate. They warned, however, that reliable data do not go back far enough to support even that estimate.[5] Labelling an event as a hundred-year storm is problematic for several reasons. First, as mentioned, the historical data needed to draw such a conclusion are not available. Second, climatological research has shown that global warming will transform hundred-year extreme weather events into fifty-year events, a shift which implies that everyone in a particular region will be hit by one of those in his or her lifetime.[6] Third, such a label is uninformative about the crucial issue for disaster preparation: the event's timing. Concepts such as a hundred-year storm, even if they were useful scientifically, are unhelpful politically because they can mislead leaders and the population into a false sense of security. They leave the erroneous impression that a hundred-year disaster will not occur for almost another hundred years. Rather, such an event means, on average, one within every hundred-year period, with tomorrow being as probable a date of recurrence as a day a hundred years from now.

FRAGILITY AND PRUDENCE

This extreme weather disaster taught these leaders an important general lesson well expressed by Mayor Bourque: "What was learned was that Quebec is fragile. We are always dependent and fragile." He was using the specific example of his own modern society, but he was implying that all societies, even the most advanced and developed ones, are fragile. They never transcend the processes of nature as much as their enlightenment and postmodern culture lead them to presume. These societies continue to interact with the dynamics of nature upon which they have been superimposed. This is true both in the beneficial sense of nature's dynamics providing essential support that humans themselves cannot construct artificially and in the threatening sense of nature's disturbances constituting hazards that can overwhelm even modern society's most robust constructions. The Montreal Urban Community's Vera Danyluk concluded, "That is the primary lesson that was learned: we are all vulnerable. There had been a tendency to compare ourselves to developing countries and conclude everything is fine."

The experience of a disastrous disturbance of nature that was not foreseen in risk assessments and that took these leaders by surprise led them to have an enhanced sense of the vulnerability of even modern, technologically advanced societies when they are confrontated with nature's hazards. It encouraged them to foresee the limits of rational foresight.

This disaster prompted these leaders to have not only greater knowledge of disaster but also, and more significantly, a greater awareness of the limits of predictions of safety. They now realize that nature produces unusual disturbances and discontinuities. They are more aware that there is a great deal about nature that even science does not know.[7] It is particularly important for leaders and the population to learn this lesson as an incentive to prudence at a time when global climate change is beginning.

ACCEPTING RISK AND ACCEPTING DISASTER

There is a major difference, though, between learning a general lesson about the vulnerability of even wealthy, modern societies and willingness to invest in costly defences and mitigation. How do leaders make practical decisions concerning vulnerability in a material context with so much uncertainty? Florent Gagné answered: "Preparations are made for events that happened in the past with a certain statistical recurrence. Preparations are not made for the worst because the historical situation shows that statistically it doesn't happen. Should we always be equipped for the worst? It would without a doubt be desirable, but it would lead society to immobilize money in statistical probabilities that are very small. These are very major questions."

Investments to enable the public infrastructure to withstand rare hazards compete with needs in health, education, the military, and other areas. So how would he and the population deal with these issues in the realm of decision-making so as to ensure safety? "A minimum must be done," Mr Gagné responded. "But we also know there is a limit to protection from risk. Let's not talk about the unforeseeable yet, just everything that is predictable. Should preparations be made one hundred per cent for everything that is foreseeable? If the population were surveyed, people would not necessarily respond yes because they would very quickly see the amount of resources that would require."

Gagné gave the following example. If a security tax was proposed to prepare for emergencies, people would probably say that it should be cut in half, which would provide some protection, and the rest of the problem would be dealt with when it happened. This approach amounts to accepting risk and hence accepting disaster in the hope that it can be diminished when it strikes.

He then described the present state of preparation for hazards. "Are we better prepared? Yes, we are a bit better than we were before the ice storm. Are we sufficiently prepared? The answer is definitely no." He admitted that luck plays a role in human interaction with the dynamics of nature. To what extent should leaders and society roll the dice, and to what extent should they be precautionary? Gagné concluded that "it would be unreasonable to prepare for the maximum. It would not be justified by history. It is unfortunately necessary to take a certain risk but try to minimize it by being well-organized and having a good deployment of resources when disaster occurs. We can count on a society that is well-organized to have the necessary means, namely, the rich society in which we live." He drew his conclusion before Hurricane Katrina struck a rich society that was supposedly well-organized. Even rich societies are vulnerable if they choose to accept risk. Andre Brunelle, of the Montreal Fire Department, also insisted that "it is necessary to examine the cost-benefit relationship [of emergency preparations]."

Financially, there is a zero-sum contest with other priorities to determine what will be spent on mitigation and adaptation. Hence there is no way of evading an assessment of the reality and probability of risk claims, fallible though those assessments may be. A cost-benefit analysis leads, however, to the concept and practice of acceptable risk. This concept implies accepting disaster on the assumption that it will occur infrequently. There is also an inherent bias in cost-benefit analysis applied to disaster preparedness: the "cost is immediate and obvious, the benefit is uncertain."[8] Costs are weighed heavily because they are tangible in the present, but benefits are discounted because they lie in the future and are unsure. Refusing to pay the cost of avoiding a worst-case scenario can result in much greater costs if the risk materializes. For example, it has been estimated that it would have cost $14 billion to erect defences in New Orleans against a category 5 hurricane. This worst-case scenario was judged too costly, so now more than $100 billion in damages and reconstruction must be paid after category 4 Hurricane Katrina struck.[9] Protection could have been built seven times over and the fatalities and suffering avoided had the necessary investment been made. Similarly, global climate change can be mitigated at relatively low cost if dealt with early, as many European countries are trying to do, but if the problem is left to fester as in North America, then the costs of a belated reduction in emissions are significantly higher. Although a cost-benefit risk analysis is unavoidable, the issue is whether the assumptions underlying it will be precautionary or reckless concerning potential calamities.

Maine's Governor King argued that "you can't afford to be totally prepared for an ice storm or totally prepared for a hurricane. You have to have your preparations proportionate to the risk. But I do think there is a risk and a continuing risk, and a somewhat higher risk than it might have been twenty years ago." There is a paradox here. The ice storm exceeded even the worst-case scenario at the time. It certainly prompted decision-makers to prepare better and to envisage even worse worst-case scenarios. But it did not lead them to prepare for worst-case scenarios, only to envisage them, because such scenarios are deemed too costly and too improbable to prepare for. Although the experience of disaster prompted a heightened perception of risk, this perception followed a "decay curve" as memories faded, costs of reducing vulnerability became clear, and other priorities competed with mitigation.[10] Risk is accepted, and so also is the suffering of disasters on the expectation they will happen rarely.

Central Maine Power president David Flanagan stated that his company was not constructing stronger transmission and distribution lines since the ice storm because "that was a peculiar event. If climate change results in more ice storms, that may call for some reanalysis and some new strategy. But based on the information available to us now, I think we're adequate. Decisions should be made on the basis of objectively knowable facts and not on speculation unless there are some demonstrable trends that are relevant and documented." The difficulty consists of obtaining objectively knowable facts about the future and deciding whether priority will be given to prudence in terms of avoiding disaster or in terms of avoiding cost. Assumptions of safety are as much based on speculation as are claims of risk. The issue is: Will the burden of proof be placed on the claimant of safety or the claimant of risk? Surely we do not have to await a disaster before admitting there are objectively knowable facts of danger. Fact-based knowledge should not be reduced to disaster-based knowledge. If objectively knowable facts about danger in the future are indicative but not definitive, as is the case with the scientific evidence of climate change, will that be a reason for business as usual or for proceeding with increased caution? Since there is much uncertainty that cannot be eliminated, the crucial issue becomes whether uncertainty will be used to justify driving society full speed ahead or with caution.

Leaders and the population are stuck with the unavoidable but difficult task of distinguishing between risks that are real threats, which must be attended to promptly, and those that are fictional threats – risks in name only. The issue is not to protect against the worst case that could possibly be imagined. Rather, it is to extend protection to events against which society has hitherto had little protection and to keep protection up against the dangerous new dynamics of nature that human activities are unleashing, such as global warming.

DEFENCES: ALL HAZARDS OR SPECIFIC HAZARDS?

The Quebec scientific commission that attempted to learn lessons from the 1998 ice storm entitled its five-volume report "Confronting the unforeseeable." But how can preparations be made for that which cannot be foreseen? Worse still, the context of uncertainty has increased with the possibility of global climate change. The chronic and the acute, the slow-onset and the sudden, environmental problems and disasters, nature's hazards and socio-technological vulnerability are now coupled together. Are preparations to defend against natural disasters, technological disasters, epidemics, and acts of terrorism sufficiently similar that the same organization, lines of authority, and investments can deal with all of them? Or are these dangers so different that organizations specifically dedicated to each are necessary?

The advantages of an all-hazards organization has been described as follows: "Mitigation and response planning activities may be totally ignored in areas where they infrequently occur. While it is wise to devote limited time and other resources to those events which have the highest likelihood of occurring, other disaster agents should not be totally ignored. An all-hazards approach may provide the greatest mitigation and response service to the community one is serving."[11] André Brunelle confirmed this very pragmatic political reason favouring an all-hazards approach. "If we only had to plan for an intense, persistent ice storm," he said, "we would say that it happens about never, so we wouldn't do it."

Any one particular major hazard occurs so infrequently that society would be unwilling to invest in protections against it. Multiple specifically focused organizations would be too expensive. But taken together and combined with less serious hazards that occur more frequently, such events constitute a critical mass that provides the basis for investments that are sellable politically. Thus it is the same emergency measures organizations and often the same people who respond to large disasters and small accidents.

Maine's Governor King expressed the thinking of many of the leaders interviewed. "I definitely think you could do it with the same organization. You'd have to have different capacities. For an extreme weather incident, you'd need shelters. For an anthrax attack, you'd also need a hazardous material handling capacity. It would be crazy and very inefficient and expensive to have separate structures: here is our natural disaster emergency structure and here is our terrorist attack emergency structure. That would be an unnecessary and expensive redundancy." Another reason in support of an all-hazards organizational structure is that the lessons learned from managing one – for example, the ice storm disaster – can to a certain extent be applied to the management of other crises, even terrorism.

The goal of effectiveness is, however, elusive. The weakness of an all-hazards organization was indicated forcefully to me in an interview with a community officer within Emergency Management Ontario who preferred to remain anonymous. "Ice storm '98 and the subsequent research showed conclusively that an all-hazards approach doesn't work," this person concluded. "A strict, basic, all-hazards approach – that is, a plan and an organization that are flexible enough to deal with whatever comes – is insufficient because they don't understand what they are up against. Hence they don't know how to manage their resources, and they will fail. We were doing an all-hazards approach in 1998, and we needed to go to the next level. If you have a full understanding of what you're potentially up against, then you can take steps to mitigate or prevent."

Thus this official advocated a risk-processing system, which he called comprehensive emergency management, based on the application of resources in accordance with attempts to understand hazards. "What does work is a hazard analysis to understand fundamentally what all the hazards are and then do the risk assessments. So it depends how you define 'all hazards.' From our risk-based scenario, we look at all hazards, yes. And then we assess what the impacts are."

This approach is heavily weighted toward risk assessment, with the potential impact of what could happen having more emphasis in disaster planning than the probability of occurrence. "You have finite resources – where are you going to put them?" the official commented. "You're going to put them where they are going to do the most good. You try to bring the risk down to zero and reduce the overall impact. But if there's nothing you can do to prevent or mitigate it, you'd better put your resources into response because sooner or later it's going to have a significant impact. So there's a lot of science behind this."

The difference between an all-hazards organization and comprehensive emergency management is not merely a question of semantics. All-season tires on an automobile do not function as well as winter tires in winter or as summer tires in summer. They are simply an economical compromise. So, too, an all-hazards organization does not deal as effectively with an epidemic as one designed with an epidemic specifically in mind, and the same is true of an ice storm, hurricane, flood, earthquake, heat wave, or other emergency.

There is another problem with an all-hazards organization. Politically, all hazards are not created equal in an all-hazards structure. The particular hazard that is on the political agenda tends to dominate the structure, regardless of objective probabilities of impact or occurrence, and this factor tends to weaken preparedness for other hazards. The threat of hurricanes striking the vulnerable below-sea-level city of New Orleans was of high probability, with catastrophic potential impact, yet the all-hazards structure of Homeland Security, to which FEMA was subordinated, was weighted in favour of defending against terrorist threats, with the result that its response to Hurricane Katrina in 2005 was abysmal.

Jean-Bernard Guindon, of Montreal's Centre for Civil Security, stated that his organization attempted to find a middle ground between an all-hazards approach and organizations dedicated to specific hazards. "Our choice is to have a generalist organization to tackle almost anything. That organization must nevertheless develop specific plans with respect to particular types of disasters. Seven types of disasters have been identified for which approaches are being prepared. Emergency personnel are being trained to prepare for the worst in these scenarios. So even though the response will be taken using a generalist structure, there would nevertheless be specialized groups with specific plans to intervene, and this would endow the larger structure with specific approaches to deal with what needs to be managed."

This synthesis of seven most likely risk scenarios within a generalist organization amounts to a wise use of scarce resources and is similar to the comprehensive emergency management approach, which emphasizes assessments of the potential impact as a guide for distributing resources and, secondarily, for assessments of probability of occurrence. Its effectiveness hinges on whether occurrences will fall within the seven risks that were foreseen. Even in wealthy modern countries, disasters often involve threats that were previously unforeseen, such as this ice storm, the terrorist attack on the World Trade Center, and the extreme heat wave in Paris in 2003, when caregivers for the elderly were on holidays. Even the most sophisticated risk assessments are fallible. The comprehensive emergency management approach is only as strong as its assessments are accurate.

PROPOSING COLLECTIVE PROJECTS AS A COMPONENT OF LEADERSHIP

Mayor Bourque of Montreal pointed out an additional element of quality leadership: "It will take in all societies, leaders who propose projects, who propose to their young people to work in the Peace Corps in Latin America, or exchanges with the rest of Canada or the rest of Quebec. A society cannot live without social projects where men and women get together and meet challenges – challenges of justice, challenges of equity, challenges of the environment, challenges of beauty."

So what did he do while he was mayor? Were these just nice words, or did he put his discourse into practice? Bourque was educated as a botanist. Before entering politics, he had been director of Montreal's Botanical Gardens. As mayor, "I had a complete concept for an eco-city. This eco-city was based on neighbourhoods. I wanted two hundred volunteers in each of the neighbourhoods. I subsidized $50,000 per eco-neighbourhood. Ecological education and tools were provided. The goal was to improve the quality of life all over the city."

The eco-centres promoted the three Rs: reduce, reuse, and recycle. The development of parks and bicycle paths, diminishing urban sprawl, and in general putting nature into the city were part of this project. Bourque contended that this approach leads to a healthier environment. "There is a new Montreal that is being greened and flowered and that works well. There is a demand for that. We need to stimulate urbanites to move towards gentle activities rather than towards hard activities. A city can be humanized by nature and the participation of people."

When hazard strikes, resilience is increased where leaders can call upon pre-existing voluntary associations that have already been created to deal with environmental problems. In this vision, disaster protection and mitigating environmental problems go hand in hand. Bourque concluded that this project was successful. "I do not think there are any other solutions than to give each citizen an ethic, a form of civic responsibility. There are not a lot of things that unite us in a society, other than the environment, the quality of life. We can demand that. In Montreal we have advanced towards that goal. But for that, it takes a form of leadership; it takes projects."

He put the problem of equitable development of countries this way. "I am in favour of sustainable development. I cannot say, 'We will remain rich and the other countries will remain poor.' That is a problem of universal conscience." He added that leaders with a universal conscience "are few and far between. Each country defends its bone; each country defends its oil; each country can make war for diamonds or gold. You can not even condemn people for doing it. I don't know where that will lead us. There is not much generosity in our societies."

Asked whether the responsibility for this attitude lies with leaders or the citizens who elect the government, Bourque said that both were responsible, and he pointed to the dilemma of any ethically oriented political leader. "I was defeated because as soon as I said the word 'share,' I lost votes. I was finished." In the interview he judged that environmental activists such as himself are now on the defensive. He was right. Humanizing the city through the participation of citizens in greening it proved to be just one of many competing priorities that were weighed against one another by voters. After one term in office as mayor, Pierre Bourque was defeated in two subsequent elections for the mayoralty of Montreal: once before the interview and once after.

MITIGATION: ALTERNATIVES TO FOSSIL FUELS

Mitigation of environmental problems, such as global climate change, is an important element of protecting against acute disasters, but it, too, involves ongoing and difficult challenges. For example, greater efficiency and reduced consumption of energy are essential ways of decreasing greenhouse-gas emissions and pollution associated with the use of fossil fuels. The consumption of energy is so great in all developed, modern societies, however, that even if efficiencies and conservation were optimized, there would still be a great demand for energy. Maine's Governor King reaffirmed the need to develop sustainable forms of energy before fossil fuels are depleted "because of the environmental effects but also because it's a finite resource. All we should do [with fossil fuels] is buy time as we develop the technologies that can give us the energy we need. I'm not one who says we should put on sweaters and quit using electricity. We need to develop additional technologies in order to meet that need without fossil fuels." Here he is describing the limits of conservation as a political strategy if it is seen as involving a permanent sacrifice, such as no longer using electricity and putting on more sweaters. So conserve as much as possible, but be aware that convincing people to permanently give up those comforts will face serious resistance.

Conservation of energy runs up against the unwillingness of people to reduce their use of rapid, door-to-door, private transportation, of labour-saving devices, and of climate controls in buildings. People in wealthy countries have an expectation of entitlement to creature comforts, and the poor in developing countries aspire to those comforts. As they become more prosperous, the automobile is one of their first purchases. Populous societies such as China, India, and Brazil now have the highest rates of growth in energy consumption in the world. Improved technical efficiencies should certainly be promoted, but efficiencies have tended to result in the perverse incentive that the more energy-efficient a technology, the more consumers believe they can use it. Energy use and emissions per unit of a commodity decrease, but not total energy used or greenhouse gases emitted because more commodities are produced and used. That is one reason why intensity-based targets typically fail to reduce pollution, greenhouse-gas emissions, and energy consumption. Where will all the energy come from?

Governor King expressed the issue bluntly: "There's a major wind project proposed in western Maine, and it's being fought to the death by the Appalachian Mountain Club because – heaven forbid – a hiker walking through the mountains would see a windmill and spoil his wilderness experience. These are the same people who would bemoan SUVs and fossil fuels and we've got to get away from global warming, but don't put windmills near my hiking trail. That absolutely drives me crazy. There are choices to be made, and every choice has some consequences. If you don't want to do fossil fuels, then you got to accept windmills and dams and maybe even nuclear power." Other goals, even environmental ones to conserve pristine nature, compete with the goal of developing clean, renewable sources of energy.

Wind farms can only be constructed where there is wind, and that is usually near oceans, lakes, and other picturesque sites. Wind energy becomes expensive if the farms have to be built far from populated areas where the electricity will be used. Powerful not- in-my-backyard (NIMBY) reactions, based on noise, diminished scenic views, bird and bat deaths, apprehension about blades flying off, and an underlying fear of lower property values, are now creating serious obstacles to the development of wind energy.[12] The most high-profile NIMBY campaign against a wind farm has been led by the Kennedy clan in opposition to 130 turbines off Nantucket Sound, where it owns an estate. Hence the NIMBY opposition to wind energy could be named the Kennedy syndrome: "What's mine is mine and what's yours is negotiable as a wind farm." Nantucket Sound may be a beautiful place, but many other places are also attractive in the eyes of their inhabitants. The term "Kennedy syndrome" is particularly appropriate because the rich and powerful are well positioned to impose their definition of beauty in order to exclude wind farms from their surroundings.

Pierre Bourque argued that hydroelectric power is an environmentally friendly source of energy in the present state of technology. Quebec has an abundance of rivers in the sparsely populated far north that have been dammed and diverted to produce electricity. Government leaders in the province chose this source of energy in the early 1970s over nuclear energy. It has proven to be an economical and safe option that enabled Quebec to become more energy self-sufficient than most societies and even to sell energy to its American neighbours for a handsome profit. Storing water behind dams is one of the few ways of stockpiling electricity, and it has been well used by Quebec to generate electricity when demand is greatest and the price is highest.

Nuclear energy is another alternative to fossil fuels, and it is now being aggressively marketed by the nuclear industry as a solution to global warming. But nuclear energy does not eliminate risk; instead it exchanges one risk for other kinds. Nuclear reactors have already caused a disaster (Chernobyl) and a near disaster (Three Mile Island), have resulted in the accumulation of radioactive waste, and are closely related to the production of nuclear weapons. Maine had a major nuclear power plant that was closed down in 1997–98 during the administration of Governor King. Why did Maine phase out nuclear? King answered that "this particular plant had some problems. It was a risk assessment issue for me. The plant was right on the ocean. Our biggest industry is tourism. A serious nuclear incident on the coast of Maine would have been catastrophic for the lobster industry, tourism, the whole deal. We had to be sure it was safe. I did press to be sure it was inspected and met the highest level of inspection standards." The thorough safety audit revealed problems that would have cost US$152 million to fix. The closure was not based on anti-nuclear ideology. Instead, it was founded on a risk assessment and on a business decision concerning the improbability of turning a profit after such a major investment.

ENERGY AND EXPECTATIONS

A prominent disaster researcher explained safety versus disaster in terms of the interaction between energy and expectations.[13] If expectations about nature's energy are correct, humans can usually prepare for the hazard and minimize damage and fatalities, thereby avoiding disaster. If such expectations are erroneous, whether because of ignorance, denial, or some combination of the two, then risky decisions are inadvertently made that render communities and organizations vulnerable as they confront nature's energy; hence the potential for disaster is high. Disaster occurs at the juncture of the socio-cultural and nature's biophysical dynamics.

Energy can be either in the form of a construction of nature, such as freezing rain or hurricanes, or in the form of a human recombination of nature's dynamics, as in an electrical grid or a nuclear reactor. Disasters often involve the confrontation of, first, artificial energy[14] that humans have constructed through a recombination of nature's dynamics with, second, energy that nature has constructed. That was the case in this ice storm: the electrical grid constituted the artificial energy for heating, lighting, and production in a frigid winter climate, and it was crushed by the energy of extreme weather originating in El Niño on the other side of the North American continent. Such confrontation is also the case when hurricanes rupture dikes, as Katrina did in New Orleans, and when earthquakes fracture buildings (people die not from being shaken but by having buildings fall on them). New confrontations with nature's energies are created when humans recombine nature in novel ways, such as taking carbon from the ground and placing it in the atmosphere by using fossil fuels, thereby unleashing global warming. This is the energy and hazard side of the socio-cultural–biophysical interaction.

The other side involves expectations. Safety results from appropriate expectations about nature's energy, whereas vulnerability results from mistaken expectations. In the latter case, safety was expected, and so preparations were inadequate where danger lurked. In some instances this mistaken expectation of safety was based on an unforeseeable threat, as in this ice storm. In other cases the expectation of safety amounted to recklessness and refusal to acknowledge the threat when scientific indications were available, as in the example of hurricanes striking New Orleans. Claims of unforeseeability in the latter case amount to more an excuse than a reason for expectations of safety. I would hypothesize that unforeseeability, although it still occurs, becomes less prevalent as science develops. Hence vulnerability is increasingly related to a refusal to acknowledge risk. For example, as scientific evidence develops linking greenhouse-gas emissions with global warming, expectations of the safe use of fossil fuels become more dubious, unforeseeability is increasingly an excuse rather than a reason, and business as usual turns into irresponsible human activity.

The acute problem of disasters is connected to the chronic issue of constructing accurate expectations about nature's energy, whether primal or recombined. Developing safe, abundant sources of energy to power both wealthy and developing societies is a huge challenge. Reliance on artificial energy sources such as the electrical grid requires that they be made safe. If the expectation of safety is faulty, the result can be far more disastrous than if such artificial recombinations were not used. This difference will be examined next in a comparison between modern societies, with their sophisticated risk assessment and preparedness to avoid disasters, and some anti-modern communities that were struck by the same intense, persistent freezing rain.

CHAPTER 13

Extreme Weather without Disaster: A Reminder for Moderns

Not only was the extreme weather event of January 1998 intense and of long duration, but it also had enormous scope. In Canada the freezing rain fell from Kingston and Ottawa in the west to New Brunswick and Nova Scotia in the east. In the United States it hit northern New York State, New Hampshire, Vermont, and most of Maine. The deluge created major disruption wherever it fell, except in one group of communities. Why did those communities not experience the same disaster suffered by others when they too were hit by freezing rain of the same intensity and duration? Our understanding of disaster and of vulnerability to a hazard of nature can be deepened by investigating the answer to that question.

One might suspect that the communities little affected by this extreme weather would be those of North American Native people, but that was not the case. Although not integrated into the labour market like other Americans and Canadians, the Native American community in the Mohawk territory that spans the Canada–New York border has also become dependent on electricity for its way of life. Far from rejecting modern technologies, this society embraces and relies on them. Oil furnaces, freezers, refrigerators, electric ovens, washers, dryers, televisions, telephones, automobiles, trucks, motorboats, and snowmobiles have all become thoroughly integrated into North American Native life. The tools Native people now use are entirely modern ones. Despite the importance of tradition in their discourse, their lifestyle is modern in its use of technology. Hence the intense, persistent freezing rain that crushed the electrical system became a disaster for the Mohawk nation that depended on it just as it did in the American and Canadian societies that surround that community. Mohawks too were deprived of heat, light, and other conveniences in a dark, frigid winter. The material advantages of an electrified way of life has for them, as for everyone else, come at the cost of a specific type of vulnerability to the forces of nature. Like Canadian and American societies, the Native American communities nonetheless had the advantage of integration into powerful technological and organizational networks for emergency measures, restoration, and redistribution of costs to the wider society. This linkage mitigated the consequences of the disaster and made the Native communities resilient.

There was, nevertheless, one group of communities living in the area hit by the freezing rain whose culture set it apart from the others: the Amish. In the mid-1970s Amish families from Ohio and Ontario moved into the rural region of northern New York to acquire inexpensive farm land and a setting propitious to maintaining their traditions and raising their children. They are a conservative Christian group best known for their rejection of modern social practices and modern technology, a refusal they see as originating in biblical strictures but which many historians claim has more to do with rural life in seventeenth-century Europe. Their culture emphasizes traditional values and an ascetic lifestyle. They live their daily lives without electricity, and they heat their homes with wood stoves. They spurn telephones, radios, and television and do not own or drive automobiles, relying instead on horses and buggies. The Amish are overwhelmingly farmers. Although they refuse to use modern farm machinery such as tractors and electric milking machines, they have a reputation for being excellent farmers.[1]

FREEZING RAIN WITHOUT DISASTER

As noted earlier, one group that suffered great stress during this extreme weather event was modern farmers. Their supply of electrical energy was cut when ice-loading from the freezing rain crushed the electrical grid. As well, the low population density in their rural areas resulted in those areas being at the bottom of the priority list to have power restored. Farmers lost not only electricity for their homes but also the energy they rely upon for their livelihood. Intensive farming is dependent on electricity to power machines to milk cows, pump water, remove manure, and other chores. Sufficient labour no longer exists to do these tasks by hand. Cows became sick and died when left unmilked for long periods of time. The big herds these farmers had taken many years to build were threatened. The same was true for chicken and pig farmers. The ice storm was far more disastrous for modern farmers than for most other occupations.

By 1998 Amish farms had become well-established in northern New York State. After they were struck by the same freezing rain as other farms in this area, the surrounding modern society, believing that the Amish were not as technologically and culturally advanced as they were, assumed that the Amish farmers would be particularly traumatized by the worse disaster in the history of the region. Rescue workers were sent to monitor the needs of everyone in the vicinity, including the Amish. What did they find? "When searchers returned from Norfolk, they reported that Amish families were almost unaffected by the storm."[2] The maple trees in the sugar bush tapped by the Amish to produce syrup had suffered some damage, as had other trees on their farms, but unlike modern farmers and urban dwellers, their lives changed little during the storm.

Accounts of the ice storm in letters from the Amish themselves, published in *Die Botschaft: A Weekly Newspaper Serving Old Order Amish Communities Everywhere,* confirm that the ice-loading caused little damage and had little disruptive effect on their communities. One Amish woman wrote: "Beautiful scenery when one looks out the window. Saturday night we had freezing rain and then one inch of snow on top by Sunday morning, making every weed and wire a winter wonderland. The beautiful works of God ... Hasn't gotten warm enough since to melt it all off yet. It is mostly just on the trees still."[3] There was much written about the weather in letters to this Amish newspaper: a great deal about the beauty of the landscape that resulted from the freezing rain, but nothing from these witnesses to suggest that harm or trouble had been caused to Amish communities by the freezing rain. "We have some beautiful winter scenery all around us the last few days," another woman wrote. "The ice on the trees, shrubs, etc. is all covered with sparkling snow."[4] The ice storm did not result in the collapse of the essential infrastructures of the Amish way of life. As in periods of average weather, those families used their labour to work their horses, milk cows by hand, and pump water. They lighted their homes with kerosene lamps or candles and heated them with wood stoves, on which they also cooked their food, just as in normal weather.[5] There was some disruption, but it was minor. "Boys had intended to go to Norfolk on Sat. to get more of our hay," someone else wrote, "but called ahead and found out they had lots of ice, so they postponed the trip."[6]

The Amish observed the problems of their non-Amish neighbours, who depend on electrical power and fossil-fuel–powered vehicles. "Week before last we had freezing rains which lasted for several days," a letter to *Die Botschaft* recalled. "It got worse enough that a lot of English people were out of power. This neighborhood didn't have electric for a week, also broke down trees. People were warned to stay off the road as much as possible. If they got caught without business they got fined for a $1000 or 6 mo. in jail. According to reports several got caught."[7] "English" is the term used by the Amish in the United States to refer to non-Amish people, whether they speak English, French, Japanese, or another language. This is because the Amish learn as children to speak a dialect of German often referred to as Pennsylvania German. When the Amish realized their non-Amish neighbours were in difficulty, they began to help. "Our neighbors came over to sit by the stove and warm up and drink water," said a letter in other paper. "The next day they got our generator and Dad's oil stove. After that they could make it without having to go to the shelter."[8]

A non-Amish journalist who visited the Amish during the ice storm made the following observation: "About the only people in the Canton area totally unaffected by the outage were a hundred Amish families, … choosing to live without electricity. While people all around them were busy staving off tragedy, the Amish continued sending their children to school and helped neighboring farmers milk cows [by hand] … They pronounced the event 'nothing unusual.' Another Amish friend … summed up the ice storm like this: 'Modern conveniences turned into inconveniences. Our inconveniences remained mostly the same.'"[9]

The Amish that I met during my research in the affected region (Heuvelton-Depeyester) of New York State were mainly descendants of the early settlers of Ohio. They did not gloat that their way of life had enabled them to avoid the disastrous consequences of the freezing rain suffered by their modern neighbours.[10] On the contrary, they regretted not realizing sooner that the "English" had no heat and could not milk their cows. When they became aware of the severity of the problems afflicting the non-Amish, they tried to lend a hand in any way they could. For example, they sent their sons to help milk their neighbours' cows and do other chores. During the disaster in this region, electrical power was replaced by Amish labour power. The disposition to help others that the Amish invoke in time of disaster within their own community was quickly extended to non-Amish neighbours. "The habits of care address all sorts of needs triggered by drought, disease, death, injury, bankruptcy, and medical emergency. The community springs into spontaneous action in these moments of despair, thus articulating the deepest sentiments of Amish life."[11] The whole community constitutes an all-hazards response organization. No emergency measures bureaucracy needed there.

Although the Amish were not much affected by the freezing rain within their own community, they were affected when they depended on the English. The collection of the milk they sell and the bread ordered from them, the mail pick-up and delivery that they too use, the buses they take for long distance travel, and other activities were all impacted by the freezing rain. Since the Amish refuse to use electricity from the power grid, they can not themselves process the milk they produce. So Amish farmers sell their milk to cheese factories run by the English, and these were affected by the collapse of the power grid. For example, the Heritage Cheese House in Heuvelton, New York, turns milk from Amish farmers into cheese for sale in a store at the front of the factory and at local stores in the vicinity.[12] Electricity powers the factory's machines, and the freezing rain knocked out that energy on Thursday, 8 January. The milk was temporarily cooled by placing the containers in cold water. On Friday some cheese was made in cauldron pots on Amish farms in order to use the milk before it went bad. Electricity came back on Saturday, so more milk was picked up from Amish farmers. Then the power went off again. On Sunday the power came back on permanently. Fortunately, in winter it was easier to keep the milk from spoiling than it would have been in the summer. The Amish use modern society in ways they choose, and because of this linkage, they were affected at the interface between their practices and those of modern society. Amish communities, nevertheless, were much less disrupted by the freezing rain than the wealthy, advanced modern society that encircles them.

What is to be made of this finding? It could be referred to as the protection paradox: Amish communities, which did not construct defences against disasters, proved to be less vulnerable to this extreme weather event than modern societies that invested heavily in monitoring hazards and preparing for emergencies. I would suggest that, in a deep sense, modern society can learn about itself and about broader aspects of the challenge of disaster mitigation by comparing the consequences it suffered with the much more modest ones Amish communities faced during the same extreme weather.

The experience of the Amish demonstrates that disaster resulted not from the freezing rain per se but, rather, from the vulnerability of the infrastructure that modern society had constructed and upon which it had become dependent. A calamity only occurred where society had become reliant upon a centralized electrical grid that was prone to collapse under the weight of ice. Susceptibility to disaster was manufactured, not natural. Freezing rain that was benign for the Amish became lethal for modern society. The striking contrast between the disastrous experience of modern communities and the normal life of Amish ones when confronted by the same extreme weather demonstrates that this was not a natural disaster. A non-disastrous disturbance of nature precipitated a technological disaster because of constructed vulnerability in modern society. By eschewing modern technology, the Amish averted a disaster that resulted from that technology's vulnerability to this type of extreme weather. This protection paradox leads to an important conclusion. When society becomes dependent on new technologies, it faces new vulnerabilities to nature's forces; hence safety requires additional monitoring, preparedness, and precautions. But even this action does not guarantee safety because the new, tightly coupled technologies have the ability to produce far more disastrous consequences than the decentralized technologies of the past.

Intense, persistent freezing rain fell on Amish communities too, but they did not suffer a technological disaster. They enjoyed the same heat and light they had under normal winter weather. Their amusement did not depend on electrical lines coming into their homes even in the best weather, so no loss was felt by the absence of soap operas, Hollywood movies, fashion shows, and the Super Bowl on television. By choosing to forego the amenities of a modern lifestyle, the Amish steered clear of the infrastructures those amenities and that culture depend on. Thus they avoided the corresponding vulnerabilities, and the freezing rain was not transformed into disaster. The case of the Amish demonstrates that the construction of a powerful electrical system and dependence on it internalizes autonomous dynamics of nature into society. Modern societies apparently much more robust and better prepared than the Amish suffered from the technological disaster of deprivation of heat, light, and running water in a frigid winter setting. The Amish had all these essentials. Reliance on a centralized electrical grid had rendered modern society more susceptible to disaster from freezing rain than the counter-modern society of the Amish.

Modern societies rationally calculate the probability of specific ice-loadings and then build robustness into electrical lines to resist loadings of high probability, and they judge higher loadings of lower probability to be an acceptable risk. This approach places an extremely heavy safety burden on accurate assessments of probability and of timing of occurrence, as well as on willingness to install expensive safety measures and robust equipment. Loading did not occur only in terms of ice on electrical lines. Safety estimates were also loaded with uncertainty. In a world characterized by global climate change, already poorly based calculations of probability will be made even more uncertain.[13] Amish communities did not construct disaster-preparedness organizations or hire specialists with expert knowledge in risk analysis and disaster readiness. Instead, their everyday traditional practices reduced their vulnerability to a disaster of this sort, whereas the everyday practices of modern society, particularly its dependence on a central power grid, exacerbated vulnerability.

The reflective rationality of modern society seeking disaster reduction was no match for the everyday practices of Amish communities. Modernity cannot be equated with invulnerability, or even straightforwardly with lesser vulnerability. An unintended consequence of Amish values was protection against this kind of disaster. Robustness did not reside in resistant power lines so much as it did indirectly in cultural values, which in the case of the Amish were precautionary toward the adoption of new technologies and commodities and hence indirectly toward technologically aggravated disasters. Nevertheless, Amish communities could be vulnerable to other types of disaster, including those that are socially manufactured by the surrounding modern society, with its activities that result in global environmental change.

WHO ARE THE AMISH?

The Amish are descendants of Swiss Anabaptists who took their name from Jakob Ammann of Bern. The Anabaptist movement, advocating adult baptism, began around 1525 in Switzerland, Austria, and Holland. There are many divisions among Anabaptists, a movement that also includes the Hutterites and the Mennonites. The Anabaptists went further than the Reformers, led by Luther and Zwingli, in separating the church from the state. As a result, they brought on themselves opposition from the Catholic Church, the Reformers, and the state. Most Amish households even today have a copy of the *Martyrs' Mirror*, whose one thousand pages describe the public executions of Anabaptist martyrs in Switzerland, Holland, and Germany. The Amish fled persecution into isolated rural areas, where their attachment to a rural way of life began. They were innovators in early agricultural technology, introducing the scythe, crop rotation, meadow irrigation, and building up the soil by planting clover and alfalfa.[14] Later they migrated to northern Germany, Prussia, Russia, and eventually to the central plains of the United States.

The first Amish came to Pennsylvania around 1727. Emigration to North America increased at the end of the Napoleonic Wars in 1815. Large numbers of Bavarian and Alsatian Amish came to Ohio, Illinois, and Upper Canada (Ontario) between 1815 and 1840. The availability of land in North America meant they could live on adjacent family farms and form self-sufficient, closely-knit communities. There are now no Amish in Europe, the communities having migrated, died out, or been assimilated into other Anabaptist groups or the larger society.[15] The military draft in the United States, from World War II to the Vietnam War, led some American Amish to migrate to Ontario because Canada does not have conscription. When the American draft was abolished and the Ontario government enacted a law requiring refrigeration of milk on farms, hence electricity, this migration ceased completely and some reverse migration occurred. The largest settlements of Old Order Amish, as the more conservative are known, are found in Ohio, Pennsylvania, Indiana, Iowa, Illinois, and Kansas. About 1,500 live in southern Ontario.

Within the Amish there are divisions. In 1913 a particularly conservative group that became known as the Swartzentrubers split off from other Old Order Amish. "Of all the major Amish affiliations in North America they are the most reluctant to accept new changes in technology."[16] They have been called the Old Order within the Old Order Amish family.[17] They were the most numerous Amish group in northern New York State, where the freezing rain fell, and were the ones I visited. A second conservative schism occurred around 1950 under the leadership of Bishop Andy Weaver. Progressive New Order Amish emerged in the 1960s.[18]

The Amish are not evangelical: they do not seek to convert others. Nevertheless, their high birth and retention rates in North America, where they have not been persecuted, have resulted in a population increase from 5,000 at the beginning of the twentieth century to 180,000 at its end.[19] Amish families have from seven to twelve children on average. The most conservative Amish communities have the highest retention rates. Congregations are intentionally small-scale; when they become much larger than seventy-five baptized members, meaning adults, they divide into two congregations. "Local church districts – congregations of twenty-five to forty families – shape the heart of Amish life."[20] Power among the Amish is left at the local level, with only a very weak federation holding the various communities together. Religious values are the principal inspiration and common bond between communities. In many Amish communities the bishop or minister is chosen by lot, and in their enchanted world this is interpreted as his (only men are eligible) being chosen by God.

The Amish have become more distinct as the modern society around them developed in the twentieth century. When the century began, most North Americans lived on farms, used horses for transportation and work, heated and lit their homes with wood or coal stoves and kerosene lamps, and educated their children in one-room schools. Thus their way of life was not very different from the Amish. By the end of that century, the Amish were continuing these practices as before, but the rest of the population had begun using fossil-fuel–powered tractors, trucks, and automobiles, migrated to cities, become dependent upon a centralized electrical grid, and educated their children in large, comprehensive schools and postsecondary institutions. Amish farmers reject the technologies that modern American farmers have adopted as necessities of efficient farming, such as tractors. Nevertheless, the Amish at times have the highest net farming incomes in the United States because of their low costs of production.[21] Whereas many small family farms in North America are in financial difficulty and are being replaced by huge, mechanized factory farms, small Amish family farms are thriving.

The "Old Order Amish have consciously rejected the criterion of efficiency (as defined by the larger society) as a master value ... Whatever competitive disadvantage might have resulted from this partial resistance to the dictates of efficiency was mitigated by accepting, even idealizing, a standard of living that was grossly incompatible with the rising expectations of the rest of the population."[22] Amish cottage industries and shops have generally been profitable. "Low overhead, minimal advertising, austere management, modest wages, quality workmanship, and sheer hard work grant Amish businesses a competitive edge in the public market."[23] Specialization as well as the industrialization and commercialization of agriculture have been systematically resisted by the Amish. But this resistance is always a struggle, and its success never assured.

Amish education is limited to eight grades in a one-room schoolhouse, which they run themselves. As in many rural communities where parents themselves have little education, the children are known as "scholars." Parents discourage their children from formal education after grade eight. Despite its short duration and lack of emphasis on science, this schooling has produced entrepreneurs who manage flourishing, though not large, businesses. No MBA needed here. Artisan skills are developed and enhanced through informal learning and innovation.

The Amish refuse to collect social security in the United States or government pensions in Canada. They do not accept unemployment insurance or welfare in either country. They pay all taxes except social security and do not receive its benefits: "federal aid in the form of Social Security or Medicare would erode dependence on the church and undercut its programs of mutual aid, which the Amish have organized to help members who have suffered fire and storm damage as well as onerous medical expenses ... They have stubbornly refused direct subsidies even for agricultural programs designed for farmers in distress."[24] Because they end their schooling at the eighth grade, they do not produce medical doctors, nurses, and medical technicians. Nevertheless, they have the same medical needs as everyone else, so they too use the medical system of the "English." They pay for it themselves, aided by their church community but without the added cost of a intermediate insurance company. Many Amish in northern New York State claimed that they found Canadian hospitals less expensive and less crowded than American ones, so they often come to hospitals in Canada for their medical needs (Brockville, the Children's Hospital of Eastern Ontario in Ottawa, or Montreal). This practice will surprise Canadians, who frequently complain about hospital wait times, overcrowding, and the taxes they pay for medicare.

Resilience is an important characteristic of Amish communities. In the case of small disasters such as house fires, the house is quickly rebuilt by the self-reliant Amish community without the help of government, insurance companies, or external donors. An Amish furniture builder told me that when his furniture shop burned down, church members financed and rebuilt it within two months. "A spontaneous, humane social security springs into action in the face of fire, disability, sickness, senility, and death."[25] The Amish elevate the self-reliance of their community to a virtue. Mutual aid takes the place of disaster preparedness. The dependency cycle of disaster relief is thereby avoided. The Amish constitute a hard-working, self-reliant community that adapts to the biophysical and social dynamics in which members find themselves. They are not, however, totally self-sufficient. They depend on the wider society for the sale of their products, such as milk, for long-distance transportation between their communities, and for medical services.

Amish communities are characterized by distinctiveness, smallness, homogeneity, and self-reliance.[26] Although different church districts disagree somewhat on where to draw the line concerning technology, all agree that a line must be drawn. Badges of Old Order Amish identity include "(1) horse-and-buggy transportation, (2) the use of horses and mules for fieldwork, (3) plain dress in many variations, (4) a beard and shaven upper lip for men, (5) a prayer cap for women, (6) the Pennsylvania German dialect, (7) worship in homes, (8) eighth-grade private schooling, (9) the rejection of electricity from public utility lines, and (10) taboos on the ownership of televisions and computers."[27] Each community has its own *Ordnung*, usually unwritten, which specifies the prescriptions and proscriptions. A dress code common to the community symbolizes collective loyalty. This is the polar opposite to dress as a sign of individual expression in the surrounding modern society. Underlying these visible signs are more fundamental elements: individualism, with its rights, freedoms, and achievements, is replaced by the collective church community and obedience to it; citizenship in the nation-state, with its welfare benefits and obligations, is supplanted by collective reliance upon and contribution to the church community; consumption and leisure are superseded by a work ethic; science, technology, formal education, and (post)modern values are treated with skepticism.[28] Physical labour is seen as healthy for both soul and body. All Amish reject modernity, and those located where the ice storm struck are mainly Swartzentrubers, the strictest of all Amish communities.

The Amish participate in the whole process of feeding and slaughtering their livestock before eating them. This practice contrasts with the division of labour in modern society, which leads moderns to perceive only steak, ham, or sausage on their plate because they are detached from the animal when it is alive as well as from the stunning, slitting of the throat, and bloodletting, as it is "rendered" (or killed, to use less politically correct language). Such detachment leaves moderns without a practical understanding of the difference in suffering underlying a meal of bacon or of broccoli. The intimate practical connection of the Amish with the processes involved from barn to plate does not, however, lead them to be more in favour of animal rights.

The Amish interpret "working for wages as 'the real problem,' since it often meant working only eight hours a day, thus leading to idleness."[29] Rather than working away from home, many families turn to cottage industries. One Amish family I visited made money by assembling plastic lacrosse sticks in their living room. They were pleased with this contract from a company because it enabled them to earn money in the home without submitting to the influences of the wider society. Bonds between father, mother, and children were reinforced by working together long hours, and idleness was avoided. Amish communities are characterized by a decentralized, artisan economy. The Amish hold private property and engage in market transactions for commodities (milk, medical services, and other needs) with other Amish and with the "English." For example, auctions of retiring farmers are frequently attended. Amish production is not only sold locally. When I visited the Amish community in southern Ontario, I was told about an enterprise there to manufacture improved wood stoves that are sold to Amish, Hutterite, and Mennonite communities throughout Canada and the United States, as well as to non-Amish throughout the world who enjoy heat from a wood stove. The only major advance in the wood-fired household stove in three hundred years, this product, which has an airtight combustion compartment, was developed by two Amish brothers in Aylmer, Ontario, and is now produced in this Amish factory.[30]

Instead of the conspicuous consumption and leisure of the modern and postmodern society that encircles them,[31] the Amish strive to live simply and unostentatiously and succeed in doing so. They take pride in referring to themselves as "plain people." Too much leisure is interpreted as idleness, which promotes individualism and has corrosive effects on the community and its values. Instead they prize labour itself. Leisure, property, the market, and technology are accepted, but they are kept in their place, in line with Amish religious and community values, and not allowed to become fetishes. Needs are defined to be as basic as possible. Like everyone else, the Amish need to earn their living and meet basic health and emotional needs, but they feel no desire to see the world, keep up with the Joneses (which they view as envy), be fashionable, or corner the market. They are very innovative in their own terms, adapting and inventing technology to do what they need. They emphasize their plainness and sameness with others of their community, not the cultivation of distinctions.[32] Like-mindedness, not one-upmanship, characterizes their culture. Division of labour exists in Amish communities, but it, too, is kept under control, especially by the restriction of formal education to eight years.

The Amish do not see the *Ordnung* as a painful burden, and the vast majority of individuals choose to abide by it, as indicated by the high retention rates of community members. The Amish with whom I spoke and those who write in their various publications appear to accept the regulations of the church as useful, necessary, and satisfying guides: "Each day successfully lived within the Ordnung is akin to a job well done."[33]

The reasons the Amish give for living simply are as follows. The first involves very fundamental questions of equity. "How can I live on a level far beyond what I see my neighbor living? ... In short, how can I live in luxury when my neighbor lacks the necessities?"[34] The Amish give the example of the private automobile, which most Americans believe is a necessity. Then they give the statistic that in the United States there is one car for every two people, compared to one car for every one hundred people in Third World countries: "how would we explain to a group of 100 people in a third-world country that we can have fifty vehicles, and they can only have one?"[35] While some think the answer is to help them to have fifty vehicles too, that is a pseudo-solution. It is not possible "that the earth's resources can support a global fleet of cars so that there is one for every two persons"[36] because fuel is limited, agricultural land would be lost by paving it over, and enormous pollution would result. The only responsible solution the Amish see is to reduce the number of American vehicles. "Why do we live simply? We live simply because we are neighbors ... 'We live simply so that others may simply live.'"[37] Another reason they give for living simply is that physical work, especially work done collectively, is good for a person. Hence labour-saving devices have had the perverse effect of eliminating something beneficial and essential. "We feel that is much better for her [or his] character than being idle, or expecting others to do things for her [or him] all the time."[38]

The most basic motivation of the Amish for avoiding material baggage is clearly spiritual. "We are in enemy territory and we dare not linger. Here we are travelers, strangers, and pilgrims, just passing through in search of a better home ... If the destination is all important, it just makes sense to travel lightly."[39] The Amish have chosen to take a horse and buggy to heaven, rather than a Mercedes to sample the delights of Manhattan. Otherworldly spirituality is a powerful source of motivation for the rejection of modern Enlightenment values and postmodern relativism, of the pursuit of the dollar and the consumption it brings. Science and its technological applications are treated with suspicion and caution by the Amish, but so are eco-centric theories that raise animals to the level of humans and dismiss a personal God. Amish communities constitute not a subculture but, rather, a counterculture that rejects modern and postmodern commitments and discards the Enlightenment project, with all its ideas of science and progress.

The Amish should not be romanticized. There are aspects of Amish culture that are far from praiseworthy: strict patriarchy, shunning of members who deviate from the *Ordnung*, restriction of education to the eighth grade to prevent outside influences, and so on. One Amish mother spoke of a central feature of Amish child-rearing as follows: "If you spank them [children] a lot, you break their will and they become like you want them."[40] Thus the very high retention rates among the most conservative Amish communities has been explained in this way by a former member of the community: "Young people socialized into such a different cultural world 'really have no other option.'"[41] The rare Amish who chafe at the dress restrictions complain that even plainness can be a basis of one-upmanship and self-righteousness: "Someone wants to be plainer than someone else. Plainer than you and proud of it."[42]

TRIAGE: A MODEL OF DEVELOPMENT THAT AVOIDED DISASTER

Contrary to the stereotypical image of the Amish, they do not oppose all change. Instead they select, reject, and modify according to their values. Immersed in a modern society continuously developing new technologies and commodities, Amish communities have made triage, not rejection of everything, a crucial part of their social practices. "The screening of new behaviors and products often involves an assessment of their long-term impact on the Amish community. Changes that erode loyalty to the community, create undue temptations, foster inequality, and encourage individualism are suspect."[43] The Amish recognize that technology has power over us if we let it and if we become dependent on that technology. An editorial in the *Mennonite Weekly Review* put it this way: "The lesson from the Amish is that for the good of the community or the family, sometimes one needs to give up the good in order to avoid the bad. That's an attitude that puts technology in its place, making it the servant, not the master."[44] The Amish seek to benefit from technology's advantages but avoid its liabilities. "They try to discern the long-range effects of an innovation before deciding whether to adopt it."[45]

In my visits to Amish households, a subject of discussion was a bishop who had benefited from triple-bypass surgery.[46] The Amish will try a traditional herbal cure first, but when that does not work, they do not hesitate to purchase the best medical technology and care modern science has to offer. "Throughout the twentieth century, the Amish gradually began using modern medicines and the services of physicians and trained midwives. These practices reduced infant mortality."[47] The growth of their population has resulted precisely from the fact that they are willing to benefit from the advances in modern medical science. This acceptance demonstrates clearly that the Amish do not replicate traditional life as it was several centuries ago. They use modern technologies and organizations, just not all of them.

The Amish chose to do without specific technologies they decided were threatening when these were developed in the twentieth century. "The Amish community has traditionally forbidden the installation of private telephones, the ownership of vehicles, and the use of 110 volt electricity from public utility lines"[48] because these are seen as a threat to the community's values and solidarity. Avoiding electricity indirectly forbade a multitude of devices dependent on it. Whereas modern society seeks electrical energy in order to have labour-saving devices and creature comforts, the Amish shun them for fear of what they will lead to. The modern technological inventions of the motor vehicle, airplane, telephone, radio, television, and computer have compressed space and time, allowing the individual to escape these constraints, see the world, bring all the world's cultures to him or her, and become cosmopolitan rather than being restricted to the local. What is seen as a major achievement by moderns is perceived as a threatening peril by the Amish, a folly that would unravel tight-knit local communities and lead to moral relativism.

Ownership of automobiles is banned because it is seen as resulting in individualistic behaviour. "As one Amish bishop noted, 'The car is not such a bad thing in itself; it's what it brings along with it.'"[49] Because of their large families, the Amish have dispersed to distant places in North America in search of suitable, reasonably priced farmland. This was especially the case for the most conservative communities, in particular the Swartzentruber Amish of northern New York State whom I investigated. Travelling by horse and buggy back over a long distance to visit relatives and other Amish friends is impractical. In these cases their preferred means of transportation is a chartered van with a paid "English" driver for family members so that the trip itself builds communal bonds without any member have to concentrate on the road. Even when a modern technology is used, the particular choice is an expression of Amish values: a plain van rather than vanity limousine service, despite the fact that the Amish could afford the latter.

Some new technologies are totally rejected (televisions, centralized electricity), but those that have both benefits and perils are redeployed according to Amish values. For example, a test the Amish apply to new technologies is this: does it draw us apart (individual ownership of an automobile) or bring us together (a rented van enabling an entire family to visit distant relatives together)? The telephone was deemed an interruption of normal face-to-face family communication but was useful in emergencies and for business. So it was forbidden in the home but placed in a shanty at the end of a shared lane. "Keeping the telephone in an unheated shanty in a field, or even an outhouse, was keeping the phone in its proper place."[50] Thus not only rejection or acceptance but also the rules for use are negotiated.

The computer has been banned because it leads to contaminating influences from the Internet and video games and diminishes face-to-face collective communication in favour of individual communication with a machine. Word processors and calculators are, however, sometimes grudgingly accepted for work. The Amish do use some Internet sites to sell their traditional goods such as quilts and wooden wares. They also use gas generators to charge 12-volt batteries and then inverters to convert this power into 110-volt electricity in order to operate cash registers, copy machines, and word processors. Doing so enables the Amish to profit from some equipment judged necessary for work without becoming dependent on the centralized electric power supply. The Amish are now debating about the proper place for mobile phones. Since they can be left outside the home or turned off when others are present, Amish religious leaders, who meet twice a year, have not yet forbidden them.

Of particular interest to this study is the Amish refusal to use 110-volt electricity from public utility lines yet their acceptance of 12-volt electricity from batteries. "These and many other compromises show the ingenious ability of the Amish to tap the resources of technology without forfeiting their religious souls. Striking a host of cultural compromises, the church has selectively picked the fruits of progress while retaining a firm grip on traditional ways."[51] The Amish have prospered in the modern technological world as a result of this triage. In some cases, change is forced on the Amish from the surrounding society, but they adapt to it their own way. In the 1960s the milk industry no longer wanted to pick up small quantities of milk from farmers. It insisted on changing from small cans, which the Amish cooled in streams, to bulk refrigerated storage tanks on farms. Amish farmers had to make the change or lose their business, so they installed the tanks but powered them not with electricity from centralized utility lines but with their own diesel engines.

Members of Amish communities are hard-working and innovative and have avoided indebtedness by not purchasing expensive machinery. This practice has led them to be successful, which is true of both their farms and their shops. Increased productivity is defined by economists in terms of producing more with less labour, but for the Amish it means producing more with less machinery, less use of fossil fuels, and less land. Success usually leads to bigness. The Amish believe, however, that there are social diseconomies of scale: bigness leads entrepreneurs to make their own rules, have undue influence, and be arrogant, and it disrupts the egalitarian character of Amish communities. Hence "the church frowns when business operations become too large ... only 6 percent of the enterprises had more than six employees."[52] It more than frowns: "Some Amish have indeed been excommunicated because they refused to downsize their businesses."[53] No massive factory farms (e.g., pig farms) are to be found among the Amish. Their communities are renowned for aiding members when tragedy strikes, rallying together to help the victims financially, and offering their labour and skills to rebuild. But when one of the Amish businesses that had prospered and grown in size burned down, there was resistance to building it back to its former size. Many Amish felt it was too big and worldly in the first place and should have been split up.[54] This is another dimension of triage: there is a conscious choosing of small-scale organizations over large ones. This taboo on bigness has stimulated a proliferation of small shops, which has diffused entrepreneurial talents and energy throughout the community.

The Amish attempt to keep work whole rather than break it down into separate areas of specialization as modern technological culture does. They believe that well-shared practical knowledge promotes community cohesion, enabling members to work together in barn raisings, house building, and threshing wheat and, more generally, in community decision-making. Everyone is considered competent in the community; the modern deferral of judgment to specialists is thereby avoided.[55] The restrictions inherent in this multi-dimensional triage constitute constraints, but they have also fostered a spirit of ingenuity. "The technological restrictions in Amish life have encouraged an inventive spirit that, rather than stifling creativity, has actually spurred Amish mechanics to experiment"[56] so as to be productive within the restrictions. The Amish are technically very innovative. Although they reject some technologies, they innovate with those they accept. The example of an improved wood-fire stove manufactured in Aylmer, Ontario, has been described above. The most questionable taboo of all – no formal education beyond grade eight and therefore no possible entry into professions requiring higher education – "has lured the best and brightest into the world of business and marketing."[57]

A *Wired* magazine article concluded that the Amish are "very adaptive techno-selectives who devise remarkable technologies that fit within their self-imposed limits."[58] They are choosy about their technologies. "The

Amish consciously steer their cultural course in the sea of alternatives opened by technological advance, accepting only those that enhance their way of life. They often adopt new technologies only after a trial period in which they assess the effects."[59] Modern societies, on the contrary, launch new technologies and commodities on the market without assessing their potential social impacts. Instead, they unreflectively proceed as if the results will be favourable or at least reversible.

Stereotypes to the contrary, Amish communities do not all hold the same positions on modern technology, even concerning electric power. All Amish communities refuse electricity from public power lines, but beyond that the consensus ends. Differences emerge concerning what should be accepted or rejected in their technological triage: "knowing where to draw the appropriate line on electrical usage has not been easy. In some settlements 12 volt electric motors are used to power small equipment in shops and barns. Some affiliations permit the use of generators to power electric welders in shops and portable electrical tools at construction sites. Moreover, inverters that create 'homemade' 110 volt electricity from 12 volt batteries are also being used in some communities."[60] Diesel generators used to produce electricity have resulted in independence from public utility lines, so even those Amish communities that permitted this form of electricity were unaffected by the ice storm. Deciding which technologies and commodities to use and which to forbid has produced tension. "Because moral authority ultimately rests in the local congregations, bishops vary in their interpretation and application of rules."[61]

As a result, there is a patchwork of somewhat different decisions among the various decentralized church districts. Within a church district the debate over rejecting or adopting a technology has resulted in conflict that is framed in religious terms and has led to schisms if church leaders take different sides on the question, or to shunning and excommunication if a member does not have the support of a church leader. It is difficult for non-Amish to comprehend how the use of the telephone or a gas generator can become the basis for religious dispute so intense that it results in a schism, but it does.

Thus the Amish do not mindlessly hold to seventeenth-century technology, as is commonly thought. Nor do they unreflectively absorb every new technology that comes along as modern societies do. As one anthropologist put it, "The use of unsophisticated technology does not indicate an unself-conscious or uncritical approach to existing social institutions. In the context of the contemporary United States, its use represents the very essence of self-conscious choice."[62] The Amish critically examine new technological developments, accepting some and rejecting others according to their goals of decentralized, self-reliant communities. Similarly, commodities from the surrounding society that do not threaten their values are consumed, whereas those that are seen to endanger their basic values are vigorously avoided. The Amish "are cultural conservationists who cherish the beauty of traditional ways but who also accept technological advancements that fit within their moral order and social structure."[63]

The Amish have a way of life that mitigates environmental problems just as it prevented this disaster. These "plain people" live lightly on the planet, not because of environmental values, but indirectly because they practise technological and commodity triage for communal and religious reasons. The consumption by Amish communities contributes little to the depletion of resources and to waste accumulation because its quantity is restrained and its quality is biodegradable and renewable. Horse and buggy travel instead of automobiles and planes, laundry washed in tubs and hung on clotheslines rather than in electric washers and dryers, production based on human and animal labour, and very limited use of pesticides result in limited effects on the environment. A few biodegradable horse droppings left on the road are unlikely to burn up the planet. Amish levels of fossil-fuel emissions are microscopic compared to the modern non-Amish who drive automobiles, take planes, are connected to the electrical grid for lighting, air conditioning and information transfer, and buy numerous commodities from fossil-fuel emitting factories. The Amish have proportionately much less impact on the environment than moderns. The self-sufficiency of their local communities is the polar opposite of the fossil-fuel driven, transportation-dependent, just-in-time interdependency and hyper-specialization of modern society. They have no need for high-efficiency and dangerous technologies: no nuclear energy, no ultra-rapid jets to travel the globe. The pressure to open up Arctic nature reserves for oil and gas exploration is not coming from the Amish. It is not these communities that have produced the "risk society"[64] in the sense of constructing novel recombinations of nature's dynamics that then unleash new dynamics of nature which turn back to threaten society.

Amish communities provide a model and a successful practice of development founded on small-scale production, little division of labour, an attachment to ideals that keep consumption in check, and a purposive selection of technology based on fundamental values. In other words, these are groups that have successfully operated a triage of production and consumption, rather than accepting every new technology and commodity that has been developed with all their environmental ramifications. Theirs is an everyday praxis that does not contribute to global climate change and that is robust when confronted by at least some extreme disturbances of nature, such as the ice storm of 1998. The Amish constitute a useful point of comparison for modern societies under the threat of global climate change.

A REMINDER FOR MODERNS

Whereas moderns were left by this extreme weather event with no heat in frigid winter, without light, without power for machines, and with severely disrupted farms and businesses, the Amish had heat and light and lived as usual. To use the disaster language now fashionable, Amish communities achieved a "business continuity" that the "all-hazards approaches" of the "impact-based planning" of modern societies failed to achieve. Intense, persistent freezing rain fell on Amish communities, but it did not result in the disaster that occurred in the surrounding modern societies of Canada and the United States. The extreme weather event that proved to be devastating for modern communities was harmless for Amish ones.

Unlike the Amish, modern societies with high demands for consumption are dependent on centralized infrastructures such as the electrical power grid, on advanced technological knowledge, on specialists to assess risk, on economists to construct a model of projected costs and benefits, and on disaster experts to prepare for emergencies within cost-benefit projections. Essential infrastructures are constructed to be robust in their interaction with the broader dynamics of nature, but that robustness has its limits. These two modern societies ran up against those limits when the extreme weather exceeded expectations. If modern societies make themselves dependent on such infrastructures to meet basic needs and attain consumer goals and lifestyles, they must intentionally ensure that robustness and preparations are adequate. Serendipity cannot be counted on.

Modern societies may think they have knowledge that transcends the long-term, trial-and-error approaches of the Amish, but they have only pushed trial and error to a new level of generating uncertainty. In addition, the more modern societies unleash emergent dynamics of nature such as global climate change, the more difficult it becomes to foresee hazards. Ever greater expertise is required in this circle of technological development and monitoring of its risky side effects. If the consumption of resources and the creation of waste are at high levels, if complex, highly coupled technology employing dangerous materials of nature is used, and if population levels are high, then the best available monitoring and planning become necessary.

The comparison of the disastrous crushing of the electrical system by freezing rain with the scarcely perceptible effects of the same ice storm on the Amish communities gives additional credence to the conclusion of an environmental anthropologist: "given the emerging instability and unpredictability of environmental conditions ..., it is no longer possible to simply assume the adaptive fitness of our own social system. Quite the contrary, in fact. A number of factors, including population growth and concentration, the toxicity and volatility of some technologies, and productive forms maximizing resource exploitation, have placed that fitness in some doubt."[65]

The difference between modern society and Amish communities has been depicted as follows: "Old Order faith is less reflexive than modern culture, where thought and action continuously interact as people weigh multiple options."[66] This assessment is true in the sense that the Amish cautiously repeat traditional patterns when in doubt, rather than ceaselessly innovating novelties as modern societies do. But this portrayal needs to be qualified.[67] In another sense Amish culture is more reflexive than modern culture. The Amish thoughtfully reflect on, debate, and weigh their options at the level of local communities as they continuously assess the impact of new technologies, commodities, and forms of organization on communal solidarities and values. They then select some, modify others, and reject many. They attempt a cultural-impact assessment to foresee consequences so as to select consequences judged positive and avoid those deemed negative. Modern society rarely performs such a triage based on reflective cultural assessment, instead choosing to unleash the technologies, commodities, and gigantic organizations and then see what happens. If the metaphor of a river current is used to depict how technological and organizational changes and market forces affect society, then it could be concluded that the Amish attempt to steer their community according to their own values in this current, whereas moderns let themselves be unreflectively swept along. In this sense it can be concluded that, compared to the situation in Amish communities, disasters and environmental problems in modern society are the result of its failure of foresight.

Some observers argue that the market system has placed citizens on a treadmill of production and consumption.[68] The experience of the Amish shows that technological innovation and consumption do not have to be a treadmill. It puts the lie to the hypothesis that we cannot live in today's world at a low level of consumption of energy and resources and of waste and that we cannot select or reject particular technologies according to specific values. If moderns are on a treadmill of production, it is because they have chosen, as a result of their consumer values, to climb on and remain there. The example of the Amish demonstrates that if we stay on the treadmill of production and refuse to practise triage, it is not because we cannot get off but, rather, because we do not want to. Amish practices constitute a reminder to all communities that remaining on the treadmill of production and consumption is a collective decision, not an inevitable fate determined by some technology imperative. A technology, commodity, or form of organization does not have to be used just because it has been invented. "It is possible, the Old Orders argue, to tame technology, to control the size of things, to bridle bureaucracy, and to hold things to a humane scale."[69] Triage – selecting some technologies, commodities, and forms of organizations while rejecting others according to assessments about which are benign and which are threatening to fundamental shared values – can and, in the case of the Amish, does occur. Rather than letting loose new technologies, commodities, and organizations without examining the risk of doing so, the Amish attempt to assess in advance what those new technologies will lead to and proactively separate the bad from the good to avoid consequences that would undermine valued practices. Theirs is an early and sustained version of the precautionary principle in a social and religious, rather than an environmental, sense, but it does result in practices that are environmentally friendly as if they were environmentalists, and it mitigated this disaster better than modern communities that prepared to avoid disaster.

All this implies that moderns, too, could choose to step off the treadmill and practise triage. Technology could be steered and consumption made more selective in safety and ecological terms. An incentive structure could be constructed to promote what is ecologically good and safe, while avoiding what is ecologically bad and risky. If the selections are appropriate for the constructions of nature, then disasters and environmental degradation would be mitigated, as in the case of the Amish. Marc Olshan, an "English" anthropologist, concluded that the Amish are "a light to the modern world ... [in their effectiveness] to define what is appropriate, and to insist on the need for limits despite a lack of control over exactly where those limits will be drawn."[70] He argued that "given current conditions we need more of what the Amish have. How much more and in what form is very much problematic. For the present, however, the Amish constitute a valuable model for development."[71]

This assessment does not imply that moderns have to convert and become Amish or that all of humanity should adapt their anti-modern way of life. That would likely be impossible in a world of six billion people of different cultures and, if pushed to its logical conclusion, would imply the abandonment of important developments over the past four centuries, such as medicine. Even the Amish need the surrounding modern society to satisfy basic needs such as health care. Olshan concluded, in addition, that the Amish "stand as a reminder that the deliberate imposition of limits can have enormous costs. Beyond these two guideposts we must navigate the future on our own."[72] Amish communities provide the benefits of identity, security, meaning, and belonging at a price: the sacrifice of individualism, consumerism, leisure, and the mass media. They require the acceptance of the principle that community welfare is more important than personal freedom. Hence it is necessary to do a triage separating the best from the worse of the Amish approach, rather than choosing between the polar opposites of either wholesale acceptance or outright dismissal. Modern societies can learn from the Amish experience about the range of choice that is possible and about the dangers and costs of both their choices and ours.

The comparison of Amish communities with surrounding modern societies was undertaken to better understand the latter. It raises the issue of whether modern societies can develop their own pathways to technological, commodity, and organizational triage in order to avoid disasters and steer clear of environmental problems. The master values underlying these pathways could be ecological ones. Development could be selected that is sustainable and that prevents disaster rather than simply recovering after disaster strikes.

HARMONY BETWEEN SOCIAL PRACTICES AND NATURE'S DYNAMICS

The comparison of the Amish with the modern societies of Canada and the United States demonstrates that the issue of whether disaster and the degradation of the human-supporting natural environment are avoided does not require intentional disaster mitigation and planned environmentally friendly action. It does, however, require that everyday social practices be appropriate for the dynamics of nature. Whether the two are synchronized or out of line occurs on the material level, regardless of intentions. Disasters may be intentionally mitigated, prepared for, and hence avoided. Or they may be circumvented for reasons having nothing to do with disaster planning, as in the case of the Amish. The successful avoidance of this freezing rain disaster resulted from practices based on the religious and social beliefs of the Amish, whereas disaster mitigation, preparedness, and robustness of modern societies failed to prevent disaster. On the contrary, the everyday reliance of modern societies on an electric grid for heat, light, and other necessities transformed extreme weather that was innocuous for the Amish into a technological disaster. The expectation of safety on the part of these modern societies proved to be inapt for the forces of nature. The safety of Amish communities when confronted by intense, prolonged freezing rain was based on a culture and technology suitable for this type of disturbance of nature. In this case, substantive appropriateness proved better than formally rationalized prediction, defences, and mitigation. What the Amish avoided for religious and social reasons turned out to be precisely the technologies that made modern communities vulnerable to this extreme weather event, namely, the electrical grid. The appropriateness, whether intentional or serendipitous, of social constructions to nature's constructions determined whether communities were safe or were beset by disaster.

The calamity faced by the modern societies of Canada and the United States is typically referred to by their citizens as an "ice storm disaster," a term that portrays it as the result of nature's construction of freezing rain. It could, however, as well be defined as a "feeble electrical-grid disaster," a description that would draw attention to socially constructed vulnerability. The infrastructure upon which these modern societies depend was built on the premise of near-average weather. Assumptions about nature's dynamics underpin social order. Disruption resulted when the presupposition of near-average weather proved to be out of line with nature's dynamics. Robustness could have been built into the modern electrical grid to withstand the ice-loading that occurred. That was not done because it was judged too expensive and because such an intense ice storm was assumed to be highly improbable. Expectations about the dynamics of nature and socio-economic priorities were embedded in the essential infrastructures. These then became the foundation not only of the recombinant physical structure of these modern societies but also of their social order. But the expectations proved inadequate for this freezing rain, and the result was a disastrous disruption of the modern way of life. When intense, persistent freezing rain occurred, Amish practices were in step, whereas modern practices were awkwardly out of step. Safety results from practices that are suitable, whether such aptness be the outcome of reflective planning or serendipity.

The Amish are not intentional environmentalists, but they do live lightly on the planet and cause few, if any, environmental problems, even though they inhabit the heartland of the superpower propelling the treadmill of production. Just as their activities are not planned to mitigate a disaster, neither are they planned to avoid environmental degradation. Nevertheless, avoidance of disaster and of environmental degradation is the result of their normal activities. There has been a serendipitous harmony of their activities with the dynamics of nature that completely mitigated this disaster and greatly diminishes harmful effects on the environment. Their prevention of disaster and of environmental problems is largely based on using technologies that have been tested by trial and error for centuries. Whatever technological modification their communities undergo involves cautiously slow change, which is dramatically different from the rapid technological transformations of modern societies.

Serendipity can not, however, be relied upon, especially by modern societies that recombine nature's dynamics so extensively. But as the ice storm disaster demonstrated, modern extrapolations of safety cannot be relied upon either, when uncertainty is great in relations between social constructions and those of nature. Knowledge that is only partial, as well as discontinuities in nature's dynamics, can result in time-series projections of safety that are false, especially if human activities unleash emergent dynamics of nature that intensify or hasten its disturbances. Under these conditions a precautionary approach is particularly necessary, which is the main feature of Amish practices. In modern societies it will take quality leadership and an environmentally reflective public to transform the treadmill of consumption and production into an ecological variant of triage in order to mitigate both environmental problems and disasters.

CHAPTER 14

Survival in the New Frontier

The unexpectedly intense and persistent freezing rain of January 1998 did not harm people by falling on their heads. Rather, it damaged the infrastructures upon which modern societies and their citizens have become dependent to meet their needs, with disastrous results. Transportation was disrupted, but this disaster became more than anything else a crisis of electricity when transmission lines and towers were crushed by the ice-loading. The blackout reminded everyone that electricity has become the essential infrastructure underpinning other basic infrastructures. Even natural gas and oil furnaces require electricity to function, so they provided no heat in frigid winter weather. Gas pumps and oil refineries operate using electricity, so they did not work. Neither did automatic teller and credit card machines, and people no longer had access to their own money. Computers, fax machines, televisions, and traffic lights all need electricity to function. Water purification, filtration, and pumping ceased when the ice-loading crushed the electrical lines needed to power these plants and their pumps.

Although extreme weather seemed to be the cause of this disaster, there are compelling reasons to believe that this is an oversimplification. The disaster was brought on by nature's disturbance, but it only became disastrous where there was socially constructed vulnerability. Hence it was more of a hybrid disaster than a natural one. Modern society paradoxically manufactured a natural disaster. Moreover, the change by leaders from transparency to withholding information documented in chapter 8 as the crisis deepened constitutes another type of social vulnerability and has troubling implications for the management of disasters and environmental problems.

Disturbances of nature such as this one demonstrate that even the most scientifically advanced countries can have their climate-control systems and time-and-space-shrinking technologies disrupted by dynamics of nature. Fortunately, the two societies involved had competent leaders who were able to rush in aid, restore essential infrastructures quickly when (though only when) the disturbance ended, and redistribute the costs to the whole society. Even the low loss of life remains problematic, however, because it depended on the duration, intensity, and scope of nature's disturbance.

Modern technology and organization have brought many benefits, not the least of which is a much higher life expectancy and hence a much larger population. Even the anti-modern Amish choose to benefit from these elements of modernity. The successful calculation of some of nature's forces has enabled the manipulation of those forces and given modern societies new capacities to act. But that success in practice has become inflated in discourse. The belief that there are no longer mysterious incalculable forces and that modernity can master nature's dynamics by calculation is clearly a misleading overstatement. The intensity, duration, and scope of this extreme weather event were incalculable and unforeseeable before it occurred. Anticipation of risk and reflective mitigation of calamities are problematic because scientific knowledge of nature's dynamics is only partial, and hence the foresight of decision-makers is at best incomplete. Modern societies have recombined nature's dynamics to form essential infrastructures such as electrical grids, but this process has rendered them more vulnerable to extreme weather events of the type examined here. It confirms Weber's theory, advanced a century earlier: despite all the advantages of the intensification of modern rationality, it has in some ways inadvertently brought with it a magnification of irrationalities. Perceiving the strengths of modern society and failing to recognize its weaknesses would be naive and risky. It is important to identify fragilities in order to take action to make society less vulnerable and prevent disaster, rather than simply striving to recover after disaster strikes.

THE PROTECTION PARADOX

The world image the Amish developed out of Protestant asceticism led them to reject modern values and make a triage of new technologies. Their anti-modern communities were robust when confronted by this extreme weather event. Modern societies that heavily invested human and financial resources in monitoring nature's hazards and in emergency preparedness proved to be more vulnerable to this extreme weather than Amish communities that did not. The finding of this protection paradox leads to an important conclusion. When a society becomes dependent on new technologies, it inadvertently results in new vulnerabilities to nature's broader forces. The tightly coupled technologies have the property of propagating disastrous consequences far more profoundly and rapidly than the decentralized technologies of the past. Safety then requires the chronic task of risk assessment, mitigation, preparation, and planned response to nature's forces. These measures have tended to reduce fatalities but not property damage, and they could be overwhelmed because knowledge of nature's dynamics even under reflective modernity is only partial. This study of the 1998 ice storm demonstrates the unreliability of presumptions of safety. It showed that disasters result when risk analysis underestimates the likelihood of danger or cannot provide information about its timing. Worst-case scenarios such as this ice storm need to be taken into account,[1] even if they are assessed to be improbable, because they are disastrous and do occur and because assessments of their probability and timing have serious limitations. The disastrous-but-improbable dimension of precaution should not be ignored.

Constructed robustness in modern societies has its limits because of restrictions on the willingness to pay and the capacity to know. If nature's dynamics exceed expectations, then consequences can be worse than if the infrastructures, knowledge, and specialists had not existed, as the comparison of modern societies with Amish communities demonstrated. The concept of "acceptable risk" is based on the assumptions that disaster will occur extremely rarely and be manageable. This premise places a heavy safety burden on accurate assessments of probability, timing, and magnitude of occurrence. In the case of this ice storm, loading did not occur only in terms of ice on electrical lines. Safety estimates were also loaded with uncertainty. Poorly based calculations of probability will be made even more uncertain with the advent of global climate change. There is no avoiding the conclusion that the concept of acceptable risk implies the acceptance of disasters.

If consumption of resources and creation of waste are at high levels, if complex, highly coupled technologies employing dangerous materials of nature are used, and if population levels are high, then the best available monitoring and planning become necessary to intentionally augment safety. Ever-greater expertise is required in this circle of technological development and the monitoring of its risky side effects. The need for rationalization in the form of science, technology, and complex, expert organizations feeds off itself. Modern societies are condemned to the chronic and expensive task of making attempts, first, to foresee the frequency, timing, intensity, and magnitude of nature's destructive dynamics; second, to prepare for those hazards; and, third, to mitigate their consequences. Yet they are also condemned to recognizing that these attempts, and the extrapolations and time series behind them for inferring safety, are imperfect. Forecasting the timing of nature's disturbances is especially important for preventing disaster, yet it is particularly problematic. The more modern society unleashes emergent dynamics of nature such as global climate change, the more difficult it becomes to foresee hazards.

When pristine nature is transformed and reconstructed into technological elements upon which modern communities depend, it is crucial to perceive risk accurately and make social constructions that take into account the interaction with the broader dynamics of nature because those dynamics have been internalized into society. But there is a major problem: the tools, including conceptual ones, for the accurate assessment of risk and especially the timing of nature's disturbances do not always exist. Nature's continuing surprises demonstrate that abstract systems have not brought nature as a domain external to human knowledge to an end. Ignorance and uncertainty rather than knowledge still characterize much of the human relationship with nature, hence the more modest claim that knowledge of nature remains partial is appropriate. Partial knowledge makes technological advances possible, but it does not ensure the safety of those changes. For example, the knowledge that permitted the construction of a centralized electrical grid inadvertently enabled freezing rain to deconstruct society temporarily when its extremeness exceeded expectations derived from meteorological knowledge that was incomplete. Since this temporary deconstruction would not have occurred with previous technologies, new technologies manufactured a natural disaster.

The experience of the Amish demonstrates that communities can step off the treadmill of production and carry out technological and commodity triage, even in the bosom of the world's main consumption and fossil-fuel–burning societies. For the Amish, this has been done according to the "switchman" of Amish spiritual values, which guide their pursuit of material interests along tracks that are environmentally benign and proved robust when confronted by this extreme weather construction of nature. The consumer world image of modern society based on technological and commodity indulgence has, on the contrary, propelled the pursuit of material interests along tracks that have resulted in global warming, ocean degradation, forest dilapidation, biodiversity loss, and other deterioration. It has also made modern society dependent on an electrical grid for basic necessities, hence vulnerable to the type of extreme weather that occurred in northeastern North America in 1998. The challenge for modern society is to develop and use a more complete world image – namely, an ecological one – to direct the pursuit of interests along tracks less harmful to the human-supporting natural environment than economic rationality alone. It is to transform the treadmill of consumption and production into an ecological variant of triage in order to mitigate both environmental problems and disasters.

UNINTENDED CONSEQUENCES

There is a long tradition in the social sciences of studying unintended consequences of intentional action; for example, of Protestant asceticism inadvertently giving rise to a spirit of capitalist consumption.[2] The investigation of unintended material consequences of purposive action has become particularly important in the contemporary world because science and markets are recombining nature's only partially understood dynamics and because of the scale of human activities interacting with the processes of primal nature. The manufacture of a disaster as a result of the construction of an electrical grid was a specific example of unintended consequences in human interaction with nature's dynamics. The increase in greenhouse-gas emissions that cause global warming is an example of more generalized unintended consequences of human-nature interaction.

As disaster research has shown and as this study confirms, many of the most consequential events for humans occur because humans are not oriented to them in advance or are inappropriately oriented. This misorientation occurs either because the risk was unforeseeable in the state of risk-detection technology (as in the extreme weather disaster studied here), or because such technology was not used (tsunami detection devices existed but were not employed in the Indian Ocean prior to the 2004 tsunami), or because the severity of the risk was not acknowledged or was defined as acceptable (as in the 2005 disaster in New Orleans). Where risk has, on the other hand, been perceived and acknowledged – that is, where leaders and the population were accurately oriented to nature's actions – successful avoidance, mitigation, or at least adaptation was often carried out, as in the cases of earthquakes in Japan and the Montreal Protocol banning CFCs. Hence the social sciences need to be expanded to analyze not only social action as defined by Weber but also the non-social action of purposive human actors oriented either appropriately or incorrectly to the non-social actions of nature's actants. Whether the consequences of human actions based on the specific orientation to the actions of nature's actants are those anticipated or instead are unintended – in particular, whether they prove safe and sustainable or harmful – is typically known definitively only in hindsight.

Conceptions of the social construction of nature, social nature, technonatures, and so on only capture one aspect of these phenomena, which must not be allowed to obscure another: that nature's autonomous constructions remain constitutive of primal nature. They are also constitutive of technology and machine production wherein nature's constructions are only temporarily harnessed by social constructions. Such recombinant nature carries the threat of slipping its leash. Both intentional and unintentional transformations of primal nature bring the risk of unleashing its dangerous, poorly understood emergent forces. The so-called social construction of nature remains singularly incomplete; primal nature continues to be a forceful, autonomous constructor. Even the most technically advanced societies can only manage the unwanted consequences of nature's major disturbances through warnings, shelters, evacuations, and other such responses, and they depend on enlightened leadership to prepare for and mitigate theses consequences. In some ways, modern technologically developed society has rendered itself more vulnerable to emerging forces of nature. As modern constructions affect the self-regulating mechanisms of nature and invade virgin wilderness, emerging processes of nature invade society to operate alongside existing ones. Far from eliminating nature, modern societies have inadvertently amplified their dance with nature's movements.

CHOREOGRAPHY AND IMPROVISATION IN THE DANCE OF HUMAN AGENTS WITH NATURE'S ACTANTS

Technological innovation in terms of the construction of an electrical transmission grid was a powerful illustration of humans taking the lead in their dance with nature by manipulating its actants, transforming the gravitational force on water in the distant north into electricity for the urban south, and producing electricity by assembling a critical mass of fissionable matter in a nuclear reactor. Human actors tried to retain the lead by planning a choreography in anticipation of freezing rain, albeit lesser amounts than in fact occurred. Part of the dance with nature involved the monitoring of nature's weather dynamics by meteorologists and climatologists. The knowledge provided was, however, only partial. Everyone, including meteorologists, made sense of the available evidence up to 4 January 1998 by assuming and acting as if intense, persistent freezing rain would not occur. This scenario was wrong, sufficient robustness was not constructed into the power lines, public security was ill-prepared, and the result was disastrous. The severity, duration, and scope of the freezing rain were unforeseen, so no sequence of steps had been planned for such a hazardous dance. Plans were based on faulty assumptions about atmospheric movements and when they would occur. As in a dance, anticipation of the timing of the partner's movements was crucial. But for most extreme disturbances of nature, including this one, timing is the most difficult aspect for science to foresee. Human agents and their societies were not ready for such massive movements of nature, hence choreography of human actions with nature's movements was deficient, and the two societies were overwhelmed by nature's movements, which resulted in disaster.

The extreme weather disaster examined here was a case of nature's actants taking the lead and human agents reacting and improvising after nature's moves in this dance. During and immediately following the ice storm, small variations in precipitation, temperature, wind, and other elements resulted in major changes in the biophysical context within which humans made their social constructions. Human agents danced with every move of these non-human actants of primal nature to form social constructions as best they could after each prompt. This disaster itself was a hybrid construction, resulting from a hazard produced by nature and vulnerability constructed by human societies. When expectations about nature's dynamics are faulty, human actions are tripped up by being out of step with those of nature, which can lead to disastrous results, as in the case investigated here. These failures of foresight and mistakes in the incubation period of disaster were documented retrospectively after the event. The dance in this case was ineptly performed at the very beginning because the risk of such extreme atmospheric movements was unperceived.

As the extreme weather advanced and became clearly visible, the response and improvisation of the population and leaders became robust, and the dance was thereafter adroitly performed, preventing the calamity from becoming catastrophic. The disaster prompted people to learn lessons and attempt to take the lead in the future by planning a choreography in anticipation of nature's future moves and proacting. In this dance of human agents with the non-human actants of nature, the specific reactions or proactions of humans were social constructions shaped by expectations based on cultural predispositions, economic interests, the power of decision-makers, and prior material experiences. Expectations about nature's dynamics are real in their consequences. One crucially significant reason is that they lead to social practices which, if expectations prove to be erroneous, are inappropriate for nature's actants; this outcome can be disastrous. Socially constructed expectations of safety or risk are trial balloons that can be punctured by nature's disturbances.

The freezing rain produced by primal nature crushed not only electrical transmission lines but also discourse which claimed that it would not happen. Discourse is important precisely because it results in action that is either in harmony with nature's movements or out of step. It must be understood in its social, cultural, *and* biophysical context; that is, as one form of movement of human agents that has material consequences because of the interaction of the practices it creates with the movements of non-human actants in the dance they perform together.

Analysis in terms of the metaphor of a dance based on movements between agents and actants is a good way to bridge the nature/culture divide and transcend the limitations of one-sided approaches that abstract either cultural or nature's constructions out of the analysis. It is important to comprehend how the perception of managing both disaster and environmental problems is socially constructed through communicative action responding to prompts from nature's dynamics, how authorities are tempted to replace openness with secrecy when those prompts become particularly dangerous, and how disasters and environmental problems are used for other social constructions. An analytic approach to movements between human agents and nature's actants, like partners in a dance influenced by the other's creative movements, underscores the importance of appropriate expectations by the human dancers, as well as the adverse consequences of incompatible ones. It points to the significance of learning to move in harmony with nature's dynamics and of studying why such learning does or does not occur. The present investigation informs us about errors of expectations concerning nature's dynamics, about the material consequences of such errors, and about social barriers to learning from the prompts of nature. The study of this disaster shows that modern societies had better learn from the adverse consequences of extreme weather events already experienced to take into account the vulnerabilities of modern recombinant nature when faced with extreme forces of primal nature and to build into the choreography of humanity's dance with nature generous margins of error regarding expectations about primal nature's movements.

Conceptions of nature and risk are socially constructed, but not in a material vacuum. They are constructed by sensory beings using as prompts observations and experiences of both nature's everyday dynamics and extreme disturbances, as well as scientific discoveries about the material world. It is not enough to study only how perceptions, practices, and technologies are socially constructed in a context of socio-cultural contingencies. Material contingencies have to be included in the analysis to respond more completely to this "how" question. Thus Part Two of this book gave a detailed account not only of the interaction between human actors in extreme biophysical conditions but also and especially the interaction between multifarious constructions of nature's actants and those of human actors, both leaders and the population. It presented a thick description of the dance between nature's movements and those of human agents, thereby taking into account, rather than ignoring, the biophysical contextual contingencies that influence the social construction of discourse and practices.

DISASTER OR CATASTROPHE

Disaster specialists distinguish between accidents, disasters, and catastrophes.[3] Accidents are confined to specific sites. The event usually finishes quickly, although the consequences can last longer. Even when it lingers, it is still an accident if localized. Plane crashes, train derailments, fires, and toxic spills that affect a small, circumscribed area are accidents. A disaster, on the other hand, affects or threatens a broader community, disrupts essential infrastructures such as transportation and communication, and makes difficult the determination of what needs to be done and how to do it. It can result from a sudden impact, such as a hurricane or an earthquake, or it can gradually build to a destructive climax, as with a flood. Catastrophes are even more consequential, and they substantially diminish the capacity of a community to respond. For example, specialists conclude that Canada has had only one catastrophe: the explosion in Halifax Harbour in 1917 resulting from the collision of two wartime armament vessels, which demolished buildings, started a thousand fires, killed all the senior officers of the fire department, overwhelmed hospitals with severely injured patients, and prevented the city from returning to normal for months.

Although the concept that best describes the consequences for humans of the 1998 extreme weather in northeastern North America is a disaster of great scope, intensity, duration, and expense, the occurrence also contained some elements of a catastrophe. By knocking out electricity upon which modern society had become so dependent, the freezing rain decreased to some extent the capacity of the affected areas to respond. Extreme weather that crushed the electrical grid over such a wide area prevented help from being received from the usual neighbouring sources because they, too, had severe problems.

These conceptual distinctions between accidents, disasters, and catastrophes are useful analytical tools, but they have limitations. They leave the impression that the severity of nature's blow determines whether the outcome will be an accident, a disaster, or a catastrophe. Things are not so simple. Whether a hazard becomes disastrous or catastrophic depends a great deal on leadership before the hazard strikes and during the event. The preparations implemented by leaders and their reaction to the crisis shape whether a community has the capacity to respond and whether it is used effectively. This distinction becomes evident in comparing the disaster that resulted when unexpectedly intense, persistent freezing rain fell on northeastern North America in January 1998 with the catastrophe that occurred when Hurricane Katrina hit New Orleans a well-anticipated glancing blow in 2005.

The vulnerability of the below-sea-level city of New Orleans to a category 4 or 5 hurricane had been known for years, but the budget to reinforce the levees was cut. The weather service predicted an active hurricane season months in advance and forecast the power and path of Hurricane Katrina days before it struck, yet evacuation preparations for the poor, the elderly, and the infirm were inadequate. At the federal level President George W. Bush was slow to become involved, and Vice-President Dick Cheney did little if anything. Many of the troops and equipment of the National Guards of Louisiana, Mississippi, Alabama, and other states were out of the country fighting in Iraq and Afghanistan. Federal authorities required states to specify their needs in detail before federal resources were deployed, as opposed to a more general appeal for help; so help was not sent. Technologies believed to defend against disaster were inadequate, leadership was weak, and institutions such as the Federal Emergency Management Agency (FEMA) responded in an ineffectual way after being placed under the aegis of Homeland Security. Different levels of government stepped on each other's toes and got in each other's way, and turf wars broke out. Organizational paralysis left residents particularly vulnerable.[4]

The impact of Hurricane Katrina on New Orleans was catastrophic, not because of the force of the hurricane by itself, but because of inadequate prevention, preparation, and response, all of which were FEMA's and ultimately the president's responsibility. George W. Bush said on television, "I don't think anyone anticipated the breach of the levees."[5] Yet that was precisely what had been predicted in scientific studies if New Orleans was struck by a category 4 or 5 hurricane, and meteorological studies forecast that a hurricane would likely strike the city. Calling it a "natural" disaster and unexpected is often the excuse for inadequate preparation and response. Some disasters are surprising, but not when the hazard involves hurricanes as frequent as those in the Gulf of Mexico and when vulnerability is as well-known as that of a city lying below sea level with deficient levees. What was called the "ostrich defence" in litigation against the directors of Enron and WorldCom and against Conrad Black has been repeatedly rejected by judges as a legitimate excuse. It is just as unacceptable for leaders to hide their heads in the sand to avoid knowing about imminent disasters such as hurricanes on the coast of the Gulf of Mexico and environmental problems such as global warming.

In contrast, the consequences of the 1998 ice storm in northeastern North America were mitigated because they were well managed, despite the fact that the intensity, duration, and extent of the freezing rain were unexpected. Emergency plans that had been choreographed and practised were put into effect. Improvisation, which is always necessary in a disaster, was prompt and efficient. In Quebec Premier Lucien Bouchard and the CEO of the electrical utility appeared daily on television to inform the population about the latest movements of nature and their attempts to normalize the situation and to advise people what they should do. Governor Angus King did the same in Maine. Many people no longer had television service, but they received this news via battery-powered radio and telephone calls from friends with electricity. Mayor Pierre Bourque of Montreal was on vacation in a distant region of China, but he rushed back to deal with the crisis when told of its gravity. Vice-President Al Gore came to Maine and offered any help he could provide, including the federal chequebook. The freezing rain struck an enormous territory, so electrical power crews had to be brought from afar to help with repairs. Central Maine Power arranged with North and South Carolina to borrow crews and bucket trucks, but it would take too long to drive them to Maine. Governor King asked Gore for massive military cargo planes to transport them. Within twenty-four hours, the airlift had been organized and the planes were landing in Maine with the crews and trucks. This was the golden age of FEMA, and it responded effectively. The Maine National Guard was available and was sent into action to clear debris, support the electrical workers, provide technical assistance, and do search and rescue.[6] In Quebec Premier Bouchard requested the aid of the Canadian army from Prime Minister Jean Chrétien, and 12,000 troops were promptly placed at his disposal. Despite being on opposite sides of the secessionist-federalist fault line in Canada, the two worked well together during the disaster. There was no evidence that the response to the ice storm was hampered by an attempt by either side to gain political advantage.

Managing a disaster is often a decisive test in the careers of political leaders. Bill Clinton's commanding, empathic visits to the Midwest during its flood of 1993 and to Oklahoma City after its bombing in 1995 solidified his presidency.[7] Hurricane Hugo in 1989 and Hurricane Andrew in 1992 became, on the contrary, sources of criticism for George Bush Sr because of slow, weak, bureaucratic responses. Emergency management leaders, too, face a decisive moment in their leadership when a hazard tests whether their partly choreographed, partly improvised responses are adequate. FEMA had been criticized for a poor response to Hurricane Andrew in Miami. A post-mortem was carried out, lessons were learned, and a more competent leadership was put in place. The agency's subsequent responses to the flood of 1993 and to the 1998 ice storm were much improved and praised. As a result of 9/11, however, priority was given to security from terrorists, rather than from nature's hazards, FEMA was subordinated to Homeland Security, and the leadership of the agency was weakened.[8] The catastrophic consequences in New Orleans indicate that one form of security was chosen over another instead of attempting to achieve both. FEMA director Michael Brown failed the test when Hurricane Katrina struck the Gulf Coast in 2005. Lessons learned from Hurricane Andrew were unlearned, with the result that the preparation for hurricanes and the response to Katrina were appalling. Learning lessons from disasters[9] is much like learning foreign languages: they can be learned, but they have to be maintained; otherwise they will atrophy and be unlearned. Such atrophy does not merely involve a failure of corporate memory and forgetfulness. As indicated by the inadequate preparation for Hurricane Katrina, it consists, at a deeper level, of other priorities that push aside the lessons learned from previous disasters.

It is easy to interpret the visits of political leaders to a disaster area and their use of the media cynically as a mere show of compassion whose motivation springs from their desire for voter support. But tarring every leader's performance with the same brush hides more than it reveals. Although leaders tend to put the best face on their performance and even hire "spin" specialists to help them accomplish this goal, evaluations of response also have to do with concrete problems that people face and whether they are being dealt with or left to fester despite high-sounding words and promises. Some performances of leaders in these critical periods amount to best practices; others constitute worst practices. The difference between the two must not be obscured. Overall, the responses of leaders to the 1998 ice storm and to Hurricane Katrina illustrate best practices and worst practices of leadership respectively.

In both Canada and the United States a bottom-up response to disaster has been choreographed: the municipality is in charge, aided by the province or state, but if the municipality becomes overwhelmed, the latter takes over, aided by the federal government. During the ice storm, the Canadian army followed the orders of the governments of Quebec and Ontario, and the Maine National Guard obeyed the commands of the state of Maine. In both cases the chain of command was clear. The provincial and state governments were decisional, and the federal governments in Canada and the United States played supportive roles. No debilitating turf battle broke out between leaders. This structure worked well for both countries when they confronted the ice storm and would likely have worked well in the case of Hurricane Katrina had the vulnerability of New Orleans to hurricanes been taken as seriously as the Iraq war by the federal government of the United States. The weakness in preparation, response, and recovery in the case of Hurricane Katrina was not in the federal government's playing a supportive role. Rather, it was that the federal government failed to play the role adequately. Suggestions that disaster management should be placed under the command-and-control hierarchy of the federal army are unwarranted and would likely result in an inferior response because it would not be as effective in utilizing knowledge, resources, and responsibilities that are local.

During the ice storm, the Canadian army and the Maine National Guard were present to keep order, and their proximity reassured the population. But they were hardly used for that purpose because the crime rate dropped dramatically. Disaster researchers have found that people do not usually panic or steal when told bad news calmly and clearly in a disaster.[10] They follow instructions, to the advantage of all, rather than acting in their own self-interest in ways that harm everyone else. The panic and looting that occurred in New Orleans are the exceptions. Such behaviour takes place where the living conditions of some groups are disastrous in normal weather and when decision-makers themselves panic and respond chaotically.

RISK PERCEPTIONS AND BIOPHYSICAL RISK

The word "nature" is a shorthand label for an enormous multiplicity of autonomous biophysical dynamics. The more those dynamics are manipulated, the more it is necessary to reflectively attempt to foresee how this manipulation will interact with broader dynamics of nature, to mitigate adverse consequences, and to adapt appropriately. Such attempts are fallible, difficult, and expensive. If society does not get it right, then it will be vulnerable to nature's emergent hazards, to serious environmental consequences, and perhaps to disasters.

Whether environmental problems and disasters are fostered or mitigated is determined by the fit between human activities and nature's dynamics. Practices that degrade the environment and/or result in disasters can be risks carelessly run or can be the unforeseen, perverse consequences of human activities. No one except terrorists wants a disaster, and natural and technological disasters result from either self-delusion or ignorance. Some disasters occur when social practices interact with nature's dynamics in dangerous ways that were foreseeable but leaders failed to acknowledge the risk and deal with it – for example, the failure of leaders to heed scientific warnings about the risk of hurricanes for New Orleans. This lack of response runs parallel to the failure of those same leaders to heed scientific warnings about greenhouse-gas emissions causing global climate change. Other disasters occur when nature's dynamics are unforeseeable beforehand and risk is only perceived in hindsight, as in the unexpectedly intense, persistent freezing rain in 1998 that crushed the electrical grid of inadequate robustness upon which Canada and the United States had become reliant.

Serendipity can not be counted on when the scale of modernity's manipulation of nature's processes is so grand that all of the planet and its surroundings are being affected by human activities. Learning in hindsight from the experience of disasters and environmental problems is needed to expand the foreseeable and diminish vulnerability to future hazards. The challenge is to reflectively ensure harmony in advance in order to avoid being out of step with nature's dynamics; this is a particularly onerous challenge because appropriate choreography needs to be designed for specific interactions of social practices with nature's movements. The risk society brings not only the never-ending duty of reflective anticipation of danger but also the dilemma of knowing that such anticipation is fallible. Better to confront these twin problems than to seek solace in an illusory extrapolation of normality into the distant future. Experiencing disaster or irreversible environmental degradation is a particularly unpleasant method of gaining knowledge about risk.

Risk Talk or Material Risk?

There is so much talk about risk, and even the wealthiest societies do not have the resources to defend themselves against it completely. Belief in fictional threats, such as panic and looting during disasters, leads to inappropriate preparations and poorly chosen actions, as documented in chapter 8. Modern societies are both constructing novel biophysical risks and talking more about risk. Hence a difficult question has to be answered at least provisionally in each case: does talk about safety or risk in the biophysical world have a material referent, or is it just talk? Risk and safety are not always what they are said to be. Political leaders, emergency management experts, and the population have the difficult and unremitting dilemma of distinguishing between risks that are real threats and must be attended to promptly and those that are imaginary threats – risk in word only. Similarly, they must distinguish between wishful thinking about safety and material safety.

Discourse Environmentalists and Material Environmentalists

Talk does not interact with nature's dynamics, but activities and practices do. Meanings are important for the natural environment only to the extent that they shape practices and activities which then affect how we materially relate to the biophysical world. In many cases, discourse is relatively autonomous from practices. Hence a distinction must be made between discourse environmentalists and material environmentalists, between mitigation talk and mitigation action. There are many discourse environmentalists who agree that human activities are causing climate change. There are, however, fewer material environmentalists who vote for political leaders who increase the tax on gasoline to support public transit, who agree to place windmills where there is wind, who require the sequestration of carbon as a prerequisite for the use of coal and for extraction of oil from tar sands, and who consent to implement a cap-and-trade emissions and pollution reduction system for industry that would increase the price of commodities which harm the environment.

In a January 2007 national survey, 83 per cent of Canadians agreed that global warming has the potential to harm future generations, and 93 per cent stated they were willing to make sacrifices to solve the problem, but 64 per cent admitted they were not ready to pay significantly more for gasoline.[11] Political leaders know that there is a discrepancy between the population's environmental talk and its willingness to materially change its lifestyle. Hence even when the discourse of political leaders has been green, their policies have often been brown for fear of voter retaliation. Alongside greenhouse-gas emissions is emitted green propaganda, which has been referred to as "greenwashing." Therefore physical emissions need to be kept in view as a reference point for assessing discourse. Admittedly, it takes a daring leader to challenge the North American population to become material environmentalists and put its practices and its money where its mouth is.

LEADERSHIP OF THE RISK SPECIES

The very interdependencies that make the modern system so efficient under the normal dynamics of nature render it vulnerable to nature's extreme disturbances which exceed expectations and to any incapacity of leaders to, first, make accurate sense of biophysical dynamics and then act in the best interests of society.[12] This is as true for slow-onset environmental risks such as global warming as it is for sudden disasters triggered by extreme weather. To err is human, but the consequences are all the more catastrophic if a leader errs when directing a highly coupled social system that has deployed dangerous dynamics of nature or that lies in their path. By deploying complex interdependent technologies, modern societies have replaced individual self-reliance with dependence on leaders to coordinate specialized organizations and lead wisely. The population has always relied on leaders concerning relations between societies and between people. But now that human activities are affecting the whole biosphere, dependence on leaders has been extended to include steering human activities so that recombinant and primal nature do not strike back against society. Leadership involves not only social relations but also social-natural relations – that is, relations between socio-cultural practices and biophysical dynamics.

Leaders can be either careless or careful concerning hazards in this dance with nature. They can be attentive to dangers and vigilant in the interaction with nature's movements, hence diligent in avoiding harm and act with prudence. Or they can be lax – inattentive to environmental risks and the mitigation of disasters. They can behave irresponsibly and recklessly, even with negligence, when judged by the standard of mitigating disaster and environmental degradation. A tightly coupled sociotechnical system manipulating dangerous dynamics of nature within an overpowering biophysical system requires care on the part of leaders, and hence due vigilance on the part of the population in choosing leaders. Successful leadership with respect to social-natural relations entails a sensitivity to prompts indicating that the culturally received ways of doing things could bring harmful consequences, hence an ability to make sense of novel situations and foresee danger. Where the means do not exist to foresee or where unawareness exists, leadership demands a precautionary approach to steer clear of foisting risk on others. A leader who is a risk-taker in terms of risking his or her own money, career, or even life to accomplish goals is very different from one who is a risk-maker imposing danger on others.

Recombining nature's dynamics in innovative technologies and eliminating pristine nature results in the chronic burden of accuracy and caution in human relations with nature's dynamics. The possibility that adverse effects from the technological redeployment of nature's processes could be outweighed by benefits is conditional on accurate assessments of danger, on willingness to take action to avoid danger, and on caution where assessments are uncertain. The awareness that our knowledge of nature is partial – namely, an awareness that nature's dynamics have repeatedly objected to what is said about them – warrants a precautionary principle as a necessary guide to human interaction with nature.

Humans are indeed the risk species: the only species to develop science and technology and the only species to produce the risks associated with these developments. Those special capacities also empower humans to deal with ecological risks through reflective action. Biological evolution has equipped the human species both to produce and to deal with risks associated with the manipulation of nature. Social and cultural constructions, for their part, determine whether this potential is actualized or not. They shape meanings and ideologies that equip humans with institutions, leadership, and material practices which determine whether the dangerous consequences of hazards will be acknowledged or denied, minimized or maximized. The risk species has produced the risk society, and social and cultural dynamics determine whether it will be a minimal-risk society or a risk-maximizing one. Humans are the only species that makes in large part its own planetary future in the anthropocene period, however fraught with uncertainty that may be.

SURVIVAL IN THE NEW FRONTIER

In 1972 the celebrated author Margaret Atwood argued that the central symbol of Canada is survival, as manifested in its literature – "bare survival in the face of hostile elements and/or natives: carving out a place and a way of keeping alive. But the word can also suggest survival of a crisis or disaster, like a hurricane or a wreck ... For French Canada after the English took over it became cultural survival, ... [and] in English Canada now while the Americans are taking over it is acquiring a similar meaning ... Bare Survival isn't a central theme by accident, and neither is the victim motif; the land *was* hard."[13] The harshness of nature in the space on the planet where the Canadian nation has been constructed influences the experiences of its people and the way they conceive of nature. It also results in differences between the Canadian view of nature and the American conception. "Pretending that Nature is the all-good Divine Mother ... can't really stand up very long against the Canadian climate and the Canadian terrain, measured against which Wordsworth's Lake District – Divine Mother country – is merely a smallish lukewarm pimple."[14] The Canadian conception brings with it the danger of the individual and the nation degenerating into a passive victim of nature: "you might wonder, in a snowstorm-kills-man story, whether the snowstorm is an adequate explanation for the misery of the characters, or whether the author has displaced the source of the misery in their world and is blaming the snowstorm when he ought to be blaming something else ... it expresses a premature resignation and a misplaced willingness to see one's victimization as unchangeable."[15] The challenge of the Canadian conception is for the individual and the country to become a creative non-victim in this harsh context.

The vast majority of Americans live farther south and experience a less harsh variant of nature than Canadians, as indicated by the flight of Canadian tourists, popularly called "snowbirds," to American locations to escape the chronic cold of Canadian winters. The romance of nature is not attenuated to the same extent in the United States. Atwood agrees with others who contend that the unifying symbol at the core of American culture is the Frontier. This symbol suggests a new place where the old order is discarded, "a line that is always expanding, taking in or conquering ever-fresh virgin territory (be it The West, the rest of the world, outer space, Poverty, or The Regions of the Mind); it holds out a hope, never fulfilled but always promised, of Utopia, the perfect human society."[16] Since Atwood first published her book *Survival*, the attempt to conquer ever-fresh virgin territory now includes the genetic basis of life. The frontier functions as an ideal and must not be naively mistaken for reality. "Most twentieth century American literature is about the gap between the promise and the actuality."[17] The American conception views nature as territory to be conquered, yet it consists of the all-good Divine Mother; so it encompasses a tension between the two (after all, why would anyone want to conquer an all-good Divine Mother?). This conception is based on the premise that nature is either good or perpetually plastic and can be remoulded at will by human reason, the ultimate resource.[18] Both representations avoid the possibility that nature can be unconquerably destructive.

The goal of conquering nature has to different degrees and in different ways come to prominence in many other countries, notably in Europe, in Japan, and recently in China. One of its main spokespersons is now a European.[19] Swiss scholars write doctoral dissertations and books about the "Americanization of Switzerland," even though it was neutral and prosperous during World War II and was not occupied or reconstructed by the Americans thereafter.[20] They do not mean Americanization in the sense of American corporations buying up Switzerland or its companies – the Swiss are very competitive in the market – but, rather, that the Swiss themselves have internalized what were previously American values: desire for an automobile, a detached home, playing at a golf course, and so on. These scholars conclude that Americanization is in many ways a synonym for modernization. Hence it would not be incorrect to deduce that all modern countries have become American in the sense of opening the frontier of conquering nature in order to attain consumer goals.

The American conception suffers, however, from premature triumphalism because its premise of plastic nature conquered by reason is most certainly wrong, as demonstrated by disastrous disturbances of nature that have overpowered modern societies, such as the United States itself, which have rationally attempted to master nature. The challenge for the American conception is to recognize the risk that the all-good Divine Mother could be tipped into a new steady state, becoming harsh and victimizing her self-proclaimed conquerors if the attempt is made to expand the frontier beyond what the human-sustaining processes of nature can tolerate.

Differences between countries in both the experience of nature's dynamics and the cultural representation of nature must not, however, be exaggerated. There are many similarities between Canada and the United States in both experience and beliefs. The United States experiences harsh nature too. Hurricanes, tornadoes, wildfires, floods, earthquakes, and even freezing rainstorms are frequent. Not only Alaska but also Maine is farther north than most of the metropolitan centres in Canada. The 1998 ice storm struck a larger land area in the United States than in Canada. Where it hit determined its impact on American culture: the sparsely populated state of Maine does not define American culture the way the larger state of Massachusetts, the metropolis of Boston, Harvard University, and the Massachusetts Institute of Technology do. It was only by good fortune that the intense, persistent freezing rain barely missed Boston and its avant-garde, rational institutions. The peoples of both countries interact with nature's processes and are threatened by its disturbances.

Atwood argued that in the war against nature, it was assumed that if man "won he would be rewarded: he could conquer and enslave Nature, and, in practical terms, exploit her resources."[21] The United States and Canada thought they were conquering nature by building valuable constructions such as electrical grids and cities dependent on them, but these were vulnerable and had been placed in what turned out to be the path of nature's overpowering disturbance of freezing rain. The vision of conquering and enslaving nature proved to be a mirage, a social daydream accurate only under average dynamics of nature. The extreme forces of nature underlying natural and technological disasters provide a lesson that nature has only been manipulated by technological development, not socially reconstructed and mastered. It is fitting that one of the main metaphors used on both sides of the American-Canadian border to describe this extreme weather disaster was of nature conquering modern society. Many people interpreted the freezing rain as if nature were waging war on these two societies, with its ice bombers flattening electrical grids and well-planned urban trees, orchards, timber forests, and sugar bushes.

Nature can be hostile, as the ice storm demonstrated, but seeing it only that way is too one-sided. It is also beneficial, providing humans with air, water, sunlight, and other resources they need. The main danger is winning the fight against nature and enslaving it by, for example, clear-cutting a forest, paving over the land, fishing out the oceans, and combusting all the oil on the planet. Then nature's capacity of regeneration would be lost, and humans would have to undertake all the creation themselves, a costly and impossible task. There is an appreciation by at least some leaders and part of the population in both countries that if humans fight against nature in a destructive way and win, they will lose. Pushing the frontier to the extreme of taking in all the previously virgin territory on the planet and reconstructing it as human space may produce something very inferior to the perfect human society. It could even expose humans to dangerous new forces of nature. Nature thought to be harnessed could then slip its leash and turn back against its self-proclaimed human masters. There is evidence this process is now beginning because of the use of the atmosphere as a dump for greenhouse gases, thereby unleashing global climate change. Learning to distinguish what would better be left intact in nature from what could be advantageously changed, as well as a safe rate of change, is a difficult but crucial responsibility. It will involve technological and commodity triage based on ecological values.

CULTURE-NATURE INTERACTION

Through its many processes, nature supplies us with what we need, but it is also dangerous. It constitutes the absolute beneath cultural representations of safety or risk, confirming or refuting expectations of well-being or disaster. The demonstration in this study that modern societies – and not just developing societies or Western societies in the past – can be overwhelmed by the forces of nature precisely because of their successes contributes to placing technological advance in the context of continuing vulnerability to the forces of nature. The development of an electrical grid that rendered modern societies more susceptible to disruption from freezing rain was the case documented in this book. A counterweight is thereby provided against two misleading oversimplifications in the cultures of modern societies: the wishful thinking of a blind faith in technology and the romanticized representation of nature. The modern dance of humans with nature involves the art of balancing different dynamics: recombining nature's materials and forces to gratify human desires, yet relying on processes of primal nature to satisfy human needs. Understanding the achievements and the limitations of modern societies can contribute to this balance and avoid tipping nature's dynamics into a state that may well be less satisfying for humans.

Wealthy modern societies cannot presume their invulnerability. The chronic and the acute, the slow-onset and the sudden, environmental problems and disasters, nature's hazards and socio-technological vulnerability are now coupled together in troubling new ways. As modern societies recombine nature's processes in novel forms and inadvertently modify the autonomous forces of primal nature, they face the unforeseeable.

Their need to anticipate the severity and timing of nature's hazards accurately in order to reflectively ensure safety and sustainability, together with their incapacity to do this in cases of true technological innovations involving recombinant nature and of the unleashing of new constructions of primal nature, is one of the great dilemmas for modern societies. It could be called the foresight insufficiency of modernity.

Although this investigation examined only dependence on an electrical grid, it raises questions concerning reliance on many other technological infrastructures and on specific resources that nature has taken millennia to construct. Modern societies have made themselves dependent on other recombinations of nature's dynamics, such as those based on fossil fuels, in particular oil. This reliance has left those societies vulnerable over the long run to possible scarcity of these resources, which have become essential, as well as to the pollution and degradation of nature's broader support system, such as the atmosphere necessary for human life. Can nature be socially reconstructed safely by technology and the market in a timely manner in order to satisfy human consumption desires? Is there, on the contrary, a treadmill of production and economic pressures that will prevent the construction of a steering mechanism and brakes to control the direction and speed of modernization? Will reflective and ecological modernization anticipate and acknowledge risks and deal with them sufficiently to avoid disasters and environmental calamities? Will the answers to those questions differ from one modern society to another, a difference that seems to be emerging between northern Europe and North America? Such broad questions cannot be definitively answered by a single investigation into the management of the disastrous collapse of an electrical grid struck by extreme weather, but this analysis does strive to contribute to elucidating the parameters within which answers may be sought.

TOWARD ENVIRONMENTAL TRIAGE IN A MARKET SOCIETY

The Amish have accomplished technological and commodity triage based on resolutely anti-modern and otherworldly values and a rejection of the big organizations, lifestyle, and consumption practices of the modern world. But few people convert to the Amish faith now, and few in the future would choose their labour-intensive lifestyle and plainness. How can technological and commodity triage to mitigate environmental problems and perhaps disasters be accomplished in populous modern market societies? One possibility consists of using the tax system to stimulate a triage between economic goods and ecological bads.

In North America taxes on fossil fuels are among the lowest in the world, encouraging fuel-inefficient vehicles and urban sprawl. For example, thc United States and Canada have the lowest rates of taxation on gasoline among developed countries (ten and five times lower respectively than the United Kingdom) and have very few road tolls.[22] Subsidies and tax concessions have been given to the automobile industry to build manufacturing plants for gas-guzzling vehicles in Ontario and in American states and to the oil industry to extract oil from tar sands in Alberta and construct pipelines. Coal-fired electrical plants are allowed to profitably dump carbon into the atmosphere. All this, then, makes the prosperity of these regions dependent on emitting greenhouse gases and locks them into a high-carbon economy.[23] In turn, it causes harm to other countries and to North America itself: permafrost and glaciers are already melting, pine beetle infestations are occurring, and increased drought, wildfires, extreme weather, and flooding of coastal cities are expected in the future. These are indeed immense costs against which the present low level of fossil-fuel taxes in North America pales in comparison. Polluters do not have to pay the full cost of their fossil-fuel emissions because the harm done by emissions is not included in the price of fossil fuels. Dumping emissions in the atmosphere is cost-free to the polluter – much as dumping effluents in waterways was free for polluting factories in the past – with the effects to be paid by someone else in the future in terms of living in a degraded environment. This perverse incentive structure has resulted in Canada and the United States having the world's highest per capita rates of greenhouse-gas emissions among developed countries.

Economists refer to the above as "market failure": the market competition has been structured in a way that generates destructive externalities. Organizations and consumers have market incentives to force unwanted costs onto others.[24] The operational principle is that of the polluter profits rather than pays. North Americans have become free riders refusing to pay the cost of their destructive carbon emissions and leaving it up to other countries to suffer the consequences and deal with the problem. In the early industrial revolution, the tragedy of the commons was that polluters were free to unburden themselves of their waste into rivers and the surrounding air. That problem was largely solved through tall smokestacks, regulations, and the change from dirty coal to sweet oil as an energy source, but the remaining, more subtle tragedy of the commons is that polluters are still free to use the global atmosphere as a waste dump.

Technological improvements to solve the problem of greenhouse-gas emissions can not be counted on to occur automatically in a timely fashion when needed. Why would a company bother with the expense of developing technological innovations to reduce emissions when it and its competitors can dump their pollution into the atmosphere at no cost? Market competition makes ecological saints rare indeed. Comparatively low prices of fossil fuels, based on unpaid pollution costs, have blocked development of alternative energy sources and of more efficient use of fossil fuels. In a market system, technological alternatives to pollution emerge and harmful practices are diminished when there is an environmentally appropriate system of incentives and disincentives, as was constructed in the cap-and-trade system to deal with the problem of acid rain, or regulations, as implemented to solve the problems of airborne lead and ozone-layer–depleting CFCs, or carbon taxes, as exist in Europe. The unfettered market got modern society into the problem of greenhouse-gas emissions provoking climate change, hence a reconfiguration of the market will be necessary to get it out.

Resetting the rules of the market contest so as to promote an environmentally friendly triage of technologies and commodities can be done through regulation and/or taxation. Economists have shown that the least costly solution for society is to increase taxes on activities that produce pollution (burning gasoline and coal, extracting oil from tar sands) and reduce taxes on personal and corporate income correspondingly.[25] The overall tax bill would not be increased, but taxes that target pollution would stimulate technological innovations and behaviour in favour of fuel efficiency. Lower taxes on commodities and on production processes that are good for the environment and higher taxes for those that are bad would push the market to develop and install low-emission technologies because that is where profitability would lie.

The challenge is to integrate the complete cost of industrial production and of commodities into their price instead of externalizing part of the cost in terms of environmental degradation. This polluter-pays solution to global warming is readily available and could be implemented immediately. But it involves shifting taxes and, by implication, a transformation of cultural and lifestyle habits and structures that in turn depends on the public's trust in their leaders not to make a tax grab and on leaders' trust in the population not to vote them out of office for making a reasonable proposal. Will there be quality leadership to correct market failure by implementing a polluter-pays taxation shift? And will the population accept and push for such leadership, or will it vote for short-sighted leadership that maintains externalities whose cost will have to be paid in the future? The answers to these questions remain open in North America, and in many other countries as well. If a system of technological and commodity triage based on ecological values is not implemented, it is because modern societies do not want to practise it and not because a technical imperative prevents leaders and citizens from doing so. This issue involves not only what happens in North America but also setting an example of social structures promoting environmentally friendly triage for rapidly developing countries such as China.

AN EPISTEMOLOGICAL EXPANSION OF THE SOCIAL SCIENCES

This book, including the literature reviewed in chapter 1, constitutes a detailed refutation of the premise underlying too much (but not all) social science that nature is a constant and therefore can be ignored by social science.[26] Nature is, on the contrary, dynamic. It has movement. Nature's processes within and outside the human body constitute action that affects human experiences, practices, and beliefs. Humans give verbal representations to nature's dynamics, but as Latour so aptly stated in metaphorical language, nature often objects to what humans say about it.[27] Nature's dynamics can contradict discourse about them both in the laboratory and in a field setting, with unanticipated natural and technological disasters being examples of nature's contradictions of claims of safety. Material consequences are a testing grounds for expectations and discourse. They can lead to a de-representation of what had been said about nature's dynamics and result in a revised representation. Hence social science would be strengthened if the misleading premise that the biophysical world is a constant, which can be disregarded as far as social constructions are concerned, was corrected. Instead of the social sciences being restricted to one side of the culture/nature divide that resulted from the disciplinary specialization between the social and natural sciences, such a correction would extend the social sciences to the study of the interaction between the socio-cultural and the biophysical. The book documents in detail the need for such an epistemological expansion. It demonstrates that social constructions should be understood in their material context of nature's dynamics, rather than treated as free-floating socio-cultural entities. A theoretical framework that takes into account the interactive effect of socio-cultural and biophysical dynamics is necessary – that is, a true complexity theory.

To better understand the interaction between social constructions and nature's constructions, this book has suggested from a social science perspective the following concepts to facilitate research across the nature/culture divide. It proposes that the social action of purposive human agents be analyzed in its biophysical context of the non-social action of nature's actants, as well as in the context of the non-social action of human agents oriented to those actants. Specifically, it suggests exploring the utility of the metaphor of a dance between human movements and those of nature while capturing the specificity of each. It advances concepts such as the three types of nature's prompts – extreme, everyday, and scientific – to analyze how the biophysical influences but does not determine the socio-cultural. Whether the influence of one type of prompt is reinforced by another type or diminished is particularly important. Expectations and practices constructed by winning the rhetorical struggle of sense-making are trial balloons that can be burst by their interaction with nature's dynamics. Technology is reconceptualized in this approach as recombinant nature to emphasize that nature's dynamics embedded in technological constructions at times slip their leash. Biophysical processes are viewed as primal nature, which retains its force in human-natural interaction even as pristine nature wanes on our planet. It includes anthropogenic primal nature, unleashed by human activities and superimposed upon naturogenic primal nature, which continues its forceful existence.

This approach theorizes the internalization of nature rather than the end of nature as modern societies develop and population grows. It argues that the anthropocene epoch is characterized by an extremely intense interaction between human constructions and nature's constructions. Not only has nature been societized, but society has also been naturized. The book has documented the advantages of appropriate choreography and of technological and commodity triage to diminish unintended, perverse boomerang effects resulting from human recombinations of nature's dynamics. The perspective views technological and societal development as transforming the global biophysical environment into a medium through which some groups and countries inadvertently or knowingly harm vulnerable groups and countries distant in space and time (future generations). Such a perspective promotes sensitivity to the difference between the socio-culturally plausible and the biophysically accurate, to interpretative flexibility yet inflexibility of material consequences, and to the incompleteness of knowledge in knowledge societies.

APPENDIX ONE

Methodology: Doing Interviews at the Top and Listening to Plain Folk

Before I undertook this study, I had written several books about social theory and debates in the field of sociology.[1] My discipline prides itself on taking into account the broader context in which social and cultural life occurs, rather than specializing narrowly in one area like economics or political science. I had come to the conclusion that sociology and a large part of the social sciences had, however, ignored and abstracted out an important set of contextual influences on social and cultural life. It had, to use its own language, "put in parentheses," "bracketed," and "suspended" the effects of the biophysical context. That was a mistake because humans are beings embedded in biophysical dynamics. Although they can culturally construct any beliefs and social practices they want, these have material consequences because they interact with the constructions of the biophysical world. It is important to take into account material consequences and the interaction between the socio-cultural and the biophysical, rather than restricting investigations to discourse alone.

Expectations and beliefs do not exist in a biophysical vacuum. Humans are not pure socio-cultural spirits. Culture mediates material experience, but we must not lose sight of the fact that there are material dynamics of nature within and outside of the human body to mediate. When biophysical phenomena are experienced, material consequences can affect cultural beliefs and social practices. Such beliefs and practices can not be explained only on the basis of prior beliefs and practices. Humans should not be assumed by the social sciences to be blind and senseless. Beliefs about the material world are fallible, some more false than others. Since beliefs shape practices, mistaken expectations about the physical world can lead to particularly harmful consequences. The appropriateness or fit between our cultural representations of the dynamics of the biophysical world and their material referents have to be included in the analysis, even it this can only be done tentatively or retrospectively.

INDICATING THE REFERENTS OF DISCOURSE

The terms "nature" and "risk" are cultural constructs, but they have biophysical referents. Investigations miss an important dimension if they proceed as if those referents do not exist. The term "nature" is admittedly an umbrella concept, much like the word "sick." It refers to a complex aggregate of biophysical dynamics. Like many other words, it is used in this book as a shorthand designation to prevent the writing from becoming too heavy.

This book, like all such studies, is necessarily socially constructed discourse. The material upon which it is based is discourse: social science theory, empirical research studies, interviews, media reports, and other sources. But the book does not treat this discourse as if it had no referents, as if the world and, in particular, the extreme weather disaster were fiction existing only in the minds of the authors or speakers. So how can research document the referents? Does it have to be agnostic about the existence of disasters, extreme weather, and environmental problems? That approach would not be very convincing. It would ignore important phenomena and would have the same social consequence as denying risk and environmental problems. It would be limited to the question of how discourse is constructed and give only a partial answer to even that question by neglecting biophysical contextual contingencies that affect the social construction of discourse.

The best way of indicating the referent of discourse is to take into account in the analysis the best available evidence about the phenomena referred to and to use triangulation. Concerning the dynamics of nature, this means using rather than ignoring the results of natural science research in the social sciences, that is, taking interdisciplinarity seriously. Triangulation, in turn, means getting at the referent from various sources and angles. This process solidifies documentation of the phenomena under investigation, but it can lead to repetition in the writing of a book. Hence a balance was sought, with a great deal of empirical material being shaved off in the final revision to minimize repetition. The result will not completely satisfy everyone; some readers will conclude that there is still too much repetition and others that the referents should be documented further.

A RESEARCH PROJECT FALLS FROM THE SKY

By January 1998 I was looking for an empirical research project that would not be limited to the socio-cultural but would instead examine the socio-cultural in its context of the natural world in which we live. I had concluded that this approach was crucial for the study of many areas where the interaction between the socio-cultural and the biophysical is central: science, technology, risk, health, aging, and so on. In fact, it is difficult to find an area where that interaction is not important and can be ignored. With all these concerns in mind, I had become particularly interested in studying environmental problems resulting from the interaction between social constructions and nature's constructions, rather than examining environmental issues as solely socio-cultural discourse.

One morning during the week of 5 January 1998 I was teaching a graduate course on environmental sociology at the University of Ottawa when we were told that the university was being shut down because the head of the regional government, Bob Chiarelli, had declared a state of emergency. We were instructed to go home. At the time I found this decision strange. True, there was freezing rain, but the area experiences that phenomenon about a dozen times a year. I must admit my first thought was that Canadians had become wimps, so preoccupied with comfort they could not withstand what their hardy ancestors would have seen as a minor inconvenience.

But I was wrong and Mr Chiarelli was right. Our ancestors were not dependent on a centralized electrical grid to supply heat, light, energy for production, entertainment, and other activities. They did not teach and learn in a windowless classroom dependent on heat exchangers, air recyclers, and florescent lighting. Our modern society had inadvertently made itself more vulnerable to nature's infrequent but massive disturbances. The scope, intensity, and duration of the freezing rain that crushed electrical transmission and distribution lines and disorganized modern society demonstrated the power of nature's autonomous dynamics, with which we humans interact. The unexpectedly intense, persistent freezing rain falling on societies that had become reliant on a centralized electricity network resulted in the most expensive disaster in the history of Canada and the state of Maine. This danger, which was not perceived in advance, showed that risk cannot be reduced to threats that are foreseen and debated. Its autonomy reminded everyone – and, it is hoped, social scientists – that only our symbolic representations of nature but not the referents are social constructions. There is a major difference between, on the one hand, assuming that nature's independent dynamics can be abstracted out of the analysis of the social and, on the other, making material reality conform to that premise. I had found my research project: namely, integrating research on environmental problems and on disasters through an investigation of the management of this extreme weather disaster in northeastern North America.

I read up on research about disasters and found it fascinating. It had developed intriguing concepts such as "failure of foresight," "disasters by design," "man-made disasters," and the "sociology of mistakes" that are particularly relevant for the study of the consequences of socially constructed discourse when it encounters dynamics of autonomous nature which cannot be reduced to discourse about it or, stated more simply, "when nature objects to what has been said about it."[2] Most disaster sociology had seen that social practices based on discourse have material consequences even if judgments have to be made post facto. Hence determinations were made that one discourse is correct (there is an iceberg in the path of the *Titanic*) and that the opposite discourse is wrong (clear sailing for the *Titanic*). Once we admit that mistakes have been made rather than assume a relativist equivalence of all discourses, learning can begin.

PUBLISHED MATERIAL

I wrote up a project to investigate governance during and following this extreme weather disaster and to study learning by key leaders from their experience. It was awarded a grant from the Social Sciences and Humanities Research Council of Canada in a national competition, which made this research possible, and I am very grateful for that funding. The first part of the project involved collecting and analyzing all the written material on the subject. Much information had appeared in newspapers and magazines, books of stories and photos had been published, a Quebec scientific commission had issued a five-volume report based on its research,[3] meteorological studies had been done, and insurance companies had undertaken analyses from their perspective. This material provided the documentation for Part Two.

DOING INTERVIEWS AT THE TOP

The next and main part of the research consisted of interviews with leaders who had had key roles in the management of their society during its interaction with this extreme weather. The interview material forms the basis of chapters 7 to 12. Leadership during a crisis is crucial for determining its outcome and is particularly revealing about how a society is organized. Who would be interviewed for the disaster? I decided to begin at the top. Since Quebec was most affected, I would try to interview Premier Lucien Bouchard, Mayor Pierre Bourque of Montreal, and the president of the electricity company, André Caillé, because from all accounts they played crucial public roles in managing this crisis. If they were unavailable for whatever reason, I would ask them for other sources and work down the hierarchy. A colleague who knew more than I did about all three men informed me that Mayor Bourque would likely be the most approachable, Premier Bouchard, who had retired from politics, might agree to an interview, and the electrical company head, Mr Caillé, would probably not want to be interviewed. I wrote to all three inviting them to be interviewed and asking them to communicate with me if they accepted. If they did not communicate within several weeks, I telephoned them. Since they were so highly placed, I foresaw that I would have to do the interviews with these leaders myself to increase my chances of getting an interview. I could not leave the interviews to a research assistant.

I began with Mayor Bourque, who graciously and quickly agreed to an interview. That first interview took place in Montreal on 16 June 2004. At its end, Mayor Bourque suggested that I interview the head of Montreal's Fire Department, who was much involved in managing the crisis, as well as the president of the Montreal Urban Community, Vera Danyluk. What is called a snowball approach to getting interviews had begun. I then sent invitations to these people. Premier Bouchard responded that he could not do an interview because, though he found the project interesting, he is solicited for so many interviews that if he set a precedent, he would spend all his days giving interviews. He had been the charismatic leader of the secessionist party in Quebec and now had an important position in a major law firm. He did generously give me the name of his chief of staff when he was in office, Hubert Thibault, as a knowledgeable person to interview. Mr Thibault is now the vice-president of international affairs for the Mouvement Desjardins, the largest cooperative bank in Quebec. I interviewed him in Quebec City on 13 July 2004. When I was there, I also visited the civil security ministry seeking more interviews. Mr Sylvain Tremblay, who is now chief of the Quebec National Centre of Public Security, granted me an interview on the spot that same day. Back in Montreal I had an appointment to interview the head of the fire department, but he was unavailable and assigned his assistant, André Brunelle, who was also the interim coordinator of emergency measures for the city of Montreal.

After each interview, I would put the names of the respondents already interviewed on my next invitation. So I was building up a rather impressive list of interviewees to entice others to accept an interview. The deputy minister of public security in Quebec in 1998, Florent Gagné, now deputy minister of transportation, accepted my invitation for an interview. It took place on 3 November 2004 in his office atop the tallest building in Quebec City, with a majestic view of the St Lawrence River. The president of the Montreal Urban Community in 1998, Vera Danyluk, was the next person interviewed, in Montreal on 8 November 2004. At the time of that interview the metropolitan government of the Montreal Urban Community had been disbanded after the amalgamation of communities into a greater city of Montreal, but there was still much conflict about amalgamation and de-amalgamation. Ms Danyluk also arranged an interview for me that same day in Montreal with her former director of civil security, Jean-Bernard Guindon, who after amalgamation became the director of civil security for the enlarged city of Montreal.

Meteorological reports demonstrated that the freezing rain had resulted in the greatest natural disaster in the recorded history of the state of Maine. So I wrote to the governor at that time, Angus King, explaining my project and requesting an interview. He not only gave me an interview in the living room of his home in Brunswick, Maine, on 15 December 2004, but also arranged interviews for me with three other key decision-makers in the state that same day: David Flanagan, who was president and chief executive officer of Central Maine Power; Earl Adams, commissioner of the Department of Defence, Veterans, and Emergency Management; and John Falona, the vice-president of transmission and distribution operations at Central Maine Power.

Opportunism can be of help in seeking interviews. I was invited to a conference on the topic of the SARS outbreak in Toronto. A director for Emergency Management Ontario gave a presentation in which he alluded to work he had done during the ice storm. So I invited him to be interviewed. He arranged for another person at Emergency Management Ontario to be interviewed at the same time. The interview was done in Toronto on 19 July 2005. I also requested an interview with Mayor Bob Chiarelli of the city of Ottawa, whose declaration of a state of emergency had triggered my interest in governance during this disaster. Because of his busy schedule, he granted me two half-interviews, which took place on 7 October 2005 and 10 January 2006.

In short, I received remarkable cooperation from these key decision-makers and community leaders. I owe them an enormous debt of gratitude. I did not have the impression they were concealing important information from me, nor were they unduly concerned about their image. They engaged in a frank and honest discussion in response to my questions and were willing to admit errors where they saw them. Everyone of them concluded that, overall, they did a good job in avoiding fatalities and minimizing devastation and disorganization when confronted with this unexpected, intense, persistent, and, indeed, overpowering ice-loading from freezing rain. My own research and independent assessments of the management of this disaster[4] lead me to the conclusion that this evaluation was not self-congratulation on their part. Despite the problems that I describe in the book, which should be avoided in the management of future disasters, the disaster response of these leaders to the extreme weather event was excellent, and they deserve to be proud of their achievements.

THE INTERVIEW GUIDE

I drafted an interview guide of important questions based on the information I had collected. This was a way of beginning a conversation. These people knew many things that I ignored at that point, and I wanted them to talk and reveal issues that I had never imagined and that could not be found in the published literature. Thus the interviews were open-ended, welcoming new information that came up during the course of the discussion and seeking to obtain elaborations of it. I was very sensitive to the possibility of an "interviewer effect" and tried to minimize it by being as objective as possible in tone and substance. At the same time I sought to elicit not only information but also judgments, assessments, and conclusions from the respondents after their experience with disaster. The interview guide had a core of questions for everyone that I had formulated from reading the literature, but the remaining questions varied so as to be appropriate for the specific people being interviewed and the problems they encountered. The interview guide also evolved in the light of the previous interviews. The guide used for the chair of the Regional Municipality of Ottawa-Carleton is included in appendix 2. The interviews were almost all conducted one-on-one in the office of the leader. I would begin by requesting that the individual read and sign the ethics consent form while I was setting up my tape recorder.

LIMITATIONS OF THE INTERVIEW MATERIAL

One limitation of this study is that it is based on interviews with only thirteen key leaders. There was one refusal to be interviewed, and I made the decision that another interview was not usable because the interviewee, unlike the other leaders cited in the book, was much too concerned about the image of the institution and tended to bury problems rather than learn from them for the future. When that interviewee insisted on having the institution's lawyer and its director of public relations approve my manuscript before I submitted it for publication, I decided not to use the interview at all.

It is easy to enumerate other important decision-makers who would have been appropriate to interview as well. The main reason for not doing additional interviews was a lack of time and resources. I decided that I had to interview these people myself because of their key positions as leaders in the management of this disaster in their respective societies, and unfortunately, there is a limit to what one person could do, especially since I was chair of my department when the interviews began and had administrative as well as teaching duties. Despite their limited number, the leaders interviewed consist of a reasonable cross-section of political and emergency-management leaders: the governor of the state of Maine, who was independent of both major political parties in the United States and who previously worked as a lawyer in the energy sector; a general in the Maine National Guard, who was also head of the state's emergency management; the president of a private electrical company in Maine and a high-ranking official in that company; the chief of staff for the premier of Quebec; the deputy minister of public security in the Quebec provincial government; the mayor of the city of Montreal, who had previously been director of a botanical garden; the appointed president of the metropolitan government of Montreal; the director of emergency management for that metropolitan government; a deputy fire chief of Montreal; an official in the public security ministry of Quebec; the elected chair of the national capital of Canada; and an official in the Emergency Management Organization of the province of Ontario. Political and operational elements overlapped in this disaster, as they do in most extreme events; hence it was important to deal with both.

The interviews upon which Parts Three and Four are based were done six and seven years after the ice storm. This interval has the obvious disadvantage that the key decision-makers interviewed might have forgotten some details about the event and their management of it. It does, on the other hand, have the advantage of giving them time to reflect on their decisions and actions. Were decisions that were made rapidly in the heat of the action later judged by these leaders to be the right ones in light of the way the crisis unfolded? Could better actions have been taken? The advantage of a retrospective assessment outweighs the above-mentioned disadvantage for a study, such as the present one, that focuses on broad issues of leadership rather than on narrow administrative and technical details.

Having done the interviews, I doubt that the "interviewer effect" – my own beliefs and my construction of a line of questioning – had much influence on the answers given by these well-informed, independent-minded leaders, who are by their positions often required to answer questions. I tried to be as neutral as possible during the interviews and not to influence the answers, but I do give my conclusions on a range of issues throughout the book.

The next task involved transcribing the interviews. I paid transcribers to do that work – transcribers whose schooling was in French if that was the language of the interview and transcribers whose schooling was in English if that was the interview language. Since I was the only researcher present at the interviews, I checked the work of the transcribers myself, a time-consuming task. I next sent the transcribed texts to the interviewees for their verification, invited them to revise them if they so chose, and requested signed approval. In doing so, I warned the interviewees about the difference between spoken speech and written text. The spoken word often contains many repetitions, "you know's," and ungrammatical talk as speakers formulate their thoughts while speaking. At first I tried to be faithful in the transcription not only to the content but also to the way things were said in the interview, but then I received a message from one interviewee who was shocked at seeing spoken speech in writing. I did not want to insult these leaders who can write very well. Since my purpose was not a linguistic analysis of the way things were said but, rather, an understanding of the content of what was said, I subsequently cleaned up ungrammatical speech and eliminated repetitions before sending the transcriptions to interviewees. When I selected sentences to quote, I did a more thorough cleaning up. Admittedly, this process involved some judgment calls on my part. My intention was to capture the meaning of what was said as closely as possible, and I think I achieved that goal while eliminating the "noise" in speech. The interviews concerning Quebec were done in French, and the quotations from the transcripts were translated into English by me. Although English is my mother tongue, I read, teach, and work in French as well. The original tapes can be reviewed to determine if the transcriptions and translations were well done.

Since I was analyzing a relatively small number of in-depth interviews with key leaders that I had done myself and since I had a set of broad themes and questions that I sought to investigate, I decided not to use a computerized program of coding, retrieval, and analysis. Instead, I examined the four hundred plus pages of single-spaced transcribed material many times over to cull the most salient information relevant to the particular subject being investigated. This was obviously a very time-consuming task, and the reader can judge its value. This procedure may seem old-fashioned in a computer age, but I am convinced that this open-ended methodology was the best way to proceed in a case such as this. In presenting the interviews, I sought to let the key leader speak for himself or herself to the extent possible in a book that analyzes specific issues. Of course, the overall presentation, selection of material, premises, discussion, and conclusions are mine alone. There is no guarantee that these independent-minded leaders would agree with all the conclusions I have drawn, although I suspect that in most cases they would.

LISTENING TO PLAIN FOLK

From meteorological information, I discovered that the freezing rain also fell on the United States – in fact, on a larger geographical area than in Canada, but not on a metropolitan region. I decided to drive to one of the areas that was struck, namely, upper New York State. I collected books and newspaper articles that many communities had produced on the effect of the storm in their area. In one book about the ice storm in Watertown, I made an intriguing discovery: there were Amish communities living in the area. Search parties had been sent out to help these anti-modern farming families, but were told by the Amish that they had not experienced a disaster or even any problems except for a few falling branches. The contrast with the disastrous effect on modern communities and especially on modern farmers so dependent on electricity was stunning. I wanted to interview the Amish. So I first read extensively about this group.

By this time I had done some of the very informative interviews listed above. I drove back to northern New York State with a plan to use the same techniques that had already worked so well: written invitation, ethics consent forms to sign, tape recorder and note pad in hand, and so on. I drove around the Norfolk-Heuvelton area of New York State, where I saw the horses and buggies, the children dressed in black walking home from school, and laundry on the clotheslines frozen solid in these cold winter days. This was Amish country. In the village of Heuvelton, I noticed an Amish man descending from his buggy, tying up his horse, and going into a bank. When he came out, I approached him and explained my project to him. He confirmed that the freezing rain had had little effect on his community. I requested an interview with him or any other community leader. Not with that tape recorder, he replied. It became evident that he would not sign consent forms either. The research protocols required by granting councils that are appropriate for modern societies were rejected by the Amish, even though those protocols were intended to protect the subjects of research. It was obvious I needed more experience and insight into the problems of gathering data about the Amish than I possessed at the time.

I returned home and wrote to researchers who had written about the Amish. They put me in contact with Dr Karen Johnson-Weiner, an anthropologist at the State University of New York in Potsdam, who had been studying the Amish in the northern New York State region for over a decade. I owe her an enormous debt of gratitude for making me more sensitive to the values of the group I was studying; these Amish would not sign forms, did not want to speak to a recording machine, and would be seriously put off by my taking notes while talking with them. Face-to-face conversation must not be interrupted. She argued that if consent forms or tape recorders were required by universities and granting councils, much anthropological research would be stopped in its tracks, particularly concerning groups such as the Amish. The issue is not just the impossibility of doing research but, more importantly, the fact that these groups have a principled opposition to that kind of protection being accorded to them.

Dr Johnson-Weiner kindly offered to take me with her on one of her regular visits to Amish families with whom she had established a long-standing relationship. She introduced me as someone who was interested in hearing their experiences of the ice storm in order to write about it. In their living rooms I explained what I was doing, and we had a chat about what they had experienced during that ice storm. I then made notes after the visits. This technique gave me important background information but no formal interview material or written documentation.

I did discover a few newspapers in which Amish families regularly write about their experiences for one another. I used their first-hand, eye-witness accounts in those newspapers for documentation. I was also given the name and address of an Amish man, David Luthy, who publishes and collects Amish material. I wrote him and then visited him on his farm in southern Ontario. Unlike people who grow up in Amish communities and leave school after eighth grade, Mr Luthy had been a Roman Catholic who attended Notre Dame University before converting to the Amish version of Christianity. He has transformed one of the buildings on his farm into an archive holding everything that appears in print about the Amish. Like all Amish, he refuses to be connected to the centralized electrical grid, so there are no electrically powered temperature and humidity controls in his archives. Nonetheless, he has his own system which seems to work. His collection also includes some well-preserved Christian texts that predate the development of Amish communities themselves.

Mr Luthy allowed me to search for anything that seemed relevant to my research on the ice storm. He helped me locate such materials because he has no computerized retrieval system. These documentary materials as well as published research on the Amish provides the empirical basis for chapter 13. One major problem exists in his system: it is based solely on his own knowledge of it. What will happen to his collection when he is no longer there to manage it? Mr Luthy introduced me to his family, fed me, and even let me sleep for the night in a house he had built for one of his children that was not yet occupied.

Thus in this research project I went from an interview atop the tallest building in Quebec City to stumbling my way in the middle of the night to find the bathroom in an unfamiliar Amish house without electricity. This project took me from the key leaders at the top of the hierarchy of modern society to the plain folk (as they call themselves) who reject modernity and postmodernity and insist on keeping their feet on the ground. When I told Mr Luthy about my nocturnal bathroom meanderings, he laughed, saying he forgot to tell me that he had left a flashlight on the counter. The Amish accept the flashlight because it is a modern technology that helps their decentralized community remain self-reliant, but they reject the electrical grid, which would render them dependent on the centralized structure of modern society. I had come close to experiencing first-hand the technological triage of the Amish.

SENSE-MAKING AND DECISION-MAKING

There are two ways of interpreting what the key leaders said in the interviews: either as reasons (rationality) or as rationalizations (after-the-fact sense-making). The content of the interviews could be understood in terms of the leaders describing the motivations and reasons that led them to make one decision rather than another and to act in specific ways. This is the rational paradigm for understanding the interviews: they thought and then they acted, and in the interview they are describing this process. Alternatively, the interviews could be viewed in terms of an act-then-think schema. According to this interpretation, these leaders are, six years later, making sense of their actions, which took place in an urgent crisis that did not allow time to think through the issues. They acted during the crisis and during the interview are reflecting on why they acted that way. The interviews consist essentially of justifications and rationalizations of their actions. This second way of understanding the interviews involves the sense-making paradigm:[5] leaders are *ex posteriori* making sense of their actions.

It is incorrect to read these interviews as only one or the other of these two polar opposites. Rather, both processes were involved to different degrees for various affirmations by particular leaders. They did think before they made decisions and acted, even in the heat of the action. But they also developed a more elaborate understanding of why they acted that way when they had more time to reflect on it. This reminder constitutes a warning to the reader to be aware that both reasons and rationalizations are in play. Arriving at a definitive separation of *post facto* sense-making from at-the-time reasoning is not the goal of this study. Instead, the objective is to examine and learn from the thought processes of these key leaders at the time of the interview.

HINDSIGHT

A prominent disaster researcher, Barry Turner, found that technological disasters which are acute and sudden usually have an "incubation period" during which incorrect expectations of safety are extrapolated from previous experience of well-being and signs of impending disaster are missed. He characterized decisions in the pre-disaster situation as a "failure of foresight."[6] Social actions continue that are inappropriate for dealing with the dynamics of nature. Hence he described what resulted as "man-made disasters" because they could have been avoided if signs of danger had been taken seriously and appropriate action followed during the incubation period.

Critics of Turner's work contend, however, that he judged expectations and the missing of signs in hindsight after the disaster and that appropriateness or inappropriateness were not at all straightforward before the event. It may have been unforeseeable before it occurred. Hence retrospective classification of expectations and actions prior to disaster in terms of "failure" can be problematic. We must be wary of doing only studies that work backwards once a disaster has occurred and that label prior perceptions of safety as "failures." Hindsight after disaster can show that prior expectations were mistaken, but it is an additional empirical issue to determine whether this constitutes avoidable failure. Furthermore, hindsight is not necessarily 20/20. If the outcome is disastrous, inquiries look for mistakes; if the outcome is advantageous, brilliant moves are searched for. Hindsight is itself open to contested interpretations from conflicting interest positions. There can be plural hindsights from different perspectives. Authorities often interpret failures in ways that protect themselves and their organizations from blame. After Hurricane Katrina the Bush administration, the governor of Louisiana, the mayor of New Orleans, the head of FEMA, and the director of Homeland Security all had different interpretations of who should be blamed. If all their accounts are believed, then readers would be led to conclude that everyone in authority was blameless for the inadequate response to Hurricane Katrina in New Orleans. Even different inquiries reached divergent conclusions to some extent. Lessons learned vary according to beliefs about what is desirable and feasible.

However, interpretations are not free-floating in pure rhetorical struggle. They are themselves affected by biophysical contingencies because humans are sensory beings embedded in a biophysical world. Looking for something in an inquiry does not necessarily mean that it will be found. The material outcome (disaster or safety) affects conclusions, and rightly so. Hindsight after the biophysical prompt of the outcome is based on additional relevant information, namely, the outcome itself and data gathered after the outcome. In the rhetorical disputes concerning interpretations, the disaster itself is a material contingency that renders some interpretations more credible and others less plausible.

An investigation of the Challenger Space Shuttle calamity arrived at the conclusion that the launch occurred, not because production pressures led to an acceptance of great risk, as many people assumed, but because decision-makers mistakenly believed that it was safe to launch the shuttle. Even the scientists who were worried about the O-rings never imagined the problem would lead to the destruction of the shuttle: "no one thought that a complete ring burn-through was possible."[7] Hence the researcher, Diane Vaughan, described her analysis of the incubation of disaster as "the sociology of mistake." In hindsight it was determined that expectations about launch dynamics, which were based on extrapolations from testing and from twenty-five successful launches, were erroneous and that inappropriate actions were taken. But she did not conclude that in this case the decision-makers were to be blamed for failing to see the signs beforehand.

The studies of disasters by Turner and Vaughan deal with the poles of a continuum of possibilities: from clearly visible biophysical signs of risk that are dismissed for socio-economic reasons, and hence the conclusion of "failure" is warranted, to unperceived risk where the state of knowledge and tools of forecasting do not permit adequate foresight, and therefore it would be excessive to blame decision-makers for failure. As the knowledge base of societies develops, harmful outcomes can change from inadvertent consequences of human activities to reckless denial of danger and risks intentionally run. There is a continuum of disasters and environmental problems from unavoidable to mistakes to failure. The conclusion shared by the two studies is that socially constructed expectations about the dynamics of nature led to decisions and actions that proved to be inappropriate for the dynamics embedded in these technologies, which resulted in disaster, and this conclusion can be seen in hindsight.

It is in hindsight that one can most accurately determine the material consequences that social practices have led to in their interaction with nature's constructions, precisely because the outcome is known. It is is for this reason that disasters are typically followed by post-mortems and lists of lessons learned. Thus there is a practical recognition that hindsight can be an incitement to foresight. Disastrous material consequences are significant stimuli for learning. Critics are right, though, to remind us that disaster does not determine that lessons will necessarily be learned. Whether improvements are made depends on many socio-economic factors. They are not the automatic product of a disaster. The relationship between hindsight and foresight is problematic rather than straightforward. Moreover, lessons can be unlearned as well as be learned.

Retrospection can be applied to a specific sequence of events, such as the ice storm disaster, leading to particular lessons learned, or it can be applied to event categories, such as global warming, and result in generic lessons. This distinction can be combined with that between decision-making, which involves the attempt to solve well-defined problems, and sense-making, which attempts to determine what the problem is in ill-defined situations. This hindsight study after a disaster was divided into different sections. Part Two describes the dance between nature's constructions and social constructions. It examines the interaction of the movements of sensory, embodied humans and those of nature's autonomous biophysical dynamics in which they are embedded. Part Three consists of retrospective sense-making (by leaders) of the disaster unleashed by this extreme weather event and of the decisions that were made during the crisis. Part Four deals with retrospective sense-making for broader climate change and environmental issues subsequent to the experience of this disaster. It is hoped that all these sections and the book as a whole will lead to generic lessons learned.

APPENDIX TWO

The Interview Guide

Note: The following guide was prepared for the interview with Bob Chiarelli, chair of the Regional Municipality of Ottawa-Carleton.

- Could you give your name and the position you held at the time of the ice storm of January 1998.
- Did you and the other decision-makers around you expect freezing rain of the intensity and duration of the one that occurred in January 1998?
- Could you describe the damage and the disruption that followed the freezing rain.
- Could you describe your participation in the decision-making and in the management of the consequences of the ice storm.
- If I remember correctly, one of your first acts in power was to declare a state of emergency. Was that a difficult decision? Could you describe how a state of emergency is declared in Ontario and in Ottawa and what it implies. Was the federal government shut down? For how long?
- Could you describe the communication of information between political leaders like yourself and the population. Did you appear on TV and radio in news conferences to give the population the latest information? Were there any problems in this exchange of information? Is there anything that you feel should be improved?
- Could you describe communication between political leaders like yourself and emergency management planners and technical experts. Are there any examples of friction or tensions that you could describe? Is there anything that you feel should be improved?
- What was the worst moment of the ice storm for you in your position, and what was the most difficult decision you had to make? What were the best

decisions and what were the worst ones that should be made otherwise in the future? Were there any decisions you would now with the benefit of hindsight make differently? Were you worried about panic or looting?

- In Montreal the mayor and directors of emergency measures told me that their worst moment occurred when two major water-filtration plants lost their power, water pressure decreased rapidly, and the risk of fire and water contamination increased. Was there anything equivalent here? When the filtration plants went down in Montreal, however, those decision-makers chose to hide it from the population and even from mayors in affected areas for fear that people would fill their bathtubs. Was there anything that decision-makers here withheld from the population?
- In Montreal the political leaders and emergency measures leaders were examining worst-case scenarios, including evacuation of the island city of Montreal, but they decided this was not feasible even in the worst case. Were you planning worst-case scenarios and what were they? Were they feasible?
- Was there an emergency management plan prepared in advance? Was it followed? Could you give me examples of improvisations that were necessary.
- What were the strong points and the weak points in the relations between municipal, provincial, and federal governments in the management of this crisis?
- Were you in contact with the premier of Ontario during the disaster, and if so, was there anything in your joint effort that could be improved?
- Were you in contact with the prime minister of Canada during the disaster? Did the fact that you were the head of the regional government in Canada's capital make your work easier or more difficult during this disaster? In what way?
- Did you and your organization receive all the expected cooperation from other organizations: federal and provincial governments, police, fire department, power companies, etc.?
- Did your perception of risk change as the storm developed, and if so, how did it change?
- Did the experience of this ice storm lead to improvements in the management of such crises? Which improvements?
- Could you elaborate on your views about (a) what was learned from this disaster as well as (b) what should have been learned but does not seem to have yet been learned.
- Before the ice storm, was there a culture of security here where leaders and the population were willing to invest money and time in emergency preparedness? Did the ice storm have an impact in promoting such a culture? Did 9/11? How do the two compare in the promotion of such a culture?

- Several informants told me there are cycles of readiness: after a disaster leaders and the population are willing to invest money and time in security; but as time passes, this willingness decreases as the previous crisis becomes more remote. Was that true here?
- One informant concluded that emergency planners learn much from a disaster, and this is institutionalized in improved readiness preparations, but that the population and political leaders learn little that is lasting (because new political leaders replace those who had the experience of disaster). Do you share that conclusion?
- What are your present expectations of future occurrences of freezing rain of the intensity and duration of that which occurred in January 1998? What are your expectations of other extreme weather events (floods, hurricanes, etc.)?
- One informant spoke of a reflex not to prepare for the worst case. Such preparation is costly and the worst case occurs rarely. Do you think we should prepare for the worst or for the probable? How can a balance be found between these in the uncertain situation of global climate change?
- Another line of thinking is that a disaster leads to media attention, alarmism, overestimates of recurrence, and unwarranted expenditures. Do you agree? How does a decision-maker find a balance between this argument and the argument that disaster demonstrated that improvements in preparation are necessary?
- Political decision-makers such as yourself have to make certain assumptions about the future in order to make emergency preparations and budgetary plans. What do you assume? Do you expect that extreme weather events will become more frequent and intense because of global climate change? Should we deal with these by preparing for disasters or by preventing them from occurring (supporting the Kyoto Protocol) or both? Do you think that common activities such as fossil-fuel emissions in rich countries will provoke disasters such as flooding due to ocean-level rise and extreme weather events? It is difficult enough for rich countries like Canada and the United States to prepare for these emergencies. Do we have an obligation to poor countries like Bangladesh because of our emissions?
- The *New York Times* reported that the Inuit and small island nations are suing the United States over destruction of their way of life as a result of global climate change caused by greenhouse-gas emissions (following the example of lawsuits against tobacco companies). They frame the question as a human rights issue. Do you agree?
- In Quebec several informants complained that the power utility, Hydro-Québec, was slow to give out information and their communication even with emergency officials was poor, although some informants qualified

their criticism because of the nature of the difficult situation. Was there anything equivalent here with the local or provincial power company?

- At the time, power generation and transmission was in the hands of a public utility, Ontario Hydro, but with local public utilities (e.g., Ottawa Hydro, etc.). Was coordination in a time of disaster a problem? Do you think it makes a difference in times of disaster whether the power utility is public or private, or one big company or many smaller companies?
- You know this disaster situation better than I do. Are there any omissions in my questions, important elements that I didn't ask about, etc.?
- Do you know of anyone else that it would be important for me to interview?
- Thank you for giving me this useful information.

NOTES

INTRODUCTION

1 Erikson, *A New Species of Trouble*, 22.
2 Lecomte, Pang, and Russell, *Ice Storm '98*, 8.

CHAPTER ONE

1 See Murphy, *Rationality and Nature*.
2 Weber, *The Protestant Ethic and the Spirit of Capitalism*, 181.
3 Ibid., 181–2.
4 Albrow, *Max Weber's Construction of Social Theory*, 196.
5 Weber, *The Protestant Ethic and the Spirit of Capitalism*o 181.
6 See Murphy, *Rationality and Nature*, chapter 2.
7 Weber, *From Max Weber*, 139.
8 Weber, *The Protestant Ethic and the Spirit of Capitalism*, 180–1.
9 Ibid., 182.
10 Ibid., 183.
11 Weber, *From Max Weber*, 280.
12 Ibid., 147–8.
13 Mommsen, "Personal Conduct and Societal Change," 41.
14 Beck, *Risk Society*; Beck, *Ecological Enlightenment*; Beck, *Ecological Politics in an Age of Risk*; Beck, *Democracy without Enemies*.
15 Beck, *Risk Society*.
16 Ibid., 154.
17 Ibid., 55.
18 Ibid., 154.
19 Ibid., 22.
20 Ibid., 183.
21 Beck, *Ecological Politics in an Age of Risk*, 160.
22 Beck, *Risk Society*, 180.

23 Ibid., 177.
24 Beck, *Ecological Politics in an Age of Risk*, 84.
25 Ibid., 50–1.
26 Ibid., 99.
27 Beck, *Risk Society*, 45.
28 Beck, *Ecological Politics in an Age of Risk*, 168.
29 Giddens, *Beyond Left and Right*, 195–6.
30 Beck, *Ecological Politics in an Age of Risk*, 123.
31 Ibid., 103.
32 Beck, *Risk Society*, 214.
33 Beck, *Ecological Politics in an Age of Risk*, 48–9.
34 Beck, *Risk Society*, 54.
35 Beck, *Ecological Politics in an Age of Risk*, 127.
36 Ibid., 117.
37 Beck, *Risk Society*, 167.
38 Dickens, "Changing Our Environment, Changing Ourselves"; Dickens, *Society & Nature*.
39 Dickens, "Changing Our Environment, Changing Ourselves," 104.
40 Simon, *The Ultimate Resource*; Simon, *The Ultimate Resource 2*.
41 Julian Simon, "What Does the Future Hold?" in Simon, ed., *The State of Humanity*, 642.
42 Lomborg, *The Skeptical Environmentalist*, 290.
43 Ibid., 348–50.
44 Ferry, *Le nouvel ordre ecologique*; Ferry, Review of *Entre la nature et l'homme, je choissis l'homme*.
45 Kahn and Weiner, *The Year 2000*; Clark, *Starvation or Plenty?*; Maddox, *The Doomsday Syndrome*; Hamilton, *Technology, Man and the Environment*; Krieger, "What's Wrong with Plastic Trees?"; Beckerman, *Two Cheers for the Affluent Society*; Clark, *Population Growth*; Vayk, *Doomsday Has Been Cancelled*; Smith, ed., *Scarcity and Growth Reconsidered*; Simon, *The Ultimate Source*; Simon, *Immigration;* Simon, *The State of Humanity*; Simon, *The Ultimate Source 2*; Simon and Kahn, *The Resourceful Earth*; Sjoberg, *Risk and Society*; Ray and Guzzo, *Trashing the Planet*; Maduro and Schauerhammer, *The Holes in the Ozone Are Scarce*; Bailey, *Eco-Scam*; Ray and Guzzo, *Environmental Overkill*; Easterbrook, *A Moment on the Earth*; Beckerman, *Small Is Stupid*; Huber, *Hard Green*.
46 Limbaugh, *The Way Things Ought to Be*.
47 Schnaiberg, *The Environment*; Schnaiberg and Gould, *Environment and Society*.
48 Gould, Schnaiberg, and Weinberg, *Local Environmental Struggles*, 18.
49 Novek and Kampen, "Sustainable or Unsustainable Development?"
50 Redclift, "Redefining the Environmental 'Crisis' in the South."
51 Gould et al., *Local Environmental Struggles*, 7.
52 Ibid., 8.
53 Mol, *The Refinement of Production*; Mol, *Globalization and Environmental Reform*; Mol and Sonnenfeld, eds., *Ecological Modernization around the World*.
54 World Commission on Environment and Development, *Our Common Future*.
55 Mol, *Globalization and Environmental Reform*, 211.

56 Ibid., 114.
57 Hannigan, *Environmental Sociology*, 64.
58 Hajer, *The Politics of Environmental Discourse*, 275.
59 Bell and Carolan, *An Invitation to Environmental Sociology*, 287.
60 Yearley, "Green Ambivalence about Science."
61 Ibid.
62 Hajer, *The Politics of Environmental Discourse*, 264.
63 Kingdom, *Agendas, Alternatives, and Public Policies*, 213.
64 Ungar, "The Rise and (Relative) Decline of Global Warming as a Social Problem."
65 Diamond, *Collapse*.
66 Stern, *The Economics of Climate Change*; Jaccard, *Sustainable Fossil Fuels*; Drummond, Caranci, and Tulk, "Market-Based Solutions to Protect the Environment."
67 Drummond, Caranci, and Tulk, "Market-Based Solutions to Protect the Environment."
68 Beck, *Ecological Politics in an Age of Risk*, 166.
69 Mol, *Globalization and Environmental Reform*, 203, 205.
70 Ibid., 220.
71 Hawken, *The Ecology of Commerce*; McDonough and Braungart, *Cradle to Cradle*.
72 Murphy, "Unperceived Risk."
73 Perrow, *Normal Accidents*, 75.
74 Vaughan, *The Challenger Launch Decision*.
75 Turner and Pidgeon, *Man-Made Disasters*.
76 United States, Department of Regional Development and Environment, *Disaster, Planning and Development*, 7.
77 Worster, *Nature's Economy*.
78 Klinenberg, *Heat Wave*.
79 Aptekar, *Environmental Disasters in Global Perspective*, 7.
80 Zebrowski, *Perils of a Restless Planet*.
81 Tenner, *Why Things Bite Back*.
82 Freudenberg, "Contamination, Corrosion and the Social Order."
83 Turner and Pidgeon, *Man-Made Disasters*.
84 Clarke and Short, "Social Organization and Risk"; Sagan, *The Limits of Safety*.
85 Perrow, *Normal Accidents*.
86 LaPorte, "Highly Reliable Organizations"; Weick, Sutcliffe, and Obstfeld, "Organizing for High Reliability."
87 Aptekar, *Environmental Disasters in Global Perspective*, 12–19.
88 Erikson, *A New Species of Trouble*, 22.
89 Karl, Nicholls, and Ghazi, eds., *Weather and Climate Extremes*; IPCC, *Climate Change 2001*.
90 Webster, Holland, Curry, and Chang, "Changes in Tropical Cyclone Number, Duration, and Intensity in a Warming Environment."
91 Turner and Pidgeon, *Man-Made Disasters*; Vaughan, *The Challenger Launch Decision*.
92 Cuny, *Disasters and Development*, esp. chapters 2 and 3; United States, Depart-

ment of Regional Development and Environment, *Disaster, Planning and Development.*

93 Perrow, *Normal Accidents.*
94 Lewis, *Development in Disaster-Prone Places*, 8.
95 Hewitt, "Excluded Perspectives in the Social Construction of Disaster"; Middleton and O'Keefe, *Disaster and Development.*
96 Lewis, *Development in Disaster-Prone Places*, 27.
97 Klinenberg, *Heat Wave*, 36.
98 Davis, *Ecology of Fear.*
99 Davis, *Late Victorian Holocausts.*
100 Hewitt, "Excluded Perspectives in the Social Construction of Disaster," 81.
101 Wisner, *Power and Need in Africa*, 16.
102 Lewis, *Development in Disaster-Prone Places*, xiii.
103 Hewitt, "Excluded Perspectives in the Social Construction of Disaster," 81.
104 Dynes, Quarantelli, and Wenger, *Individual and Organizational Response to the 1985 Earthquake in Mexico City, Mexico.*
105 Lewis, *Development in Disaster-Prone Places*, 30.
106 Quarantelli, "Epilogue," 262.
107 Dynes, "Coming to Terms with Community Disaster," 113.
108 Quarantelli, "Epilogue," 259.
109 Kroll-Smith and Gunter, "Legislators, Interpreters, and Disasters," 170.
110 Gilbert, "Studying Disaster," 17.
111 Thomas and Thomas, *The Child in America*, 572.
112 Oliver-Smith, "Global Changes and the Definition of Disaster"; Oliver-Smith, "Theorizing Disasters."
113 Blaikie, Cannon, Davis, and Wisner, *At Risk*, 12.
114 Lewis, *Development in Disaster Prone Places*, 40.
115 Dombrowsky, "Again and Again," 29.
116 Oliver-Smith, "Global Changes and the Definition of Disaster," 178.
117 Ibid., 188–9.
118 Lewis, *Development in Disaster-Prone Places*, xiv.
119 Sylves and Waugh, *Disaster Management in the U.S. and Canada*; Quarantelli, *What Is a Disaster?*
120 Mileti, *Disasters by Design.*
121 Platt, *Disasters and Democracy.*
122 Abramovitz, *Unnatural Disasters.*
123 Mileti, *Disasters by Design*, 18.
124 ISDR (UN), *Disaster Reduction and Sustainable Development.*
125 Handmer, "Sustainable Development Is about Disaster Reduction," 121–3.
126 Lewis, *Development in Disaster-Prone Places*, 146.
127 IDNDR Technical Committee, *Washington Declaration.*

CHAPTER TWO

1 Giddens, *Modernity and Self-Identity*, 224.
2 Beck, *Ecological Politics in an Age of Risk*, 37–8.
3 Ferry, *Le nouvel ordre ecologique.*

4 Worster, *Nature's Economy*, 350.
5 Eder, *The Social Construction of Nature*.
6 Bluhdorn, "A Theory of Post-ecologist Politics."
7 Collins, "Stages in the Empirical Programme of Relativism."
8 Merchant, *The Death of Nature*.
9 McKibben, *The End of Nature*, 55.
10 Murphy, "The Internalization of Autonomous Nature into Society."
11 Olson, *The Politics of Earthquake Prediction*.
12 Murphy, "Unperceived Risk."
13 Gross refers to this as *nichtwissen*. See Gross, "Sociologists of the Unexpected."
14 Turner and Pidgeon, *Man-Made Disasters*.
15 Beck, *Risk Society*.
16 Latour, *Nous n'avons jamais été modernes*; Benton, "Environmental Sociology"; Benton, "Why Are Sociologists Naturephobes?"; Weiland, "The Power of Nature and the Nature of Power."
17 Beck, *Risk Society*.
18 Reviews of the literature can be found in Murphy, *Sociology and Nature*; Murphy, "The Internalization of Autonomous Nature into Society"; and Murphy, "Thinking across the Culture/Nature Divide."
19 Dunlap and Catton, "Environmental Sociology"; Dunlap and Marshall, "Environmental Sociology."
20 Moreover, more complete studies of Durkheim's and Weber's writings show that they, too, took into consideration the biophysical context. See Boudes, "L'environnement, domaine sociologique," and West, "Max Weber's Human Ecology of Historical Societies."
21 Weber, *Economy and Society*, 1375.
22 Ibid.
23 Ibid., 22.
24 Ibid., 4.
25 Goffman, *The Presentation of Self in Everyday Life*; Goffman, *Interaction Ritual*.
26 Murphy, "Disaster or Sustainability."
27 Perrow, *Normal Accidents*; Vaughan, *The Challenger Launch Decision*.
28 Turner and Pidgeon, *Man-Made Disasters*..
29 Vaughan, *The Challenger Launch Decision*.
30 Diamond, *Collapse*.
31 Murphy, "Disaster or Sustainability."
32 Latour, "When Things Strike Back."
33 Vaughan, *The Challenger Launch Decision*, 402.
34 Callon, Lascoumes, and Barthe, *Agir dans un monde incertain*, 355.
35 Jackson, "Sept. 11, 2001."
36 Tester, *Animals and Society*.
37 Diamond, *Collapse*, 519–20.
38 Quoted in "Dutch Seawall a Model for Flood Protection," *Ottawa Citizen*, 2 September 2005, A4.
39 David Sellers, of Water Management Consultants of Vancouver, quoted in "It

Might Happen Only Once Every 700 years, but Winnipeg Will Be Ready," *Ottawa Citizen*, 10 September 2005, B1.

40 Diamond, *Collapse*.

41 Weick, *Sensemaking in Organization*; Weick, *Making Sense of the Organization*.

42 Turner and Pidgeon, *Man-Made Disasters*.

43 Geertz, "Thick Description."

CHAPTER THREE

1 Preece, *Animals and Nature*, 223.

2 LaRue, *1998 Ice Storm Almanac*, 9, 41.

3 Environment Canada, *Ice Storm '98* (1999), 2.

4 Lecomte, Pang, and Russell, *Ice Storm '98*, 9.

5 Milton and Bourque, *A Climatological Account of the January 1998 Ice Storm in Quebec*, 33.

6 FEMA, *Region I, Interagency Hazard Mitigation Team Report … State of Maine*, 7.

7 "The Great Ice Storm of January 1998," *Ottawa Citizen*, 15 January 1998, G12.

8 LaRue, *1998 Ice Storm Almanac*, 37, 3.

9 "The Great Ice Storm," *Ottawa Citizen*, 15 January 1998, G12.

10 LaRue, *1998 Ice Storm Almanac*, 38.

11 Milton and Bourque, *A Climatological Account of the January 1998 Ice Storm in Quebec*, 7.

12 Ibid., 81.

13 Quebec, Commission scientifique et technique, *Pour affronter l'imprévisible* (my translation).

14 FEMA, *Region I, Interagency Hazard Mitigation Team Report … State of Maine*, 19.

15 Unless otherwise indicated, the information in the following section was taken from reports that appeared in newspapers (*Gazette*, *La Presse*, *Le Devoir*), magazines (*L'Actualité*), and books on the subject: Phillips, *Blame It on the Weather*; Sperandio, *L'Enfer de glace*.

16 "The Great Ice Storm," *Ottawa Citizen*, 15 January 1998, G5.

17 Scanlon, *Ottawa-Carleton and the 1998 Ice Storm*, 13.

18 Abley, *The Ice Storm*, 29.

19 Quebec, Commission scientifique et technique, *Pour affronter l'imprévisible*, 257.

20 LaRue, *1998 Ice Storm Almanac*, 39.

21 Ibid., 13.

22 *Ice Storm 98: When Maine Froze Over*, 7.

23 *Ice Storm '98* (Denton Publications), 55.

24 Abley, *The Ice Storm*, 150.

25 Ibid., 162–3.

26 "The Great Ice Storm," *Ottawa Citizen*, 15 January 1998, G5.

27 "Hydro System Was 'Hanging by a Thread,'" *Ottawa Citizen*, 18 January 1998, A1–2.

28 Scanlon, *Ottawa-Carleton and the 1998 Ice Storm*, 14.
29 Ibid., 40.
30 Ibid., 14–15.
31 "The Great Ice Storm," *Ottawa Citizen*, 15 January 1998, G12.
32 Environment Canada, *Ice Storm '98* (1999), 2.
33 Scanlon, *Ottawa-Carleton and the 1998 Ice Storm*, 13.
34 Akwesasne Emergency Measures, *Ice Storm '98*.
35 Transmission lines are the high- and medium-voltage ones that bring power into an area. Distribution lines are lower-voltage ones that then carry the current to particular subscribers.
36 Quebec, Commission scientifique et technique, *Pour affronter l'imprévisible*, 34.
37 Ibid., 16, 33–4.
38 Environment Canada, *Ice Storm '98* (1999), 2.
39 Abley, *The Ice Storm*, 33.
40 *Ice Storm 98: When Maine Froze Over*, 46.
41 Quebec, Commission scientifique et technique, *Pour affronter l'imprévisible*, 36.
42 Abley, *The Ice Storm*, 35.
43 Quebec, Commission scientifique et technique, *Pour affronter l'imprévisible*, 257.
44 *Ice Storm 98: When Maine Froze Over*, 46.
45 Quebec, Commission scientifique et technique, *Pour affronter l'imprévisible*, 38.
46 Lecomte, Pang, and Russell, *Ice Storm '98*, 9.
47 Collins, "With a Delightful January Like This."
48 LaRue, *1998 Ice Storm Almanac*, 17–21.
49 *Ice Storm '98: A North Country Disaster*, 3–5.
50 FEMA, *Region I, Interagency Hazard Mitigation Team Report … State of Maine*, 7.
51 *Ice Storm '98: A North Country Disaster*, 12.
52 Abley, *The Ice Storm*, 59.
53 Quebec, Commission scientifique et technique, *Pour affronter l'imprévisible*, 51.
54 "The Great Ice Storm," *Ottawa Citizen*, 15 January 1998, G6.
55 *Ice Storm '98: A North Country Disaster*, 9.
56 Ibid., 11.
57 "The Great Ice Storm," *Ottawa Citizen*, 15 January 1998, G5.
58 "Montreal Stands Glazed and Confused," *Burlington Free Press*, 11 January 1998, 1A and 9A.
59 *Time Magazine*, 19 January 1998, 12–14.
60 "A Lethal Beauty," *Canadian Geographic*, March/April 1998, 36–45.
61 *Ice Storm '98: North Country Disaster*, 18.
62 Ibid., 12.
63 Abley, *The Ice Storm*.
64 "The Great Ice Storm," *Ottawa Citizen*, 15 January 1998, G12.
65 Scanlon, *Ottawa-Carleton and the 1998 Ice Storm*, 13.
66 "The Great Ice Storm," *Ottawa Citizen*, 15 January 1998, G6.
67 Ibid.
68 Scanlon, *Ottawa-Carleton and the 1998 Ice Storm*, 14.

69 Ibid.,15.
70 "The Great Ice Storm," *Ottawa Citizen*, 15 January 1998, G4.
71 Scanlon, *Ottawa-Carleton and the 1998 Ice Storm*, 16.
72 Ibid., 9, 27.
73 Quebec, Commission scientifique et technique, *Pour affronter l'imprévisible*, 34.
74 Environment Canada, *Ice Storm '98* (1999), 2–3.
75 Quebec, Commission scientifique et technique, *Pour affronter l'imprévisible*, 257.
76 *New York Times*, 12 January 1998, B6.
77 LaRue, *1998 Ice Storm Almanac*, 27.
78 *Ice Storm '98: A North Country Disaster*, 10.
79 LaRue, *1998 Ice Storm Almanac*, 29.
80 *Ice Storm '98: A North Country Disaster*, 7, 10.
81 *Ice Storm 98: When Maine Froze Over*, 3.
82 Ibid., 7.
83 Ibid, 5.
84 Ibid.
85 Ibid.
86 Ibid., 7.
87 "The Great Ice Storm," *Ottawa Citizen*, 15 January 1998, G12.
88 Ibid., G6.
89 Scanlon, *Ottawa-Carleton and the 1998 Ice Storm*, 26.
90 Ibid., 12.
91 Ibid., 16.
92 Ibid., 16–17.
93 Ibid., 9, 12.
94 Abley, *The Ice Storm*, 35.
95 Scanlon, *Ottawa-Carleton and the 1998 Ice Storm*, 28.
96 "Hydro Gridlock: A Diary of Devastation," *Ottawa Citizen*, 24 January, 1998, C3.
97 Ibid.
98 "Hydro System Was 'Hanging by a Thread,'" *Ottawa Citizen*, 18 January 1998, A1–2.
99 "The Great Ice Storm," *Ottawa Citizen*, 15 January 1998, G7.
100 Scanlon, *Ottawa-Carleton and the 1998 Ice Storm*, 41.
101 Ibid., 51.
102 "The Great Ice Storm," *Ottawa Citizen*, 15 January 1998, G7.
103 Scanlon, *Ottawa-Carleton and the 1998 Ice Storm*, 54.
104 Ibid., 21.
105 Ibid., 48.
106 Rheingold, *The Virtual Community*, 19
107 "The Great Ice Storm," *Ottawa Citizen*, 15 January 1998, G7.
108 Akwesasne Emergency Measures, *Ice Storm '98*, A7.
109 "The Great Ice Storm," *Ottawa Citizen*, 15 January 1998, G6–7.
110 Ibid., G15.
111 *Ice Storm 98: When Maine Froze Over*, 45–6.

112 Quebec, Commission scientifique et technique, *Pour affronter l'imprévisible*, 34.
113 "The Great Ice Storm," *Ottawa Citizen*, 15 January 1998, G17.
114 Quebec, Commission scientifique et technique, *Pour affronter l'imprévisible*, 234.
115 Environment Canada, *Ice Storm '98* (1999), 3.
116 Abley, *The Ice Storm*, 26–7.
117 Quebec, Commission scientifique et technique, *Pour affronter l'imprévisible*, 221.
118 Ibid., 259.
119 Ibid., 135.
120 *New York Times*, 9 January 1998, A12.
121 LaRue, *1998 Ice Storm Almanac*, 43.
122 *Ice Storm '98: A North Country Disaster*, 12.
123 *Ice Storm '98* (Denton Publications), 8.
124 *Ice Storm '98: A North Country Disaster*, 16.
125 Ibid., 11.
126 Ibid., 13.
127 LaRue, *1998 Ice Storm Almanac*, 54.
128 Ibid., A5.
129 Ibid., 44.
130 Ibid., 45.
131 Ibid., 58.
132 Ibid., 146.
133 Ibid., 52–3.
134 *Ice Storm '98* (Denton Publications), 6.
135 *Ice Storm '98: A North Country Disaster*, 61.

CHAPTER FOUR

1 Environment Canada, *Ice Storm '98* (1999), 3.
2 "The Great Ice Storm," *Ottawa Citizen*, 15 January 1998, G7, G12.
3 Ibid., G15.
4 Scanlon, *Ottawa-Carleton and 1998 Ice Storm*, 28–30.
5 "The Great Ice Storm," *Ottawa Citizen*, 15 January 1998, G7.
6 Scanlon, *Ottawa-Carleton and 1998 Ice Storm*, 17.
7 "The Great Ice Storm," *Ottawa Citizen*, 15 January 1998, G16.
8 Ibid., G13.
9 "Electricity: The 20th-Century Farm Labourer," *Ottawa Citizen*, 20 January, 1998, A1.
10 Scanlon, *Ottawa-Carleton and 1998 Ice Storm*, 26.
11 "The Great Ice Storm," *Ottawa Citizen*, 15 January 1998, G7.
12 Scanlon, *Ottawa-Carleton and 1998 Ice Storm*, 13.
13 Abley, *The Ice Storm*, 121, 122.
14 Scanlon, *Ottawa-Carleton and 1998 Ice Storm*, 66.
15 Ibid., 54.
16 Akwesasne Emergency Measures, *Ice Storm '98*.
17 "The Great Ice Storm," *Ottawa Citizen*, 15 January 1998, G7.

105 *New York Times*, 11 January 1998, 21.
106 *New York Times*, 12 January 1998, B6.
107 Ibid.
108 LaRue, *1998 Ice Storm Almanac*, A-36.
109 Ibid., 90–1.
110 Ibid., 106.
111 Ibid., 118.
112 *New York Times*, 12 January 1998, B6.
113 *Ice Storm 98: When Maine Froze Over*, 12.
114 Ibid.
115 *New York Times*, 13 January 1998, A14.
116 Ibid.
117 *Ice Storm 98: When Maine Froze Over*, 5.
118 Ibid., 15.
119 FEMA, *Region I, Interagency Hazard Mitigation Team ... State of Maine*, 19.
120 *New York Times*, 13 January 1998, A14.
121 *New York Times*, 12 January 1998, B6.
122 FEMA, *Region I, Interagency Hazard Mitigation Team ... State of Maine*, 22.
123 "Mammoth Cleanup Begins," *Burlington Free Press*, 11 January 1998, 8A.
124 "Restoration Pace Rattles Residents," *Burlington Free Press*, 11 January 1998, 8A.

CHAPTER FIVE

1 "The Great Ice Storm," *Ottawa Citizen*, 15 January 1998, G9.
2 Ibid.
3 Scanlon, *Ottawa-Carleton and the 1998 Ice Storm*, 23.
4 Abley, *The Ice Storm*, 62.
5 Ibid., 64.
6 "The Great Ice Storm," *Ottawa Citizen*, 15 January 1998, G9.
7 Scanlon, *Ottawa-Carleton and the 1998 Ice Storm*, 31–2.
8 Lecomte, Pang, and Russell, *Ice Storm '98*, 9–13.
9 Scanlon, *Ottawa-Carleton and the 1998 Ice Storm*, 42.
10 Ibid., 41.
11 Ibid., 21.
12 Ibid., 43.
13 Ibid., 40.
14 Ibid., 44; see also 65.
15 Ibid., 46–8.
16 "Ottawa Will Never Look the Same," *Ottawa Citizen*, 12 January 1998, A1. See also Scanlon, *Ottawa-Carleton and the 1998 Ice Storm*, 9.
17 Quebec, Commission scientifique et technique, *Pour affronter l'imprévible*, 54, 57.
18 *New York Times*, 13 January 1998, A1.
19 Ibid.
20 Abley, *The Ice Storm*, 64.
21 *New York Times*, 13 January 1998, A14.

22 Quebec, Commission scientifique et technique, *Pour affronter l'imprévible*, 37, 258.
23 *Ice Storm 98: When Maine Froze Over*, 46.
24 Ibid.
25 *New York Times*, 13 January 13, 1998, A14.
26 "The Great Ice Storm," *Ottawa Citizen*, 15 January 1998, G17.
27 Quebec, Commission scientifique et technique, *Pour affronter l'imprévible*, 238–9.
28 Abley, *The Ice Storm*, 83.
29 Ibid., 43.
30 Ibid., 99.
31 Lecomte, Pang, and Russell, *Ice Storm '98*, 13.
32 Abley, *The Ice Storm*, 174–5.
33 Quebec, Commission scientifique et technique, *Pour affronter l'imprévible*, 216.
34 Abley, *The Ice Storm*, 179.
35 "Un électricien cloué au pilori par J.E. réclame 3,4 millions," *La Presse*, 5 January 1999, 36.
36 Quebec, Commission scientifique et technique, *Pour affronter l'imprévible*, 86–7.
37 Abley, *The Ice Storm*, 124.
38 Quebec, Commission scientifique et technique, *Pour affronter l'imprévible*, 351, 370–1.
39 *New York Times*, 15 January 1998, A6.
40 Abley, *The Ice Storm*, 122.
41 See http://www.cprs.ca/news/e_06Jun15_Rennie.pdf.
42 Abley, *The Ice Storm*, 118.
43 "The Great Ice Storm," *Ottawa Citizen*, 15 January 1998, G17.
44 Quebec, Commission scientifique et technique, *Pour affronter l'imprévisible*, 129.
45 Ibid., 73.
46 Ibid., 158–9, 185.
47 Ibid., 240.
48 Ibid., 96–9.
49 *Ice Storm '98* (Denton Publications), 58.
50 LaRue, *1998 Ice Storm Almanac*, 150.
51 Ibid., 154.
52 *Ice Storm '98: A North Country Disaster*, 75.
53 LaRue, *1998 Ice Storm Almanac*, 136.
54 Ibid., 153.
55 *Ice Storm '98: A North Country Disaster*, 60, 64.
56 Ibid., 70.
57 LaRue, *1998 Ice Storm Almanac*, 123–4.
58 Ibid., 137.
59 Ibid., A60.
60 *Ice Storm '98: A North Country Disaster*, 71.
61 Abley, ed., *Stories from the Ice Storm*, 56, 59–60.
62 Ibid., 58–9
63 FEMA, *Region I, Interagency Hazard Mitigation Team ... State of Maine*, 4.

64 LaRue, *1998 Ice Storm Almanac*, 124.
65 Ibid., 141.
66 *Ice Storm '98: A North Country Disaster*, 80.
67 Ibid., 80, 81, 85.
68 LaRue, *1998 Ice Storm Almanac*, A-54.
69 *Ice Storm '98: A North Country Disaster*, 50; see also 53.
70 Ibid., 75.
71 Ibid.
72 LaRue, *1998 Ice Storm Almanac*, 122.
73 Ibid., 177.
74 FEMA, *Region I, Interagency Hazard Mitigation Team ... State of Maine*, 6; see also 1.
75 LaRue, *1998 Ice Storm Almanac*, 187.
76 FEMA, *Interagency Hazard Mitigation Strategy Report: New Hampshire*, 7.
77 *Ice Storm 98: When Maine Froze Over*, 31.
78 *New York Times*, 13 January 1998, A14.
79 *Ice Storm 98: When Maine Froze Over*, 2.
80 Ibid., 22.
81 *New York Times*, 13 January 1998, A14.
82 *Ice Storm 98: When Maine Froze Over*, 4.
83 Ibid., 44.
84 Ibid., 4.
85 Ibid., 5.
86 Ibid., 29, 31.
87 Ibid., 4, 32.
88 Ibid., 4.
89 Ibid., 28.
90 Ibid., 41.
91 Ibid., 33.
92 Ibid., 35.
93 Ibid., 37.
94 Ibid., 35–6.
95 Lecomte, Pang, and Russell, *Ice Storm '98*, 2.
96 Ibid., 37.
97 Ibid., 33.
98 Quebec, Commission scientifique et technique, *Pour affronter l'imprévisible*, 46.
99 Ibid., 47–8.
100 Scanlon, *Ottawa-Carleton and the 1998 Ice Storm*, 10.
101 Quebec, Commission scientifique et technique, *Pour affronter l'imprévisible*, 51.
102 *Canadian Geographic*, "Lethal Beauty," 37.
103 *Ice Storm 98: When Maine Froze Over*, 46.
104 Jones and Mulherin, *An Evaluation of the Severity of the January 1998 Ice Storm in Northern New England*, 1.
105 President's Long-Term Recovery Task Force, *A Call for Collaboration*, ii.

CHAPTER SIX

1 Wuerthner, "The Healing Woods," 60.
2 Abley, *The Ice Storm*, 128.
3 Scanlon, *Ottawa-Carleton and the 1998 Ice Storm*, 72.
4 Abley, *The Ice Storm*, 128.
5 Ibid., 180.
6 *New York Times*, 14 January 1998, A16.
7 Ibid.
8 Jennifer Robinson, in Acknowledgements to Abley, *The Ice Storm*, 11.
9 LaRue, *1998 Ice Storm Almanac*, 188–90.
10 FEMA, *Region I, Interagency Hazard Mitigation Team … State of Maine*, 14.
11 *Ice Storm 98: When Maine Froze Over*, 5.
12 Larue, *1998 Ice Storm Almanac*, 21.
13 "The Great Ice Storm," *Ottawa Citizen*, 15 January 1998, G15.
14 Ibid.
15 Callon, Lascoumes, and Barthe, *Agir dans un monde incertain*, refers to such collaboration as a "hybrid forum."
16 Quebec, Commission scientifique et technique, *Pour affronter l'imprévisible*, xiii–xv.
17 Ibid., iii, 107.
18 LaRue, *1998 Ice Storm Almanac*, 107.
19 FEMA, "January 1998 New York Ice Storm, Region II," 246, 263.
20 Quebec, Commission scientifique et technique, *Pour affronter l'imprévisible*, 18–19.
21 Ibid., 37–8.
22 Ibid., 37.
23 Scanlon, *Ottawa-Carleton and the 1998 Ice Storm*, 33.
24 Ibid.
25 Ibid., 63.
26 "Hydro Workers Get Charge Out of Torture Chamber," *Ottawa Citizen*, 14 March 1998, 27.
27 FEMA, *Interagency Hazard Mitigation Team Strategy Report …* Vermont, xiii.
28 Abley, *The Ice Storm*, 82.
29 "The Great Ice Storm," *Ottawa Citizen*, 15 January 1998, G7.
30 Quebec, Commission scientifique et technique, *Pour affronter l'imprévisible*, 34.
31 LaRue, *1998 Ice Storm Almanac*, 136.
32 Jones and Mulherin, *An Evaluation of the Severity of the January 1998 Ice Storm in Northern New England*, 22.
33 FEMA, "January 1998 New York Ice Storm, Region II," 17–18.
34 FEMA, *Interagency Hazard Mitigation Strategy Report: New Hampshire*, 13.
35 Jones and Mulherin, *An Evaluation of the Severity of the January 1998 Ice Storm in Northern New England*, ii, 45.
36 USDA Forest Service, *Ice Storm 1998*, 17, and FEMA, *Interagency Hazard Mitigation Strategy Report: New Hampshire*, 22.
37 FEMA, *Region I, Interagency Hazard Mitigation Team … State of Maine*, 27–28.
38 Quebec, Commission scientifique et technique, *Pour affronter l'imprévisible*, 62.

39 FEMA, *Interagency Hazard Mitigation Strategy Report: New Hampshire*, 19.
40 Birmingham, *Ecological Effects of the 1998 Ice Storm in New York State*, 3–4.
41 USDA Forest Service, *Ice Storm 1998*, 17.
42 Essman, "A Ice Storm: Nature Adapts."
43 Abley, *The Ice Storm*, 59, 14.
44 Birmingham, *Ecological Effects of the 1998 Ice Storm in New York State*, 8.
45 USDA Forest Service, *Ice Storm 1998*, 15.
46 Darby, *Economic Impact of the 1998 Ice Storm.*
47 Quebec, Commission scientifique et technique, *Pour affronter l'imprévisible*, 88–90, 94–6.
48 Darby, *Economic Impact of the 1998 Ice Storm* 1.
49 Quebec, Commission scientifique et technique, *Pour affronter l'imprévisible*, 248–9.
50 "FEMA Sends $3M in Ice Storm Relief," *Burlington Free Press*, 24 April 1998, 1A, 4A.
51 FEMA, *Interagency Hazard Mitigation Strategy Report: New Hampshire*, 7.
52 *New York Times*, 26 October 1998, B1.
53 Ibid.
54 Ibid. The 2008 bailout of Wall Street undermined this argument.
55 Quebec, Commission scientifique et technique, *Pour affronter l'imprévisible*, 245–7.
56 "Hydro: Ice Buildup on Wires Was beyond Any Simulation," *Ottawa Citizen*, 1998, A3.
57 FEMA, "January 1998 New York Ice Storm, Region II," 6.
58 Jones and Mulherin, *An Evaluation of the Severity of the January 1998 Ice Storm in Northern New England*, 44.
59 Quebec, Commission scientifique et technique, *Pour affronter l'imprévisible*, 255–6.
60 Ibid., 256–7.
61 Ibid., 58–9.
62 FEMA, "January 1998 New York Ice Storm, Region II," 10.
63 Quebec, Commission scientifique et technique, *Pour affronter l'imprévisible*, 37.
64 Ibid., 260–3.
65 Lecomte, Pang, and Russell, *Ice Storm '98*, 37.
66 Ibid., 33.
67 Ibid., 35.
68 Ibid., 36.
69 Ibid., 32.
70 LaRue, *1998 Ice Storm Almanac*, 24–5.
71 Quebec, Commission scientifique et technique, *Pour affronter l'imprévisible*, 47.
72 *Ice Storm 98: When Maine Froze Over*, 7.
73 LaRue, *1998 Ice Storm Almanac*, 138–9.
74 Quebec, Commission scientifique et technique, *Pour affronter l'imprévisible*, 46.
75 *Ice Storm 98: When Maine Froze Over*, 46.
76 Quebec, Commission scientifique et technique, *Pour affronter l'imprévisible*, 215.
77 Scanlon, *Ottawa-Carleton and the 1998 Ice Storm*, 12.

78 Ibid., 41.
79 Ibid., 74.
80 Quebec, Commission scientifique et technique, *Pour affronter l'imprévisible*, 275–6, 283.

CHAPTER SEVEN

1 Fischer, *Response to Disaster*, 20.
2 *Ice Storm 98: When Main Froze Over*, 7.
3 Turner and Pidgeon, *Man-Made Disasters*.
4 In this part of the book, unless otherwise indicated, quotations from leaders were taken from an interview I did with the person specified. Minor editing has been done to avoid repetition and to make oral speech more readable. For interviews in French, I translated the quotation myself. More details about the methodology can be found in appendix 1.
5 Maisonneuve, Saouter, and Char, eds., "Annexe 1," 317.
6 Bourque, *Ma passion pour Montréal*, 59–60
7 Maisonneuve, Saouter and Char, eds., "Annexe 1," 319.
8 Scanlon, *Ottawa-Carleton and the 1998 Ice Storm*, 65.
9 Quebec, Commission scientifique et technique, *Pour affronter l'imprévisible.*
10 Scanlon, *Ottawa-Carleton and the 1998 Ice Storm*, 19, 65.
11 Quebec, Commission scientifique et technique, *Pour affronter l'imprévisible.*

CHAPTER EIGHT

1 Fischer, *Response to Disaster*, 18. Fischer gives an excellent review of the research literature on the question of whether there is panic and increased crime during a disaster. Disaster researchers have been documenting since 1954 that panic and increased crime do not usually occur during a disaster, and where they occur, they only do so under very specific conditions. See Quarantelli, "The Nature and Conditions of Panic"; Genevie et al., "Predictors of Looting in Selected Neighborhoods in New York City during the Blackout of 1977."
2 Dynes and Quarantelli, *The Role of Local Civil Defense in Disaster Planning.*
3 Koring, "Katrina Didn't Discriminate, New Orleans Death Toll Shows." The data reported by Koring also show that age was the strongest determining factor of fatalities: those over sixty years of age constituted 64 per cent of the dead but only 15.1 per cent of the populace.
4 Ibid.
5 See Fischer *Response to Disaster*; Dynes and Quarantelli, "The Absence of Community Conflict in the Early Phases of Natural Disasters," 200–4; and Takuma, "Human Behavior in the Event of Earthquakes," 171.
6 Fischer, *Response to Disaster*, 18.
7 Dynes, Quarantelli, and Kreps, *A Perspective on Disaster Planning*, 31.
8 Fischer, *Response to Disaster*, 197. See also LaPorte, "Highly Reliable Organizations."
9 Fischer, *Response to Disaster*, 90.
10 Ibid., 197.
11 Ibid., 25.

12 Ibid., 195.
13 Menard, "La gestation de crises," 214–15.
14 Quebec, Commission scientifique et technique, *Pour affronter l'imprévisible*, 260 (my translation).
15 Bourque, *Ma passion pour Montréal*, 63–4.
16 Ibid., 64.
17 Ibid., 65.
18 Analyses of the 2005 Hurricane Katrina–New Orleans disaster by leading American disaster researchers can be found at the Web site http://understandingkatrina.ssrc.org/.
19 Goff, "Thousands Drank 'Lethal' Water," and Ching, "China Shows Accountability in Wake of Toxic Spill."
20 See Gusse, "Médias, information et démocratie en temps de crise," 177.
21 Maisonneuve, Saouter, and Char, "Annexe 1," 318–19.
22 Gusse, "Médias, information et démocratie en temps de crise," 163–87, esp. 177–8.

CHAPTER NINE

1 Guindon, Pauchant, Doré, and Therrien, *Rapport sur les mesures de sécurite civile de la Communauté urbaine de Montréal face à la tempête de verglas de janvier 1998.*
2 Pear, "National Guard Drained by Iraq, Governors Say."

CHAPTER TEN

1 Quebec, Commission scientifique et technique, *Pour affronter l'imprévisible*; Lecomte, Pang, and Russell, *Ice Storm '98*; FEMA, *Region I, Interagency Hazard Mitigation Team ... State of Maine*; Scanlon, *Ottawa-Carleton and the 1998 Ice Storm.*

CHAPTER ELEVEN

1 Lecomte, Pang, and Russell, *Ice Storm '98*, 8.
2 Webster, Holland, Curry, and Chang, "Changes in Tropical Cyclone Number, Duration, and Intensity in a Warming Environment."
3 Ray and Guzzo, *Environmental Overkill*, 27.
4 Kuhn, *The Structure of Scientific Revolutions.*
5 Or "habitus," as it is called by Bourdieu in *Distinction.*
6 Dickens, "Changing Our Environment, Changing Ourselves"; Dickens, *Society & Nature.*
7 Tierney, "The Courage to Vote 'No' on Ice Aid."
8 Bourdieu, *Distinction.*
9 Jaccard, *Sustainable Fossil Fuels.*
10 Nye, *Consuming Power.*
11 IPCC, *Climate Change 2001.*
12 Schindler and Smol, "Remember the Environmental Imbalance."
13 Limbaugh, *The Way Things Ought to Be.*
14 "Satirist's 'Truthiness' Tag Named Word of the Year," *Ottawa Citizen*, 9 December 2006, A6.

15 McCright and Dunlap, "Challenging Global Warming as a Social Problem"; McCright and Dunlap, "Defeating Kyoto"; Rosa and Dietz, "Climate Change and Society."
16 Freudenburg and Pastor, "Public Responses to Technological Risks"; Freudenburg and Gramling, *Oil in Troubled Waters*; Freudenburg, "Social Constructions and Social Constrictions."
17 Hawken, *The Ecology of Commerce.*
18 Drummond, Caranci, and Tulk, "Market-Based Solutions to Protect the Environment."
19 De Souza, "Green's Call for Carbon Tax Draws Attacks."
20 Patterson, "China Grasps for Green."
21 *New York Times*, 15 December 2004.
22 Allman, "Conférence de Montréal," 30–2.
23 Grundmann, *Transnational Environmental Policy.*
24 Broecker, "Thermohaline Circulation, the Achilles Heel of Our Climate System"; Broecker and Bard, eds., *The Last Deglaciation*; Quadfasel, "The Atlantic Heat Conveyor Slows," 565; Bryden, Longworth, and Cunningham, "Slowing of the Atlantic Meridional Overturning Circulation at 25° N," 655.
25 Webster, Holland, Curry, and Chang, "Changes in Tropical Cyclone Number, Duration, and Intensity in a Warming Environment."
26 MacNeill, "Re: Climate Accord 'Important First Step,' June 8."
27 Golden, "Canada's Major Cities Need Special Attention."
28 IPCC, *Climate Change 2001.*
29 Ibid.
30 Veizer, Godderis, and François, "Evidence for Decoupling of Atmospheric CO_2 and Global Climate during the Phanerozoic Eon"; Veizer, "Celestial Climate Driver."
31 Macintyre, "Polar Klondike."
32 Hansen, "Is There Still Time to Avoid Dangerous Anthropogenic Interference with Global Climate?"
33 Ibid.
34 See Festinger, Riecken, and Schacter, *When Prophecy Fails.*
35 Diamond, *Collapse.*

CHAPTER TWELVE

1 Erikson, *A New Species of Trouble*, 22.
2 Tenner, *Why Things Bite Back.*
3 Quebec, Commission scientific et technique, *Pour affronter l'imprévisible.*
4 Ibid.
5 Milton and Bourque, *A Climatological Account of the January 1998 Ice Storm in Quebec*, 84.
6 Schindler and Smol, "Remember the Environmental Imbalance."
7 Gross, "Sociologists of the Unexpected."
8 Fischer, *Response to Disaster*, 145.
9 See the Web site http://understandingkatrina .ssrc.org/.
10 Fischer, *Response to Disaster*, 128; see also Drabek, *Human System Responses to Disaster.*

11 Fischer, *Response to Disaster*, 140.
12 Lindgren, "'We Won't Have Enough Power,' Minister Says."
13 Turner and Pidgeon, *Man-Made Disasters*.
14 In the sense of an artifact that is made by human skill or labour.

CHAPTER THIRTEEN

1 *The New Encyclopaedia Britannica*, 15th ed., 1: 343. See also esp. Faber, *The Amish*, and Hostetler, *Amish Society*, 4th ed.
2 LaRue, *1998 Ice Storm Almanac*, 124.
3 Mrs Abram Ebersol, Cameron Mills, NY, in *Die Botschaft* 23, no. 34 (28 January 1998): 42.
4 Mrs Jonas Shrock, Woodhull, NY, South Dist., ibid., 21.
5 Johnson-Weiner, "An Essay on an Amish Woman's Life in the North Country"; Johnson-Weiner, "The Plain People among Us."
6 Mrs Ruth Wengerd, Clyde, NY, in *Die Botschaft* 23, no. 34 (28 January 1998): 3.
7 Rudy Mast, Rens Falls, NY, ibid., 17–18.
8 Letter in *The Diary* (Gordonville, PA), February 1998, 19, from a person in Norfolk, NY.
9 Leslie, "Powerless," 10.
10 Because there had been little documentation of the Amish experience of this ice storm, I went to the affected church districts to gather data myself. The following information has been gleaned from observing and listening to the available members of seven Amish families who were in the area during the ice storm. I was introduced to them by Dr Karen Johnson-Weiner, an anthropologist at SUNY (Potsdam) who has worked for over twenty years with the Amish in northern New York State.
11 Kraybill and Bowman, *On the Backroad to Heaven*, 113.
12 The following information is based on an interview I did with the non-Amish owner.
13 Milton and Bourque, *A Climatological Account of the January 1998 Ice Storm*.
14 Brende, "Technology Amish Style."
15 Hostetler, *Amish Society*, 4th ed., 2.
16 Kraybill, "Plotting Social Change across Four Affiliations," 55.
17 Kraybill and Bowman, *On the Backroad to Heaven*, 239.
18 Kraybill, "Plotting Social Change across Four Affiliations," 4.
19 Kraybill and Bowman, *On the Backroad to Heaven*, 103.
20 Ibid.
21 Hostetler, *Amish Society*, rev. ed., 144
22 Olshan, "Amish Cottage Industries as Trojan Horse," 145.
23 Kraybill and Bowman, *On the Backroad to Heaven*, xvii.
24 Ibid., 127–8.
25 Ibid., 278.
26 Hostetler, *Amish Society*, 4th ed., 21.
27 Kraybill and Bowman, *On the Backroad to Heaven*, 105–6.
28 Ibid., 279.
29 Olshan, "Amish Cottage Industries as Trojan Horse," 158.

30 Brende, "Technology Amish Style."
31 Veblen, *The Theory of the Leisure Class.*
32 Bourdieu, *Outline of a Theory of Practice.*
33 Olshan, "The Old Order Amish as a Model for Development," 195.
34 Stoll, "Strangers and Pilgrims Part One," 7.
35 Ibid.
36 Ibid.
37 Ibid.
38 Ibid., 8
39 Ibid.
40 Kraybill and Bowman, *On the Backroad to Heaven*, 188.
41 Ibid., 135.
42 Lewis, "For Amish, Computers Can Be a Bane – and a Boon."
43 Kraybill, "War against Progress," 46.
44 "An Amish Lesson about Technology," *Mennonite Weekly Review*, 10 June 1999, 4.
45 Scott and Pellman, *Living without Electricity*, 4–5.
46 See also Kraybill and Bowman, *On the Backroad to Heaven*, 117–18.
47 Ibid., 133.
48 Kraybill and Nolt, "The Rise of Microenterprises," 160.
49 Kraybill and Bowman, *On the Backroad to Heaven*, 249.
50 Rheingold, "Look Who's Talking," 163.
51 Kraybill and Bowman, *On the Backroad to Heaven*, 46.
52 Kraybill and Nolt, "The Rise of Microenterprises," 158.
53 Kraybill and Bowman, *On the Backroad to Heaven*, 197.
54 Reported by David Luthy, chief archivist for the Amish.
55 Brende, "Technology Amish Style."
56 Kraybill and Nolt, "The Rise of Microenterprises," 158.
57 Ibid.
58 Rheingold, "Look Who's Talking," 131.
59 Brende, "Technology Amish Style," abstract.
60 Kraybill, "War against Progress," 36.
61 Kraybill and Bowman, *On the Backroad to Heaven*, 132.
62 Olshan, "The Old Order Amish as a Model for Development," abstract.
63 Kraybill and Bowman, *On the Backroad to Heaven*, 17.
64 Beck, *Risk Society.*
65 Oliver-Smith, "Disasters, Social Change, and Adaptive Systems," 231–3.
66 Kraybill and Bowman, *On the Backroad to Heaven*, 210.
67 Modern practices are less reflective than usually assumed. They are founded on predispositions based on previous experiences and repeated past actions – what has been called a "habitus" by Pierre Bourdieu in *Outline of a Theory of Practice*, A18. It is erroneous to assume that the Amish have a habitus whereas moderns do not. It is more valid to state that the Amish and moderns have different habituses.
68 Schnaiberg, *The Environment*; Schnaiberg and Gould, *Environment and Society.*
69 Kraybill and Bowman, *On the Backroad to Heaven*, 279.

70 Olshan, "Conclusion," 240.
71 Olshan, "The Old Order Amish as a Model for Development," 199.
72 Olshan, "Conclusion," 240.

CHAPTER FOURTEEN

1 Clarke, *Mission Improbable*; Clark, *Worst Cases*.
2 Weber, *The Protestant Work Ethic and the Spirit of Capitalism* (1930).
3 Quarantelli, "Just as a Disaster Is Not Simply a Big Accident," 68–71; Scanlon, *Ottawa-Carleton and the 1998 Ice Storm*, 11–12.
4 Online reference available at: http://understandingkatrina.ssrc.org/.
5 "Anarchy: Relief Effort Called 'a National Disgrace,'" *Ottawa Citizen*, 2 September 2005.
6 Half of Maine's National Guard has been away fighting in Iraq since that war began and not available locally to cope with a disaster at home. My interview on that point reads like a premonition concerning what happened a year after the interview in Louisiana. See chapter 9.
7 Loven, "Disasters Force Presidents to Prove Their Mettle."
8 Tierney, "Recent Developments in U.S. Homeland Security Policies and Their Implications for the Management of Extreme Events"; Buck, Trainor, and Aguirre, "A Critical Evaluation of the Incident Command System and NIMS"; Aguirre, "Homeland Security Warnings"; Cooper and Block, *Disaster*; Brinkley, *The Great Deluge*.
9 Scanlon, *Ottawa-Carleton and the 1998 Ice Storm*.
10 Fischer, *Response to Disaster*.
11 "Canada: Most Willing to Sacrifice for Environment: Poll" (CTV News/*Globe and Mail* poll conducted by the Strategic Counsel, 26 January 2007); formerly available at: http://www.climateark.org/shared/reader/welcome.aspx?linkid=67951.
12 See Freudenburg, "Risk and Recreancy."
13 Atwood, *Survival*, 32, 41.
14 Ibid., 61.
15 Ibid., 41.
16 Ibid., 31–2.
17 Ibid., 32.
18 Simon, *The Ultimate Resource*, and Simon, *The Ultimate Resource 2*.
19 Lomborg, *The Skeptical Environmentalist*.
20 Schurti, "Reception and Function of American Culture in Switzerland after World War II."
21 Atwood, *Survival*, 60.
22 Drummond, Caranci, and Tulk, "Market-Based Solutions to Protect the Environment."
23 See Nye, *Consuming Power* for an analysis of how urban sprawl can lock communities into a pattern of energy consumption.
24 Drummond, Caranci, and Tulk, "Market-Based Solutions to Protect the Environment."
25 Ibid.

26 Spector and Kitsuse, *Constructing Social Problems*, 43; Fox, "Green Sociology," 23–4.
27 Latour, "When Things Strike Back."

APPENDIX ONE

1 Murphy, *Rationality and Nature*; Murphy, *Sociology and Nature*.
2 Latour, "When Things Strike Back," 115.
3 Quebec, Commission scientifique et technique, *Pour affronter l'imprévisible*.
4 Ibid.
5 Weick, *Sensemaking in Organizations*; Weick, *Making Sense of the Organization*.
6 Turner and Pidgeon, *Man-Made Disasters*.
7 Vaughan, *The Challenger Launch Decision*, 380.

BIBLIOGRAPHY

Abley, Mark. *The Ice Storm: An Historical Record in Photographs of January 1998*. Toronto: McClelland and Stewart, 1998.

– ed. *Stories from the Ice Storm*. Toronto: McClelland and Stewart, 1999.

Abramovitz, J. *Unnatural Disasters*. Washington: WorldWatch Paper, 2001.

Aguirre, Benigo. "Homeland Security Warnings: Lessons Learned and Unlearned." *International Journal of Mass Emergencies and Disasters* 22 (2004).

Akwesasne Emergency Measures. *Ice Storm '98: Akwesasne*. Mohawk Council of Akwesasne, 1998.

Albrow, M. *Max Weber's Construction of Social Theory*. London: Macmillan, 1990.

Allman, M. "Conférence de Montréal: succès et mobilisation. *Le toit de monde* 5 (2006).

Aptekar, L. *Environmental Disasters in Global Perspective*. New York: Macmillan, 1994.

Atwood, Margaret. *Survival*. Toronto: McClelland and Stewart, 1996.

Bailey, R. *Eco-Scam: The False Prophets of Ecological Apocalypse*. New York: St. Martin's Press, 1993.

Beck, Ulrich. *Democracy without Enemies*. Cambridge: Polity Press, 1998.

– *Ecological Enlightenment: Essays on the Politics of Risk Society*. New Jersey: Humanities, 1995.

– *Ecological Politics in an Age of Risk*. Cambridge: Polity Press, 1995.

– *Risk Society: Towards a New Modernity*. London: Sage, 1992.

Beckerman, W. *Small Is Stupid: Blowing the Whistle on the Greens*. London: Duckworth, 1995.

– *Two Cheers for the Affluent Society: A Spirited Defense of Economic Growth*. New York: Saint Martin's Press, 1974.

Bell, M.M., and M. Carolan. *An Invitation to Environmental Sociology*. Thousand Oaks, CA: Pine Forge Press, 2004.

Benton, Ted. "Environmental Sociology." *Sociologisk tidsskrift* 9 (2001).
– "Why Are Sociologists Naturephobes?" In J. Lopes and G. Potter, eds., *After Postmodernism*. London: Athlone, 2001.
Birmingham, M.J. *Ecological Effects of the 1998 Ice Storm in New York State*. New York State Department of Environmental Conservation Document. 1998.
Blaikie, P., T. Cannon, I. Davis, and B. Wisner. *At Risk*. London: Routledge, 1994.
Bluhdorn, Inglofur. "A Theory of Post-ecologist Politics." *Environmental Politics* 6 (1997).
Boudes, Philippe. "L'environnement, domaine sociologique : La sociologie francaise au risque de l'environnement." PhD thesis, University of Bordeaux II (Victor Segalen), Bordeaux, France, 2008.
Bourdieu, Pierre. *Distinction*. Cambridge, MA: Harvard University Press, 1984.
– *Outline of a Theory of Practice*. Cambridge: University of Cambridge Press, 1977.
Bourque, Pierre. *Ma passion pour Montréal*. Montréal: Méridien, 2002.
Brende, Eric. "Technology Amish Style." *Technological Review* 99 (1996).
Brinkley, Douglas. *The Great Deluge: Hurricane Katrina, New Orleans, and the Mississippi Gulf Coast*. New York: HarperCollins, 2006.
Broecker, W. "Thermohaline Circulation, the Achilles Heel of Our Climate System." *Science* 278 (1997).
Broecker, W., and E. Bard, eds. *The Last Deglaciation*. Berlin: Springer Verlag, 1992.
Bryden, H.L, H.R. Longworth, and S.A. Cunningham. "Slowing of the Atlantic Meridional Overturning Circulation at 25° N." *Nature* 438 (2005).
Buck, Dick, Joseph Trainor, and Benigno Aguirre. "A Critical Evaluation of the Incident Command System and NIMS." *Journal of Homeland Security and Emergency Management* 1 (2006).
Callon, M., P. Lascoumes, and Y. Barthe. *Agir dans un monde incertain*. Paris: Seuil, 2001.
Canadian Geographic. "Lethal Beauty." 18, no. 2 (March/April 1998).
Ching, Ching Ni. "China Shows Accountability in Wake of Toxic Spill." *Montreal Gazette*, 5 December 2005.
Clark, Colin. *Population Growth: The Advantages*. Santa Anna, CA: R.L. Sassone 1975.
– *Starvation or Plenty?* New York: Taplinger Publishing, 1970.
Clarke, L. *Mission Improbable*. Chicago: University of Chicago Press, 1999.
– *Worst Cases*. Chicago: University of Chicago Press, 2005.
Clarke, L., and Short, J.F. "Social Organization and Risk." *Annual Review of Sociology* 19 (1993).
Collins, Glenn. "With a Delightful January Like This, Who Needs Spring?" *New York Times*, 6 January 1998.
Collins, H.M. "Stages in the Empirical Programme of Relativism." *Social Studies of Science* 11 (1981).
Cooper, Christopher, and Robert Block. *Disaster: Hurricane Katrina and the Failure of Homeland Security*. New York: Times Books, 2006.
Cuny, F. *Disasters and Development*. New York: Oxford University Press, 1983.

Darby, Paul. *Economic Impact of the 1998 Ice Storm*. Ottawa: Conference Board of Canada, 1998.

Davis, M. *Ecology of Fear*. New York: Metropolitan, 1998.

– *Late Victorian Holocausts*. London: Verso, 2001.

De Souza, Mike. "Green's Call for Carbon Tax Draws Attacks." *National Post*, 6 June 2007.

Diamond, Jared. *Collapse: How Societies Choose to Fail or Succeed*. New York: Viking, 2005.

Dickens, Peter. "Changing Our Environment, Changing Ourselves." *Interdisciplinary Science Review* 28 (2003).

– *Society & Nature*. Cambridge, UK: Polity Press, 2004.

Dimas, Stavros. "The World Needs Canada." *Ottawa Citizen*, 1 February 2007.

Dombrowsky, W. "Again and Again: Is a Disaster What We Call a 'Disaster'?" In E.L. Quarantelli, ed., *What Is a Disaster?* London: Routledge, 1998.

Drabek, T.E. *Human System Responses to Disaster: An Inventory of Sociological Findings*. New York: Springer-Verlag, 1986.

Drummond, Don, Beata Caranci, and David Tulk. "Market-Based Solutions to Protect the Environment." 2007. Available at: http://www.td.com/economics.

Dunlap, Riley, and William Catton. "Environmental Sociology." *Annual Review of Sociology* 5 (1979).

Dunlap, Riley, and Brent Marshall. "Environmental Sociology." In Clifton Bryant and Dennis Peck, eds., *21st Century Sociology: A Reference Handbook*, vol. 2. Thousands Oaks, CA: Sage, 2007.

Dynes, R. "Coming to Terms with Community Disaster." In E.L. Quarantelli, ed., *What is a Disaster?* London: Routledge, 1998.

Dynes, R., and E.L. Quarantelli. "The Absence of Community Conflict in the Early Phases of Natural Disasters." In C.G. Smith, ed., *Conflict Resolution*. South Bend: University of Notre Dame Press, 1971.

– *The Role of Local Civil Defense in Disaster Planning*. Columbus: Ohio State University, 1975.

Dynes, R., E.L. Quarantelli, and G.A. Kreps. *A Perspective on Disaster Planning*. Columbus: Ohio State University, 1972.

Dynes, R., E.L. Quarantelli, and D. Wenger. *Individual and Organizational Response to the 1985 Earthquake in Mexico City, Mexico*. Newark: Disaster Research Center, University of Delaware, 1990.

Easterbrook, G. *A Moment on the Earth: The Coming Age of Environmental Optimism*. New York: Viking, 1995.

Eder, Konrad. *The Social Construction of Nature*. London: Sage, 1996.

Environment Canada. *Ice Storm '98*. Ottawa, Canada: Environment Canada, 1999.

– *Ice Storm 98 – The Meteorological Event*. Prepared for Avon Re Worldwide. Otawa: Environment Canada, 1998.

Erikson, K. *A New Species of Trouble: Explorations in Disaster, Trauma and Community*. New York: Norton, 1994.

Essman, J. "A Ice Storm: Nature Adapts." New York State Conservationist, December 1998.

Etkin, D., and S. Brun. *Coping with Natural Hazards in Canada: Scientific, Government and Insurance Industry Perspectives*. Part 2. Toronto: Insurers' Advisory Organization, [1997].

Faber, D. *The Amish*. New York: Doubleday, 1991.

Federal Emergency Management Agency (FEMA). *Interagency Hazard Mitigation Strategy Report: New Hampshire*. FEMA DR-1199-NH. 1998.

– *Interagency Hazard Mitigation Team Strategy Report*. FEMA DR-1201-Vermont. 1998.

– "January 1998 New York Ice Storm, Region II: A Ice Storm Overview." Available at: http:/www.fema.gov/reg-ii/1998/nyice2.htm; accessed 20 January 2000.

– *Region I, Interagency Hazard Mitigation Team January 1998 Ice Storm, State of Maine*. FEMA DR-1198-ME. 1998.

Ferry, Luc. *Le nouvel ordre ecologique*. Paris: Grasset, 1992.

– Review of *Entre la nature et l'homme, je choissis l'homme*, by Anne Deleves. *La Vie* 2465 (26 November).

Festinger, Leon, Henry Riecken, and Stanley Schacter. *When Prophecy Fails*. New York: Harper and Row, 1964.

Fischer, H.W., III. *Response to Disaster: Fact Versus Fiction & Its Perpetuation*. 2nd ed. Landam, MD: University Press of America, 1998.

Fox, Nick. "Green Sociology." *Network* (Newsletter of the British Sociological Association) 50 (1991).

Freudenburg, William R. "Contamination, Corrosion and the Social Order: An Overview." *Current Sociology* 45 (1997).

– "Risk and Recreancy." *Social Forces* 71 (1993).

– "Social Constructions and Social Constrictions." In G. Spaargaren, A. Mol, and F. Buttel, eds., *Environment and Global Modernity*. London: Sage, 2000.

Freudenburg, William R., and Robert Gramling. *Oil in Troubled Waters*. Albany: State University of New York Press, 1994.

– "Scientific Expertise and Natural Resource Decisions." *Social Science Quarterly* 83 (2002).

Freudenburg, William R., and Susan K. Pastor. "Public Responses to Technological Risks." *Sociological Quarterly* 33 (1992).

Geertz, Clifford. "Thick Description: Toward an Interpretive Theory of Culture." In *The Interpretation of Cultures: Selected Essays*. New York: Basic Books, 1973.

Genevie, Louise, et al. "Predictors of Looting in Selected Neighborhoods in New York City during the Blackout of 1977." *Sociology and Social Research* 71 (1987).

Giddens, Anthony. *Beyond Left and Right*. Cambridge: Polity, 1994.

– *Modernity and Self-Identity*. Cambridge: Polity Press, 1991.

Gilbert, C. "Studying Disaster." In E.L. Quarantelli, ed., *What Is a Disaster?* London: Routledge, 1998.

Goff, Peter. "Thousands Drank 'Lethal' Water." *Ottawa Citizen*, 27 November 2005.

Goffman, Erving. *Interaction Ritual: Essays on Face-to-Face Behavior*. Garden City, NY: Anchor Books, 1967.

– *The Presentation of Self in Everyday Life*. Garden City, NY: Doubleday, 1959.

Golden, Anne. "Canada's Major Cities Need Special Attention." *Ottawa Citizen*, 6 February 2007.

Gould, Alan, Allan Schnaiberg, and Adam S. Weinberg. *Local Environmental Struggles: Citizen Activism in the Treadmill of Production*. Cambridge: Cambridge University Press, 1996.

Gross, Matthias. "Sociologists of the Unexpected." *American Sociologist* 34 (2003).

Grundmann, Reiner. *Transnational Environmental Policy*. London: Routledge, 2001.

Guindon, Jean-Bernard, Thierry Pauchant, Michel Doré, and Marie-Christine Therrien. *Rapport sur les mesures de sécurité civile de la Communauté Urbaine de Montréal face à la tempête de verglas de janvier 1998*. Montréal: Centre de Sécurité civile de la Communauté urbaine de Montréal, 1998.

Gusse, Isabelle. "Médias, information et démocratie en temps de crise." In D. Maisonneuve, C. Saouter, and A. Char, eds., *Communications en temps de crise*. Sainte-Foy: Presses de l'Université du Québec, 1999.

Hajer, Maarten A. *The Politics of Environmental Discourse*. Oxford: Clarendon Press, 1995.

Hamilton, David. *Technology, Man and the Environment*. London: Faber and Faber, 1973.

Handmer, J. "Sustainable Development Is about Disaster Reduction." *IJMED* 20 (2002).

Hannigan, John A. *Environmental Sociology*. London: Routledge, 2006.

Hansen, James E. "Is There Still Time to Avoid Dangerous Anthropogenic Interference with Global Climate?" Presentation on 6 December 2005 at the American Geophysical Union, San Francisco.

Hawken, Paul. *The Ecology of Commerce: A Declaration of Sustainability*. New York: HarperCollins, 1993.

Hewitt, K. "Excluded Perspectives in the Social Construction of Disaster." In E.L. Quarantelli, ed., *What Is a Disaster?* London: Routledge, 1998.

Hostetler, John. *Amish Society*. Rev. ed. Baltimore: Johns Hopkins University Press, 1968.

– *Amish Society*. 4th ed. Baltimore: Johns Hopkins University Press, 1993.

Huber, P. *Hard Green: Saving the Environment from Environmentalists*. New York: Basic Books, 1999.

Ice Storm '98. Elizabethtown, NY: Denton Publications, 1998.

Ice Storm '98: A North Country Disaster. Plattsburgh, NY: The Press-Republican, 1998.

Ice Storm 98: When Maine Froze Over. Kennebec Journal, Morning Sentinel, Portland Press Herald, Maine Sunday Telegram, 1998.

IDNDR Technical Committee. *Washington Declaration*. Geneva, 1998.

Insurance Bureau of Canada. *Eastern Ontario Ice Storm – January 1998 Insurance Coverage*. 1998.

Intergovernmental Panel on Climate Change (IPCC). *Climate Change 2001*. Cambridge: Cambridge University Press, 2001.

ISDR (UN). *Disaster Reduction and Sustainable Development.* ISDR Background Document for the World Summit on Sustainable Development, no. 5. 2002.

Jaccard, Mark. *Sustainable Fossil Fuels.* Cambridge: Cambridge University Press, 2005.

Jackson, Jenny. "Sept. 11, 2001: Recalling the Day the 'Centre Could Not Hold.'" *Ottawa Citizen*, 11 September 2005.

Johnson-Weiner, Karen. "An Essay on an Amish Woman's Life in the North Country." In Natalia Singer and Neil Burdick, eds., *Living North Country: Essays on Life and Landscapes in Northern New York.* Utica, NY: North Country Books, 2001.

– "The Plain People among Us: The North Country Amish." *Quarterly of the St. Lawrence County Historical Association*, July 1999.

Jones, K., and N. Mulherin. *An Evaluation of the Severity of the January 1998 Ice Storm in Northern New England.* Fort Monmouth, NH: US Army Laboratory, 1998.

Kahn, H., and A.J. Weiner. *The Year 2000: A Framework for Speculation.* New York: Hudson Institute, 1967.

Karl, T.R., N. Nicholls, and A. Ghazi, eds. *Weather and Climate Extremes.* Boston: Kluwer Academic, 1999.

Kingdom, J.W. *Agendas, Alternatives, and Public Policies.* Boston: Little, Brown and Co, 1984.

Klinenberg, E. *Heat Wave: A Social Autopsy of Disaster in Chicago.* Chicago: University of Chicago Press, 2002.

Kopala, Margret. "The Questions We Should Be Asking." *Ottawa Citizen*, 28 December 2006.

Koring, Paul. "Katrina Didn't Discriminate, New Orleans Death Toll Shows." *Globe and Mail*, 1 December 2005.

Kraybill, Donald B. "Plotting Social Change across Four Affiliations." In Donald B. Kraybill and Marc A. Olshan, eds., *The Amish Struggle with Modernity.* Hanover, NH: University Press of New England, 1994.

– "War against Progress: Coping with Social Change." In Donald B. Kraybill and Marc A. Olshan, eds., *The Amish Struggle with Modernity.* Hanover, NH: University Press of New England, 1994.

Kraybill, Donald B., and Steven M. Nolt. "The Rise of Microenterprises." In Donald B. Kraybill and Marc A. Olshan, eds., *The Amish Struggle with Modernity.* Hanover, NH: University Press of New England, 1994.

Kraybill, Donald B., and Carl F. Bowman. *On the Backroad to Heaven: Old Order Hutterites, Mennonites, Amish, and Brethren.* Baltimore: Johns Hopkins University Press, 2001.

Kraybill, Donald B., and Marc A. Olshan, eds. *The Amish Struggle with Modernity.* Hanover, NH: University Press of New England, 1994.

Krieger, Martin H. "What's Wrong with Plastic Trees?" *Science* 179 (1973).

Kroll-Smith, S., and V.J. Gunter. "Legislators, Interpreters, and Disasters." In E.L. Quarantelli, ed., *What Is a Disaster?* (London: Routledge, 1998).

Kuhn, T. *The Structure of Scientific Revolutions.* Chicago: University of Chicago Press, 1982.

LaPorte, Todd R. "Highly Reliable Organizations: Unlikely, Demanding, and at Risk." *Journal of Crisis and Contingency Management* 4 (1996).
LaRue, R. *1998 Ice Storm Almanac*. Syracuse: Peerless Press, 1998.
Latour, Bruno. *Nous n'avons jamais été modernes*. Paris: La découverte, 1991.
– "When Things Strike Back." *British Journal of Sociology* 51 (2000).
Lecomte, E., A. Pang, and J. Russell. *Ice Storm 98*. Boston and Toronto: Institute for Catastrophic Loss Reduction and Institute for Business and Home Safety, 1998.
Leslie, Jacques. "Powerless." *Wired,* April 1999.
Lewis, J. *Development in Disaster-Prone Places: Studies in Vulnerability*. London: Intermediate Technology Publications, 1999.
Lewis, Larry. "For Amish, Computers Can Be a Bane – and a Boon." *Philadelphia Inquirer,* 30 July 2000, A18.
Limbaugh, Rush. *The Way Things Ought to Be*. New York: Pocket Books, 1992.
Lindgren, April. "'We Won't Have Enough Power,' Minister Says." *Ottawa Citizen*, 2 September 2006.
Lomborg, Bjorn. *The Skeptical Environmentalist: Measuring the Real State of the World*. Cambridge: Cambridge University Press, 2001.
Loven, Jennifer. "Disasters Force Presidents to Prove Their Mettle." *Ottawa Citizen*, 11 September 2005.
Macintyre, Ben. "Polar Klondike." *Ottawa Citizen*, 19 February 2006.
MacNeill, Jim. "Re: Climate Accord 'Important First Step.'" *Ottawa Citizen*, 8 June 2007.
Maddox, John. *The Doomsday Syndrome*. London: Macmillan, 1972.
Maduro, R.A., and R. Schauerhammer. *The Holes in the Ozone Are Scarce: The Scientific Evidence that the Sky Isn't Falling*. Washington, DC: 21st Century Science Associates, 1992.
Maisonneuve, D., C. Saouter, and A. Char. "Annexe I : Le Rapport Nicolet pour affronter l'imprévisible." In Maisonneuve, Saouter, and Char, eds., *Communications en temps de crise*. Sainte-Foy: Presses de l'Université du Québec, 1999.
McCright, Aaron M., and Riley E. Dunlap. "Challenging Global Warming as a Social Problem." *Social Problems* 47 (2000).
– "Defeating Kyoto: The Conservative Movement's Impact on U.S. Climate Change Policy." *Social Problems* 50 (2003).
McDonough, W., and M. Braungart. *Cradle to Cradle: Remaking the Way We Make Things*. New York: North Point Press, 2002.
McKibben, Bill. *The End of Nature*. New York: Random House, 1990.
McLuhan, Marshall. *Understanding Media*. New York: McGraw-Hill, 1965.
Menard, Serge. "La gestion de crises." In D. Maisonneuve, C. Saouter, and A. Char, eds., *Communications en temps de crise*. Sainte-Foy: Presses de l'Université du Québec, 1999.
Merchant, Carolyn. *The Death of Nature*. San Francisco: HarperCollins, 1980.
Middleton, N., and P. O'Keefe. *Disaster and Development*. London: Pluto Press, 1998.
Mileti, D. *Disasters by Design*. Washington: Joseph Henry, 1999.
– "Sustainability and Hazards." IJMED 20 (2002).

Milton, J., and A. Bourque. *A Climatological Account of the January 1998 Ice Storm in Quebec*. Ottawa: Environment Canada, 1999.

Mol, Arthur A. *Globalization and Environmental Reform: The Ecological Modernization of the Global Economy*. Cambridge, MA: MIT Press, 2001.

– *The Refinement of Production: Ecological Modernization Theory and the Chemical Industry*. Utrecht: Van Arkel, 1995.

Mol, Arthur A., and David Sonnenfeld, eds. *Ecological Modernization around the World: Perspectives and Critical Debates*. London: Frank Cass, 2000.

Mommsen, W.J. "Personal Conduct and Societal Change." In S. Lash and S. Whimster, eds., *Max Weber, Rationality and Modernity*. London: Allen & Unwin, 1987.

Murphy, Raymond. "The Challenge of Disaster Reduction." In Aaron McCright and Terry N. Clark, eds., *Community and Ecology: Dynamics of Place, Sustainability and Politics*. JAI/Elsevier, 2006.

– "Disaster or Sustainability: The Dance of Human Agents with Nature's Actants." *Canadian Review of Sociology and Anthropology* 41 (2004).

– "Environmental Realism: From Apologetics to Substance." *Nature & Culture* 1 (2006).

– "The Internalization of Autonomous Nature into Society." *Sociological Review* 50 (2002).

– *Rationality and Nature*. Boulder: Westview, 1994.

– *Social Closure*. Oxford: Oxford University Press, 1988.

– *Sociology and Nature*. Boulder: Westview, 1997.

– "Thinking across the Culture/Nature Divide: An Empirical Study of Issues for Critical Realism and Social Constructionism." In Frank Pearce and Jon Frauley, eds., *Critical Realism and the Social Sciences*. Toronto: University of Toronto Press, 2007.

– "Unperceived Risk." *Advances in Human Ecology* 8 (1999).

National Climatic Data Center. "The El Niño Winter of 97–98." In *Technical Report 98-02*. 1998.

Novek, Joel, and Karen Kampen. "Sustainable or Unsustainable Development? An Analysis of an Environmental Controversy." *Canadian Journal of Sociology / Cahiers canadiens de sociologie* 17 (1992).

Nye, D. *Consuming Power*. Cambridge, MA: MIT Press, 1998.

Oliver-Smith, A. "Disasters, Social Change, and Adaptive Systems." In E.L. Quarantelli, ed., *What Is a Disaster?* London: Routledge, 1998.

– "Global Changes and the Definition of Disaster." In E.L. Quarantelli, ed., *What Is a Disaster?* London: Routledge, 1998.

– "Theorizing Disasters: Nature, Power, and Culture." In S. Hoffman and A. Oliver-Smith, eds., *Catastrophe & Culture: The Anthropology of Disaster*. Santa Fe: School of American Research Press, 2001.

Olshan, Marc A. "Amish Cottage Industries as Trojan Horse." In Donald B. Kraybill and Marc A. Olshan, eds., *The Amish Struggle with Modernity*. Hanover, NH: University Press of New England, 1994.

– "Conclusion: What Good Are the Amish?" In Donald B. Kraybill and Marc A. Olshan, eds., *The Amish Struggle with Modernity*. Hanover, NH: University Press of New England, 1994.

– "The Old Order Amish as a Model for Development." PhD dissertation, Cornell University, 1980.

Olson, R.S. *The Politics of Earthquake Prediction*. Princeton: Princeton University Press, 1989.

Patterson, Kelly. "China Grasps for Green." *Ottawa Citizen*, 22 January 2006.

Pear, Robert. "National Guard Drained by Iraq, Governors Say." *International Herald Tribune*, 28 February 2006.

Perrow, C. *Normal Accidents*. New York: Basic, 1984.

Phillips, D. *Blame It on the Weather.* Toronto: Key Porter, 1998.

Platt, R. *Disasters and Democracy*. Washington: Island Press, 1999.

Preece, Rob. *Animals and Nature: Cultural Myths, Cultural Realities*. Vancouver: UBC Press, 1999.

President's Long-Term Recovery Task Force. *A Call for Collaboration: Final Report on the January 1998 Ice Storm*. 1998.

Quadfasel, D. "The Atlantic Heat Conveyor Slows." *Nature* 438 (2005).

Quarantelli, E.L. 1954. "The Nature and Conditions of Panic," *American Journal of Sociology* 60.

Quarantelli, E.L. "Epilogue." In E.L. Quarantelli, ed., *What Is a Disaster?* London: Routledge, 1998.

– "Just as a Disaster Is Not Simply a Big Accident, So a Catastrophe Is Not Just a Bigger Disaster." *Journal of the American Society of Professional Emergency Planners*, 1996.

– ed. *What Is a Disaster? Perspectives on the Question*. London: Routledge, 1998.

Quebec. Commission scientifique et technique chargée d'analyser les événements relatifs à la tempête de verglas survenue du 5 au 9 janvier 1998. *Pour affronter l'imprévisible*. Quebec: Les publications du Québec, 1999.

Ray, Dixie L., and Louis R. Guzzo. *Environmental Overkill: Whatever Happened to Common Sense?* New York: HarperCollins, 1993.

– *Trashing the Planet*. New York: HarperCollins, 1990.

Redclift, M. "Redefining the Environmental 'Crisis' in the South." In Joe Weston, ed., *Red and Green: A New Politics of the Environment.* London: Pluto Press, 1986.

Rheingold, Howard. "Look Who's Talking." *Wired*, January 1999.

– *The Virtual Community*. Cambridge, MA: MIT Press, 2000.

Rosa, E. "Metatheoretical Foundations for Post-normal Risk." *Journal of Risk Research* 1 (1998).

Rosa, Eugene A., and Thomas Dietz. "Climate Change and Society." *International Sociology* 13 (1998).

Sagan, S.D. *The Limits of Safety: Organizations, Accidents, and Nuclear Weapons*. Princeton: Princeton University Press, 1993.

Scanlon, Joseph. *Ottawa-Carleton and the 1998 Ice Storm: Sharing the Lessons Learned.* Ottawa: Project Ice 1998, 1998.

Schindler, D., and J. Smol. "Remember the Environmental Imbalance." *Ottawa Citizen* 6 February 2006, A15.

Schnaiberg, A. *The Environment: From Surplus to Scarcity*. New York: Oxford University Press, 1980.

Schnaiberg, A., and K.A. Gould. *Environment and Society: The Enduring Conflict*. New York: St. Martin's Press, 1994.

Schurti, Pio. "Reception and Function of American Culture in Switzerland after World War II." PhD dissertation, University of Texas at Austin, 2005.

Scott, Stephen, and Kenneth Pellman. *Living without Electricity.* Intercourse, PA: Good Books, 1990.

Sjoberg, L. *Risk and Society*. Boston: Allen & Unwin, 1987.

Simon, Julian. *Immigration: The Demographic and Economic Facts*. Washington: Cato Institute, 1995.

– *The State of Humanity*. London: Blackwell, 1995.

– *The Ultimate Resource*. Princeton: Princeton University Press, 1981.

– *The Ultimate Resource 2*. Princeton: Princeton University Press, 1996.

Simon, Julian, and Herman Kahn. *The Resourceful Earth*. Oxford: Basil Blackwell, 1984.

Smith, V. Kerry, ed. *Scarcity and Growth Reconsidered*. Baltimore: Johns Hopkins University Press for Resources for the Future, 1979.

Spector, M., and J. Kitsuse. *Constructing Social Problems*. Melo Park, CA: Cummings, 1977.

Sperandio, E.P. *L'Enfer de glace*. Montreal: Trustar, 1998.

Statistics Canada. *The St. Lawrence River Valley 1998 Ice Storm: Maps and Facts.* Ottawa, 1998.

Stehr, Nico. *Knowledge Politics*. Boulder: Paradigm Publishers, 2005.

– *Knowledge Societies*. London: Sage, 1994.

Stern, Nicholas. *The Economics of Climate Change*. Cambridge: Cambridge University Press, 2007.

Stoll, Elmo. "Strangers and Pilgrims Part One: Why We Live Simply." *Family Life*, October 2003.

Sylves, R., and W. Waugh. *Disaster Management in the U.S. and Canada*. Springfield: Charles C. Thomas, 1996.

Takuma, T. "Human Behavior in the Event of Earthquakes." In E.L. Quarantelli, ed., *Disasters: Theory and Research*. Beverly Hills, CA: Sage, 1978.

Tenner, E. *Why Things Bite Back*. New York: Vintage, 1997.

Tertzakian, Peter. *A Thousand Barrels a Second: The Coming Oil Break Point and the Challenges Facing an Energy Dependent World*. New York: McGraw-Hill, 2005.

Tester, K. *Animals and Society: The Humanity of Animal Rights*. London: Routledge, 1991.

Thomas, W., and D. Thomas. *The Child in America*. New York: Knopf, 1928.

Tierney, J. "The Courage to Vote 'No' on Ice Aid." *New York Times*, 26 October 1998, B1.

Tierney, Kathleen. "Recent Developments in U.S. Homeland Security Policies and Their Implications for the Management of Extreme Events." In Havidan Rodriguez, E.L. Quarantelli, and Russell Dynes, eds., *Handbook of Disaster Research*. New York: Springer Science + Business, 2006.

Turner, Barry, and Nick Pidgeon. *Man-Made Disasters*. London: Wykeham, 1978.

Ungar, Sheldon. "The Rise and (Relative) Decline of Global Warming as a Social Problem." *Sociological Quarterly* 33 (1992).

United States. Department of Regional Development and Environment. *Disaster, Planning and Development: Managing Natural Hazards to Reduce Loss.* Washington: Organization of American States, 1990.

USDA Forest Service. *Ice Storm 1998: Cooperating for Forest Recovery.* Status Report, Northeastern Area State & Private Forestry. 1998

Vaughan, Diane. *The Challenger Launch Decision: Risky Technology, Culture, and Deviance at NASA.* Chicago: University of Chicago Press, 1996.

Vayk, J. Peter. *Doomsday Has Been Cancelled.* Menlo Park, CA: Peace, 1978.

Veblen, Thorstein. *The Theory of the Leisure Class.* New York: Mentor, 1953.

Veizer, Jan. "Celestial Climate Driver: A Perspective from Four Billion Years of Carbon Cycle." *Geoscience Canada* 32 (2005).

Veizer, Jan, Yves Godderis, and Louis M. François, "Evidence for Decoupling of Atmospheric CO_2 and Global Climate During the Phanerozoic Eon." *Nature* 408 (2000).

Weber, Max. *Economy and Society.* 3 vols. Totowa, NJ: Bedminster Press, 1968.

– *From Max Weber: Essays in Sociology.* 1946. Ed. H.H. Gerth and C. Wright Mills. New York: Oxford University Press, 1958.

– *The Protestant Ethic and the Spirit of Capitalism.* 1904–5. Trans. Talcott Parsons. London: Unwin, 1930.

Webster, P.J., G.J. Holland, J.A. Curry, and H.-R. Chang. "Changes in Tropical Cyclone Number, Duration, and Intensity in a Warming Environment." *Science* 309 (2005).

Weick, K.E. *Making Sense of the Organization.* Oxford: Blackwell, 2001.

– *Sensemaking in Organizations.* Thousand Oaks, CA: Sage, 1995.

Weick, K.E., K. Sutcliffe, and D. Obstfeld. "Organizing for High Reliability." *Research in Organizational Behavior* 21 (1999).

Weiland, Sabine. "The Power of Nature and the Nature of Power." *Nature & Culture* 2 (2007).

West, Patrick C. "Max Weber's Human Ecology of Historical Societies." In Vatro Murvar, ed., *Theory of Liberty, Legitimacy and Power: New Directions in the Intellectual and Scientific Legacy of Max Weber.* London: Routledge & Kegan Paul, 1985.

Wisner, B. *Power and Need in Africa.* London: Earthscan, 1988.

World Commission on Environment and Development. *Our Common Future.* Oxford: Oxford University Press, 1987.

Worster, D. *Nature's Economy.* Cambridge: Cambridge University Press, 1994.

Wuerthner, George. "The Healing Woods." *Adirondack Life*, November/December 1999.

Yearley, Steven. "Green Ambivalence about Science: Legal-Rational Authority and the Scientific Legitimation of a Social Movement." *British Journal of Sociology* 3 (1992).

Zebrowski, Ernest, Jr. *Perils of a Restless Planet: Scientific Perspectives on Natural Disasters.* Cambridge: Cambridge University Press, 1997.

INDEX